Lecture Notes in Physics

The Editorial Policy for Proceedings

The series Lecture Notes in Physics reports new developments in physical research and teaching – quickly, informally, and at a high level. The proceedings to be considered for publication in this series should be limited to only a few areas of research, and these should be closely related to each other. The contributions should be of a high standard and should avoid lengthy redraftings of papers already published or about to be published elsewhere. As a whole, the proceedings should aim for a balanced presentation of the theme of the conference including a description of the techniques used and enough motivation for a broad readership. It should not be assumed that the published proceedings must reflect the conference in its entirety. (A listing or abstracts of papers presented at the meeting but not included in the proceedings could be added as an appendix.)
When applying for publication in the series Lecture Notes in Physics the volume's editor(s) should submit sufficient material to enable the series editors and their referees to make a fairly accurate evaluation (e.g. a complete list of speakers and titles of papers to be presented and abstracts). If, based on this information, the proceedings are (tentatively) accepted, the volume's editor(s), whose name(s) will appear on the title pages, should select the papers suitable for publication and have them refereed (as for a journal) when appropriate. As a rule discussions will not be accepted. The series editors and Springer-Verlag will normally not interfere with the detailed editing except in fairly obvious cases or on technical matters.
Final acceptance is expressed by the series editor in charge, in consultation with Springer-Verlag only after receiving the complete manuscript. It might help to send a copy of the authors' manuscripts in advance to the editor in charge to discuss possible revisions with him. As a general rule, the series editor will confirm his tentative acceptance if the final manuscript corresponds to the original concept discussed, if the quality of the contribution meets the requirements of the series, and if the final size of the manuscript does not greatly exceed the number of pages originally agreed upon. The manuscript should be forwarded to Springer-Verlag shortly after the meeting. In cases of extreme delay (more than six months after the conference) the series editors will check once more the timeliness of the papers. Therefore, the volume's editor(s) should establish strict deadlines, or collect the articles during the conference and have them revised on the spot. If a delay is unavoidable, one should encourage the authors to update their contributions if appropriate. The editors of proceedings are strongly advised to inform contributors about these points at an early stage.
The final manuscript should contain a table of contents and an informative introduction accessible also to readers not particularly familiar with the topic of the conference. The contributions should be in English. The volume's editor(s) should check the contributions for the correct use of language. At Springer-Verlag only the prefaces will be checked by a copy-editor for language and style. Grave linguistic or technical shortcomings may lead to the rejection of contributions by the series editors. A conference report should not exceed a total of 500 pages. Keeping the size within this bound should be achieved by a stricter selection of articles and not by imposing an upper limit to the length of the individual papers. Editors receive jointly 30 complimentary copies of their book. They are entitled to purchase further copies of their book at a reduced rate. As a rule no reprints of individual contributions can be supplied. No royalty is paid on Lecture Notes in Physics volumes. Commitment to publish is made by letter of interest rather than by signing a formal contract. Springer-Verlag secures the copyright for each volume.

The Production Process

The books are hardbound, and the publisher will select quality paper appropriate to the needs of the author(s). Publication time is about ten weeks. More than twenty years of experience guarantee authors the best possible service. To reach the goal of rapid publication at a low price the technique of photographic reproduction from a camera-ready manuscript was chosen. This process shifts the main responsibility for the technical quality considerably from the publisher to the authors. We therefore urge all authors and editors of proceedings to observe very carefully the essentials for the preparation of camera-ready manuscripts, which we will supply on request. This applies especially to the quality of figures and halftones submitted for publication. In addition, it might be useful to look at some of the volumes already published. As a special service, we offer free of charge LATEX and TEX macro packages to format the text according to Springer-Verlag's quality requirements. We strongly recommend that you make use of this offer, since the result will be a book of considerably improved technical quality. To avoid mistakes and time-consuming correspondence during the production period the conference editors should request special instructions from the publisher well before the beginning of the conference. Manuscripts not meeting the technical standard of the series will have to be returned for improvement.

For further information please contact Springer-Verlag, Physics Editorial Department II, Tiergartenstrasse 17, D-69121 Heidelberg, FRG

K. K. Bardhan B. K. Chakrabarti
A. Hansen (Eds.)

Non-Linearity and Breakdown in Soft Condensed Matter

Proceedings of a Workshop Held at Calcutta, India
1-9 December 1993

Springer-Verlag Berlin Heidelberg GmbH

Editors

Kamal K. Bardhan
Bikas K. Chakrabarti
Saha Institute of Nuclear Physics
1/AF Bidhannagar
Calcutta 700064, India

Alex Hansen
Theoretical Physics Group
Norwegian Institute of Technology
N-7034 Trondheim, Norway

ISBN 978-3-662-13958-5 ISBN 978-3-540-49037-1 (eBook)
DOI 10.1007/978-3-540-49037-1

CIP data applied for

Originally published by Springer-Verlag Berlin Heidelberg New York in 1994
Softcover reprint of the hardcover 1st edition 1994

Typesetting: Camera ready by author/editor
SPIN: 10127066 55/3140-543210 - Printed on acid-free paper

Workshop on Nonlinearity and Breakdown in Soft Condensed Matter

December 1–9, 1993

SAHA INSTITUTE OF NUCLEAR PHYSICS
1/AF, Bidhannagar, Calcutta 700064, India

Sponsored by:

INDO-FRENCH CENTRE FOR THE PROMOTION OF ADVANCED RESEARCH / CENTRE FRANCO-INDIEN POUR LA PROMOTION DE LA RECHERCHE AVANCEE

Advisory Committee

E. Brezin (Paris)
B. K. Chakrabarti (Calcutta)
B. A. Dasannacharya (Bombay)
E. S. Raja Gopal (Delhi)
E. Guyon (Paris)
S. S. Jha (Bombay)
C. K. Majumdar (Calcutta)
B. Sinha (Calcutta)
D. Stauffer (Cologne)

Local Organising Committee

K. K. Bardhan
S. Basak
D. Basu
B. K. Chakrabarti
S. N. Karmakar
R. Ranganathan
A. K. Sen
B. Sinha

Coordinators

K. K. Bardhan (India)
A. Hansen (France)

Preface

There have been considerable advances in recent times in understanding many common material processes such as nonlinear response, fracture, breakdown, earthquakes, packing, granular flow. etc., that are of immense practical importance. This has been mainly due to new applications of statistical physics, including percolation theory, fractal concepts, and self-organized criticality. It was felt that a meeting in the form of a workshop would be very useful to bring together people engaged in research in these closely allied fields.

The initial response was quite encouraging, especially from France, which has a relatively large number of scientists working in these areas. We found in the Indo-French Centre for the Promotion of Advanced Research / Centre Franco-Indien Pour la Promotion de la Recherche Avancée a willing sponsor to make the meeting a reality. In view of growing interest in these areas in India, and at the Saha Institute of Nuclear Physics in particular, the later was chosen to be the venue for the workshop.

The number of participants including students was limited to about 40. The country-wise breakup of participants is as follows: 9 from France, 4 from USA, 1 each from Hong Kong, Israel and Norway, and the rest from India. It is our pleasure gratefully to acknowledge assistance and encouragement from many people, especially E. Brezin, E. Guyon, S. S. Jha, B. Sinha, D. Stauffer, and J. Zittartz. We also thank E. Guyon for his generous offer to write an introduction to these proceedings. We believe that they have benefitted considerably from contributions by well known experts in their respective fields. We thank all the contributors for their sincere cooperation in meeting deadlines for publication of the volume. Finally we wish to sincerely thank Dr. P. G. S. Mony, the Director of the Indo-French Centre for the promptness and understanding that contributed in no small measure to smooth running of the workshop.

Calcutta and Trondheim, September 1994

K. K. Bardhan
B.K.Chakrabarti
Alex Hansen

Contents

List of Participants

M. Acharya	Calcutta	P. M. Hui	Hong Kong
A. A. Ali	Bombay	A. Jagota	Wilmington
G. Ananthakrishna	Bangalore	A. KarGupta	Calcutta
K. K. Bardhan	Calcutta	S. N. Karmakar	Calcutta
M. Barma	Bombay	S. Kumar	Agartala
S. Basak	Calcutta	P. L. Leath	Piscataway
R. Basu	Calcutta	K. J. Måløy	Oslo
M. Bekele	Bangalore	C. H. Mehta	Bombay
L. Benguigui	Haifa	A. Mookerjee	Calcutta
S. M. Bhattacharjee	Bhubaneswar	S. Mukherji	Bhubaneswar
D. Bideau	Rennes	L. Petitjean	Ridgefield
I. Bose	Calcutta	R. Rai	New Delhi
B. K. Chakrabarti	Calcutta	H. Ramachandran	Bangalore
R. Chakrabarty	Jhargram	R. Ranganathan	Calcutta
N. Chatterjee	Pune	P. Ray	Madras
D. Chowdhury	Kanpur	S. Roux	Paris
S. Douady	Paris	D. Salin	Paris
D. Dhar	Bombay	S. B. Santra	Calcutta
J. Duran	Paris	B. K. Sarma	Milwaukee
M. H. Engineer	Calcutta	A. K. Sen	Calcutta
A. Hansen	Rennes/Trondheim	P. N. Sen	Ridgefield
M. K. Hari	Bombay	H. Van Damme	Orléans
H. J. Herrmann	Jülich	J.-P. Vilotte	Paris

Introduction

Etienne Guyon

Ecole normale supérieure, 45 rue d'Ulm, F–75230 Paris Cedex 05, France

Linear response theory and perturbation treatments have long been among the main tools used by theoricians in particular in condensed matter and applied statistical physics. They have been extremely useful in dealing with quantum physics aspects and transport properties in particular. (Indeed the Schrödinger equation as well as classical treatments in the thermodynamics of irreversible processes close to equillibrium are linear.) Non-linear problems as dealt by other scientific and engineering communities for a long time (such as non-linear waves, fracture, turbulence...) were rediscovered by physicists more recently: The Kondo effect showed that a lowest order perturbation treatment could not predict the low temperature conductivity of dilute magnetic alloys. Josephson effect was an opportunity for them to rediscover solitons. Experiments following Ruelle and Takens discovery of the first scenario to chaos aroused their curiosity to study turbulence.

At the same time, the description of physics problems became more topological and geometrical. Many of us had been trained to work in the reciprocal space and to deal with statistical ensembles. (A counter example is the activity — in France around Jacques Friedel — involving solid state physicists and metallurgists and making use of the gometrical description of singularities (dislocations)). But microscopists were not as influential as scientists dealing with scattering techniques in particular when working with disorder geometries. Fractals providing a framework for the geometrical study of disorder at various scales encouraged experimentalists and numericians to look more into the "real world." Consequently they became more concerned to explore such disordered systems as dealt with by engineers of various fields and to study the interplay between geometrical effects and static as well as dynamic behavior.

The disorder taking place at a mesoscopic scale where quantum effects were not directly visible, and there was a strong encouragement for model experiments for the last 30 years. The theory of liquids benefited from the numerical work of Alder and coworkers as well as from simple hard ball packing experiments by Bernal and his school. The EM response of composites, cermet films, was connected with the recently development of percolation theory through experiments dealing with random mixtures of conducting and insulating balls. Mesoscopic examples were found of quantum disorder (localization) in such experminets as backscattering of optical waves in a heterogeneous medium of surface waves on a rough bottom; Chains of coupled pendulums with kinks mimicking the sine Gordon solution made this concept more explicit...

Without any doubt, these simple-minded projects have had a strong impact in getting physicists more concerned with mesoscopic physics using non-linear approaches when needed — and making full use of geometric characterization.

This book is here to show us that things have changed recently. It is first representative of a world-wide activity rather than a confidential one I was belonging to twenty years ago. Alex Hansen, one of the editors, has worked with American, Norwegian and French physicists on these problems. The Indian-French meeting to which these proceedings refers has associated physicists from various countries. This is clearly a first indication of the mature state of the field covered.

This field deals precisely with a number of problems in material science involving non-linear problems and consideration of random geometries at a mesoscopic scale. The various examples show indeed that the presence of disorder, albeit small, in such systems coupled with non-linearities at a local scale influence strongly the macroscopic properties.

One class room example is the mechanical response of a set of equal radius solid spheres tighly packed through applied stress. Despite the apparently homogeneous look of the assembly, the fine distribution of contacts of nearby separations control the mechanical stress deformation response. This is due to on (pressed) — off (pulled apart) non-linear law of an individual contact. The disorder in the distribution of the centers of the spheres is secondary, and lattice treatments reproduce the characteristics. Multiple scale, and non-local effects, as also stressed here in the contribution by J.P. Bouchaud can be visualized by photoelastic experiments. Clearly an attempt to get the global behavior from the local response using homogeneization techniques or local averaging are due to fail. Other examples are proposed in the book. They make use of arrays of non-linear conductors, LED, fuses or Josephson junctions. The general message is similar as for the first example above: We are dealing with complex systems.

Another extreme where local disorder and non-linearity influence the global behavior would be that of a simple crack in an otherwise homogeneous piece of material. Various statistical treatments since Weibull pioneering work have dealt with problems which are governed by extremes rather than by means. Several examples are discussed here such as propagation of cracks or (linear and non-linear) fluid invasion in a solid medium. Both the frozen-in disorder as well as the result of the non-linear amplification of noise can influence the pattern as well as its time dependent evolution. It is quite remarkable, but not yet understood why such crack patterns appear to display universal characteristics (see the discussion by S. Roux on self-affine cracks).

A large part of the book deals with problems where randomness in the time evolution is as crucial as the geometry itself. The archetype here is the sand pile model with the conjecture of Per Bak and coworkers that the time evolution of the relation of a sand pile continuously displaced from equillibrium would be a power law. This conjecture — properly generalized — is known as self organized criticality (SOC). The critical state (here critical cone angle) would be attractive, rather than repulsive as in the usual case in critical phenomenons. The self organized criticality concept so introduced triggered the interested to granular flow problem even if SOC does not apply well to this particular example.

Another "trigger" came from experiments of regular patterns observed in vibrating vertically a plate on which grains were laid. It was found afterwards that this "modern time" experiment by Evesque and Duran in Paris had already been made long ago by Faraday. This clearly does not limit the merits of the recent project which opened the recent interest in the problem.

Another problem of greatest interest, and considered here, is the stick-slip behavior of solids sliding on eachother. The noise pattern, here again, can be influenced by the local disorder in the contact distribution as well as by the chaotic aspect of the problem. The problem had first been proposed by Knopoff and Burridge as a model for seismicity and it remains very active in geophysics with attempts to connect to real problems (stress relaxation along geological faults). In this example as well as in other ones dealing with continuous mechanics, the choice of the grid size in simulations needs to be considered seriously and is generally connected with the search of natural length scales for the dynamical problem.

If the meeting involved essentially physicists rather than scientists of the various implied communities, the present action and discussions clearly show that such contacts and mutual awareness are active today. This is clearly very true with geophysicists and is becoming the case with material scientists and mechanicians. Many meetings deal with such gatherings but it was also good that physicists dealing with various problems of applied statistical physics gathered together to confront their methods.

To conclude, I would like to pledge for a particular approach widely found throughout the chapters here and which is often not taken seriously by engineering communities. It is use of "toy" and "table top" experiments connected with simple numerical simulations. An adequate project of this type implies that you guess the most crucial or pertinent parameters and that you develop adequate and as rigorous as possible analogies (as represented by dimensionless ratios). This is quite the opposite of engineering projects which try to mimic systems as closely as possible using many variables.

Only if the relevant questions dealing with correspondences are asked will these restricted projects be useful. Such questions can be what a two-dimensional model tells us on the three-dimensional real case. Is the use of a lattice model a correct description, in particular if one introduces the disorder only in the individual elements? What is the sensitivity to a particular choie of local non-linear laws? The response clearly depends on the class of problems considered.

A first objective of such simple projects is clearly to reproduce what is known otherwise. It is already useful to display the basic mechanisms and crucial parameters of a problem. It can also be of interest for teaching purposes. Interactive science museums which are well developed in India and France in particular, are making good use of such effects (*e.g.* size segregation or patterns in vibrated grains). But clearly we want to have more predictions (for example in the size segregation problem, just mentioned above, what is the role of size walls in the process?).

If we want to get beyond such toy projects — and we should — I recommend that we do not limit ourselves to the (fruitful) exchange with engineering communities. Just as us physicists, that have had to interpret the world to model it and to learn from it, research engineers also model the real world and reduce it as us. It is merely a matter of "point de vue." The science of mesoscopic systems discussed in this book appears to be one in which there is a considerable past and present technical history and know-how, but still containing usolved problems. Only through direct contacts will we be able to face the real world and not just propose our solutions without knowing the problems to which they can apply before visiting a cement plant or a mine to see what fragmentation is and what it costs, talking to field engineers in oil industry or seismologists... There is, however, clearly a balance to be kept if we want to be reasonable efficient in our basic research beyond our own model systems. There is a great opportunity and challenge to the scientists today to explore the various scales and models from the real world to the basic local laws on such problems as presented in this book.

Decompaction, Fluidisation and Segregation in a 2D Sandpile

J.Duran, T.Mazozi, J.Rajchenbach and E.Clément

Acoustique et Optique de la Matière Condensée- URA800 CNRS
Université Pierre et Marie Curie - B86
4 Place Jussieu, Paris 75252 Cedex 05

Abstract : Depending on the micromechanical properties of its components and submitted to a vibrational excitation, a 2D set of spherical particles can exhibit the whole range of behaviour of common dry particulates such as sand : Spontaneous heaping, size segregation , convection, surface fluidisation etc. Modelling sand with a 2D set of particles has the crucial advantage of allowing a direct observation of the interplay of these processes, as well as allowing theoretical investigation and making up privileged benchmarks for computer simulations. Based on our most recent experiments, this paper is devoted to the recognition of the pertinent parameters which govern the physics of these materials. As an example, we discuss the processes governing vibrational size segregation and we set-up and illustrate the basic equations which rule the progressive decompaction mechanism responsible for bulk motions. In particular we show how the interplay of both the friction at the boundaries, the aspect ratio of the container and the ability of the granulate to convert vertical into horizontal stresses, combine into a *single a-dimensional parameter* χ which governs the decompaction process.

1 Introduction:

Dry Granular materials are ubiquitous in nature as well as in many industrial processes. Due their specific behaviour and their great industrial concern they have been the goal of a lot of theoretical, experimental and computer simulation efforts[1 and ref. herein]. Nevertheless, most of their static and (or) dynamic properties such as flow, segregation and arching, are still under question. Real systems involving 3D polydisperse mixtures of dry particles of any shapes are reluctant to detailed experimental as well as simulation work . We have chosen to tackle this problem starting from a different viewpoint: First, attempting to reduce the dimensionality and the polydispersity of the particulate, then checking that the distinctive properties are preserved and finally taking profit of the possibility to "see through" well identified granular samples in order to investigate several characteristic behaviour via direct observation, image processing and measurements. We do not claim that the physics of dry granulates can be wholly investigated in reduced geometry's. As usually, we expect that the dimensionality of these systems plays a role in the detailed dynamics of the processes (scaling exponents), however we observe that, since well known distinctive properties are preserved, our model systems do contain in essence the motor of the basic properties of granulates.

2 Definition of granulate. 2D model systems

Recently published experimental investigations dealing with dry granulates have occasionally displayed apparently divergent results [e.g. 2,3 and 4,5]. It turns out that, under careful examination of the problem, most of the divergences can be related to different definitions of the granular materials themselves or to different experimental conditions, so that there is a need for reminding the reader of a few definitions and orders of magnitude involved in the dynamics of these materials. In the following we will ensure that the supposedly dry and non cohesive granulate will not interact with the surrounding gas or fluid via viscous dissipation, drag forces or air compression so that we restrict our investigation to relatively large (from about 100μ up to 3mm) and heavy particles in gases, so called *granular solids* in the sense defined by Brown and Richards [6]. Let us first remark that, under typical excitation (flow, vibrational excitation, jostling, shearing etc.) the velocity of a solid particle in a granulate is typically a few cm per second . Considering a typical weight of the order of a mg, we find that the typical kinetic energy is of the order of 10^{15} °K so that, unless otherwise defined [1], the usual temperature and brownian motion are not pertinent concepts. Dealing with dry relatively large and heavy particles in light gas (such as air) we can perform a straightforward order of magnitude calculation which shows that the kinetic energy largely prevails over the various causes of dissipation related to the fluid environnement. Dissipation solely originates in the particle-particle or particle-walls interaction via shock restitution coefficient $k < 1$ or (and) friction coefficient $f > 0$, here considered in the Coulomb's sense. Keeping within this framework and excluding all the spurious effects whicht might originate in the interaction of the particle with the fluid environnement, we searched for a significant reduction in the degree of complexity of usual polydisperse 3D particulates such as sand, grain assemblies etc. but retaining the presently known distinctive features of these granular materials as quoted in table 1.

Table 1. References reporting observation of some granular solid distinctive features

Distinctive properties of granulates	P	3D (sand or glass beads)	2D (beads (soft, rugose)
Convection (initiated at boundaries)	1	[7]	[8]
Heaping, Surface instability	2	[2,3]	[8]
Vibrational size segregation	3	[4,5,9]	[This work,10]
Progressive decompaction	4	[11]	[This work ,12]
Hour glass (constant flow)	5	[13]	[14]
Arching effect	6	[6]	[This work,10]

We reported in previous recent works [8,10] that, under specific conditions regarding micromechanical properties, 2D monodisperse assemblies of metallic spherical beads display different collective behaviour. Roughly stated, assemblies of highly elastic ($k\approx0.9$) and smooth ($f\approx0.1$) beads behave like poorly dissipative media and generally undergo ballistic motions at upper surfaces and therefore do not reproduce the behaviour of the common granular solids such as sand. On the contrary, corrugated aluminium beads which show up micromechanical properties (f ≈ 0.8 ; $k \approx 0.5$) as well as densities ($\rho \approx 2$) similar to silicate grains, have been found to reproduce in 2D most of the distinctive properties (labelled 1 to 6 in table 1) which have been observed in polydisperse common 3D granulates. Heaping which is the ability of a grain assemblies to exhibit an inclined upper surface when submitted to a geometrically uniform vertical agitation has been shown to originate in the convection initiated at the rubbing walls both in 3D and 2D, so that properties 1 and 2 are strongly correlated and show up only when an efficient friction interaction at the walls of the container exists. Property 3 stands for the ability for a shaken, sheared or jostled assembly of granular solids, to segregate its components depending on their relative sizes. When submitted to such an excitation, the larger particles generally tend to reach the outer limits of the particulate both in 3D and in 2D .Although the influence of the relative densities of the particles are not yet clarified, it turns out that the usual concept of mixing via stirring as in liquids is counter-productive in granular solids. Property 4 which refers to progressive fluidisation or, rather, decompaction (which stands for the ability of the particles to move with respect to each other, in the sense of the Reynolds dilatancy [15]), depicts the height dependence of this properties in a vertically vibrated container .It has been previously observed in 3D polydisperse sand [11]. It is a purpose of the present work to report observation and to model this property in the more tractable 2D geometry. Property 5 is mentioned here as a complement. It is meaningful that the well known property of a constant flow in a common 3D hour glass has been recovered in 2D. $a \sin\omega t$ Arching effect (Property 6) which stands for the ability of granulates to build up quasi-solid (i.e. of low mobility) chains of contacts simulating a sort of solid backbone in the bulk which supports the above lying granulate, have been both observed in 3D and 2D even if they are more likely to exist in reduced 2D than in 3D. In the following we will focus on the observation and modelling of properties 3 and 4 in 2D model systems designed according to the preceding considerations.

3 Vibrational decompaction of a 2D "sand-pile"

3.1 Experiments

Basically, the experiments consist in shaking vertically a flat container filled with a single 2D layer of beads at a reduced acceleration $\Gamma(>1)$ defined as $\Gamma = a\omega^2 / g$ where g is the gravitational acceleration and $a \sin\omega t$ the amplitude of the vertical motion of the container. The experimental set up as well as the image processing technique is similar to the one we used in [8,10]. The carefully controlled driving mechanism and the cell containing a single layer of about 3000 metallic beads have been described in our previous work. We also took profit of the computer posed photograph procedure (CPP) which allows to get a direct insight in the relative movements of the components of the packing. As an illustration, Fig.1 shows such an image obtained after several minutes of shaking at 15 Hz a regular packing of oxidised aluminium beads. It can be observed that a definite portion of the array is

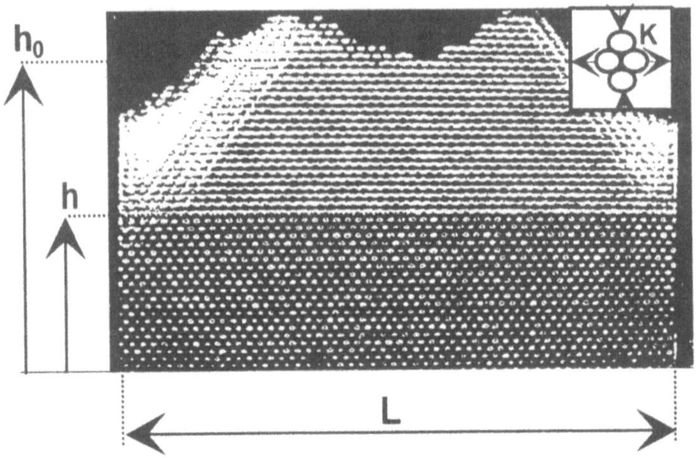

Fig.1 Experimental: Computer processed photograph (CPP) of a 2D array of aluminium beads shaken at $\Gamma = 1.4$. Insert: definition of factor K.

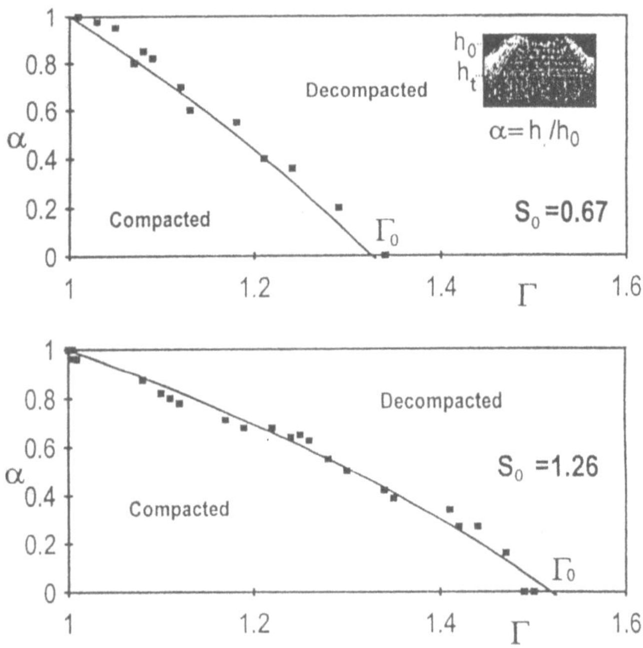

Fig.2 Compacted-decompacted phase diagram of the vibrated pile at two different aspect ratios $S_0 = h_0 / L$. Square dots are experimental while full lines have been calculated from our model using Kf=0.29.

allowed to move significantly via long range horizontal or 60° slip lines during a long lasting shaking process.

The important point here is that, starting from a rectangular perfect stacking, it is observed that the height of the decompacted portion is a constant observable, at a given acceleration. More specifically, we observed throughout series of experiments that the height of the decompacted portion $(h_0 - h)$ increases when the excitation amplitude is increased. Also and as expected, when the global decompaction of the pile is obtained at a reduced acceleration Γ_0, it is observed that the lower line of beads splits apart from the bottom of the container at every period of the sinusoidal excitation. We report in Fig. 2 experimental data plotted as $\alpha = h / h_0$ as function of Γ and obtained at two different aspect ratios of the cell.
As will be made clear in the following, it is seen that the larger the aspect ratio or the deeper the cell is, the larger acceleration is needed to allow bulk movements in the lower part of the pile.

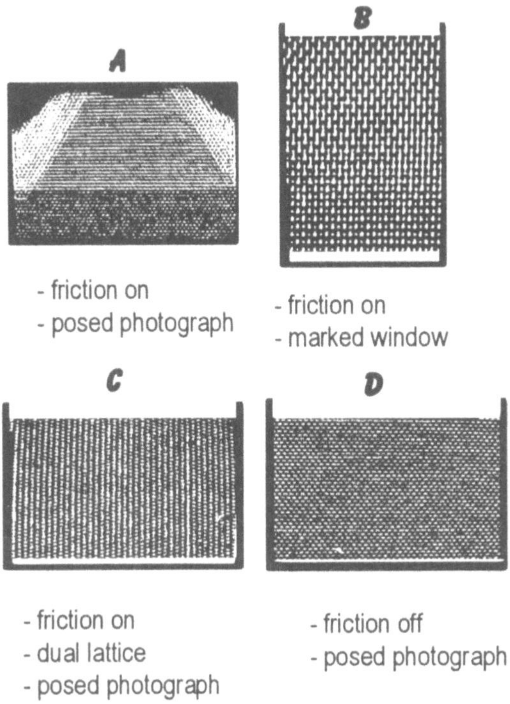

Fig. 3 Sketch of additional experiments discussed in the text

These experiments were performed in order to highlight the pertinent parameters which govern the decompaction mechanism. Fig. 3B has been obtained by letting under vibration, for a two days period, a cell ($S_0 = 1$ and $\Gamma = 2$) containing corrugated aluminium beads and exhibiting friction at the lateral boundaries. It turns out that, at the end of the experiment, the frontal windows of the glass cell have been marked by the moving beads thereby displaying a trace of the up and down motions of the successive rows of the packing. As can be observed directly on a part of the negative image reproduced in Fig.3B, the amplitude of the motion of the beads increases with altitude in the referential frame of the container. Although clearly showing the trend, this striking and simple experiment unfortunately leads to rather imprecise results. It has been reproduced in a more quantitative manner using a miniaturised CCD camera attached to the vibrating piston which supports the container. Doing this, and again working in the mobile referential frame, we observed that the distances between the beads were rapidly fluctuating in time. Maximum value of these fast and fluctuating displacements observed over a large number of oscillations are reported in Fig.4 as function of height in the cell .

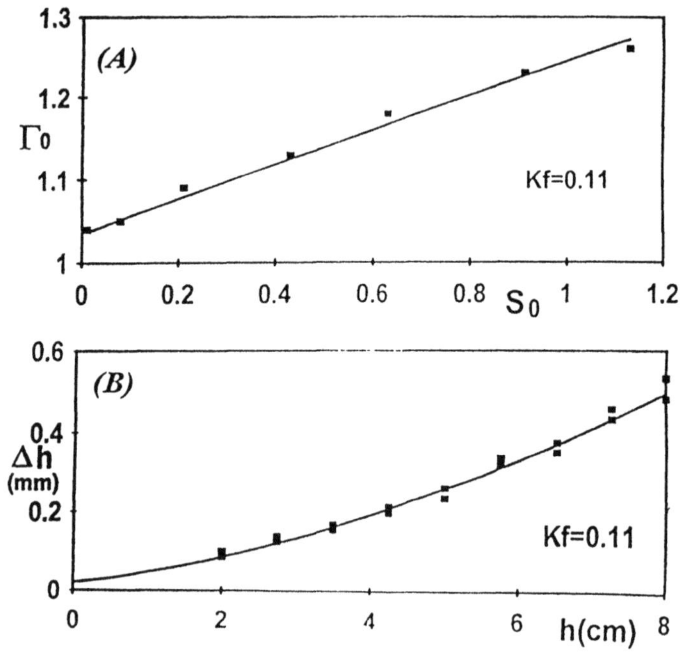

Fig. 4A Takeoff accelerations Γ_0 as function of aspect ratio . Square dots are experimental while the full line is calculated from our model (here Kf=0.11).

B Maximum elongation of the up motions of the beads as function of height in the same cell as in A. Black square dots are experimental data while the full line has been calculated from the model using again Kf=0.11.

In order to get a more global observation of these fluctuations, we recorded this fundamental feature by taking snapshots of the pile while it was vibrated and lit through by a synchronous stroboscopic light. We observed that, during short fractions of a period, the stacking undergoes fast and reversible tearing which we show and discuss elsewhere [12]. An important question arises here which concerns the various time scales of the dynamics of the 2D sand pile as it is observed through these features: It is essential to remark that computer posed photographs as in Fig.1 exhibit essentially the irreversible and slow motions of the beads in the container.

Here the time scale is of the order of tens of periods of excitation which is the time necessary for large scale and irreversible displacements to occur. In reverse, the time scale for inducing small and fast reversible beads motions involved in fluctuating cracks, is of the order of tenths of a period. Briefly stated, it turns out that the dynamics of the vibrated pile involves fast reversible and fluctuating tearing which accidentally turn out to build up constructively into irreversible block slip motions as can be seen in Fig.1 and 3A. Not being currently in a position to analyse in details the statistics of the fluctuating tearings of the packing, we remark that our experiments show that these seemingly *erratic motions combine into a smooth dependence on the altitude* in the cell as reported in Fig. 4, where we plot the results of measurements performed over a large number of clichés of the differential motion between the cell and the rows of beads. Thus, Fig.3B and 4B display the evidence that the relative amplitude of the motion of the beads increase with their altitude in the cell.

Now and in order to discriminate between the different possible mechanisms which drive the decompaction process, we performed additional experiments sketched in Fig. 3C and 3D. First and concerning Fig.3D, we performed the same experiments as in Fig 3A except using fresh and polished aluminium beads. Within the same range of excitation parameters, namely Γ ranging between 1 and 2, and according to our preceding findings [8], we observed neither convection nor decompaction of the pile which consequently behaved as a fully compact block. Moreover and using a metallic deposit on the lateral walls we observed that suppressing the lateral wall-beads friction led to the same result even if we used corrugated aluminium beads. These observations clearly exhibit once more the decisive role of the lateral bead wall friction in triggering the decompaction process. This is also consistent with our previous experiments performed in a 2D cylinder where the suppression of boundaries did suppress the heaping process whereas the introduction of a stick between the cylindrical windows restored the convection . Still more informative is the result of the experiments reported in Fig.3C, where we arranged initially the packing as a dual lattice with the rows being placed vertically in the cell. Although these experiments were performed with corrugated aluminium beads and lateral walls exhibiting bead-wall friction, we did not observe any convection nor heaping nor decompaction. Again, the pile behaved as a block. In the spirit of Fig. 3B and as expected, it left marks on the frontal windows of the cell the length of which are independent of altitude . This shows that the geometry of the arrangement is essential in determining the decompaction process. We will take all these features into account for setting up the following simple phenomenological model.

3.2 Model for vibrational decompaction of the 2D sandpile

Working within the limits of small amplitude motions and dealing with long lasting features i.e. in a quasi static limit, we may consider a simple continuum description of the 2D sandpile [12] which basically contains the three essential ingredients which we have observed all along our experiments:

i)The decisive effect of the bead wall friction as well as the influence of the aspect ratio of the container
ii)The progressive decompaction of the pile when the excitation amplitude is increased.
iii)The key role played by the geometry of the stacking.

Starting from arguments first put forward by Janssen [16], we consider the granulate as being made up of elementary slabs of height dh piled on top of the others. The vertical component P of the stress tensor is supposed to give rise to an horizontal one KP, due to the imbricated structure of the granulate stacking. Considering the balance of forces acting on an elementary slab of material at height h, we set-up the following differential equation :

$$-L\rho_0 g dh - L\frac{\partial P}{\partial h}dh + 2KfPdh = 0 \quad (1)$$

where ρ_0 stands for the mean density of the granulate and f the Coulomb friction parameter at the lateral walls. Considering the free boundary property, we integrate this equation and get an effective stress distribution in our 2D pile as :

$$P(h) = \frac{\rho_0 g L}{2Kf}\left(1 - \exp\left(\frac{2Kf}{L}(h - h_0)\right)\right) \quad (2)$$

Now we turn to the dynamical problem where the stacking is submitted to an upward launching force due to the sinusoidal excitation of the container at $\Gamma > 1$. The friction force $F(h)$ acting at lateral walls, is height dependent and opposes to this launching force. It is understandable (and can be demonstrated [12]) that the resulting force corresponds to a net upward effective reduced acceleration Γ_R such as:

$$\Gamma_R = 1 + \frac{\partial F(h)}{\rho_0 g L \partial h} \quad (3)$$

Observing that successive superimposed slabs are only interacting via the weight forces and are supposed not to be bound together via, e.g. cohesion forces, we deduce that the condition $\Gamma_R > 1$ will determine a critical height above which the 2D sandpile will be decompacted and below which it will remain solid.
This may be depicted through a parameter α and plotted against Γ (Fig.2) as a sort of phase diagram according to :

$$\alpha = \frac{h}{h_0} = 1 + (2\chi)^{-1}\ln(2 - \Gamma) \quad \text{with} \quad 1 \le \Gamma \le 2 \quad (4)$$

Which shows that the decompaction process is ruled by a *single a-dimensional parameter* $\chi = S_0Kf$, which we call "*decompaction parameter*" in the following.
An experimental test of these equation is reported in Fig.2 where it can be seen that the theoretical curves fit well our experimental measurements at two different aspect ratios of the same cell filled up to different heights with the same beads. Also, we observe that zeroing the decompaction parameter χ should supress decompaction so that the pile should behave as a compact block. Experiments reported in Fig.3C where we modified the geometry of the piling and putting K=0 and Fig.3D where we supressed the friction at the boundaries (f=0), carry out a firm support to our model.
Note that an immediate consequence of Eq.4 is that there exists a limiting acceleration Γ_0 beyond which the *whole* pile will start moving:

$$\Gamma_0 = 2 - \exp(-2\chi) \tag{5}$$

It turns out that this peculiar acceleration can be measured quite accurately by observing the takeoff acceleration of the lowest row of the 2D array. As can be seen in Fig.4A, the theoretical line obtained from Eq.5 fits well the experimental results which have been obtained in the same cell (Kf=0.11) as the progressive decompaction experiments described at the beginning of this section.

Now we push a little further our model and check its implication in the analysis of the decompaction gradient in the decompacted phase as we have observed and measured in Fig.2B and 4B. Here it is essential to realise that we are working in the referential frame of the vibrated cell .Hence we have to analyse in some details the relative motions of the cell and of the rows of beads. Using the same considerations as in the preceding and labelling t^* the time at which the slab at height h is taking off in the container referential frame, we have:

$$\omega t^* = \Gamma^{-1}\left(2 - \exp 2\chi\left(\frac{h}{h_0} - 1\right)\right) \tag{6}$$

The measured maximum Δh_{max} values for Δh which is the elongation of the motion of the beads in the referential frame of the cell are obtained via a computer solution of the following two differential equations :

$$\Delta h = -\frac{1}{2}g^*\left(t - t^*\right)^2 + a\left(\sin\omega t + \omega\left(t - t^*\right)\cos\omega t^*\right) \tag{7}$$

$$g^*\left(t_{max} - t\right) = a\omega\left(\cos\omega t_{max} + \cos\omega t^*\right) \tag{8}$$

Again, and reporting the computed Δh_{max} values in Fig.4B we find a satisfactory agreement between our data and the theoretical curve using the *same Kf parameter* obtained in the quite different preceding experiments (takeoff accelerations) performed *in the same cell* (Fig.4A).

4. Direct observation of size segregation. Convection and arching

As emphasised in section 2, 2D geometry allows via image processing (Computer posed photograph) a direct observation of the relative motion of solid particles in a granulate and, in particular, to follow the motion of an intruding particle which can be chosen of a different size.

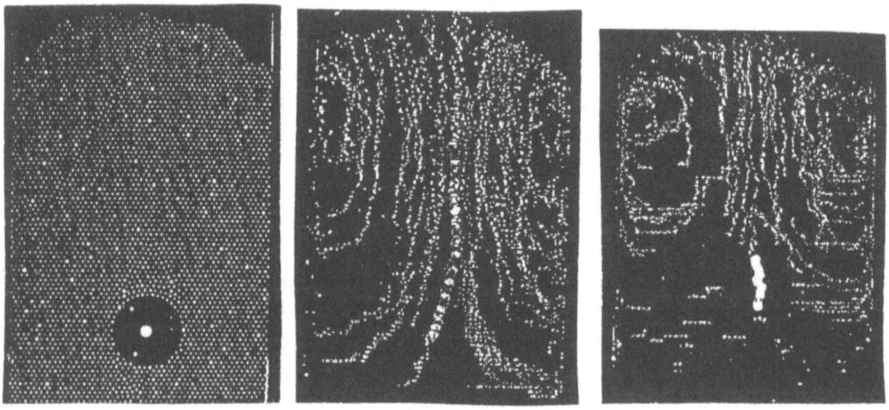

A B C

Fig.5 A) Single snapshot of the 2D container filled with aluminium beads, shiny markers and intruding disc at the beginning of the segregation experiment.

B) High amplitude excitation: (CPP). The convection driven ascent of the intruder is indicated by the large white trail in the middle of the two convective rolls. Here $\Phi = 5.3$, $\Gamma = 2$ at 15Hz.

C) Low amplitude excitation : (CPP during 10mn) The arching effect driven ascent of the intruder is clearly observed through the horizontal white trails left by the markers below the large disc. Here $\Phi = 12.9$, $\Gamma = 1.25$ at 15Hz.

Vibrational size segregation, often referred to as the "Brazil nut problem" has been recently the focus of several computer simulations as well as experimental observations [4,5,9,17] displaying apparently conflicting results. Briefly stated, the existence of a size threshold for size segregation, as well as the up motion velocity dependence on the size ratio ($\Phi = R / r$, R and r being respectively the radius of the intruder and of the particles in the granulate) seemed to be in conflict with existing experiments reported in literature or simulations. We have shown elsewhere [10] that under low amplitude of excitation which means slightly above the gravitational acceleration, there exists a size threshold Φ_C (12.9 in 2D and 2.8 in 3D) which determines the mode of ascent of the intruding disc in an assembly of smaller spheres. In short, small discs such as $\Phi < \Phi_C$ will (and do) climb up intermittently, their motion being discretized by the granularity of the particulate, whereas larger intruders ($\Phi > \Phi_C$) can move up continuously via permanent arching effects. Keeping within the limit of this paper we will restrict here to the experimental observation that vibrational segregation can occur via *two distinct processes (namely convection and arching effects*) depending on the excitation amplitude of the container

Fig.5A shows a typical 2D container built for observing size segregation. As in the preceding section, the cell is made of two rectangular glass plates (10x15 cm^2) separated by rectangular plastic wedges, the width is adapted to accommodate, as accurately and freely as possible, the monodisperse set of 1.5 mm-diameter aluminium beads. The beads we use are oxidised aluminium beads (no segregation nor convection was ever observed for a set of *polished* aluminium beads) which have high solid friction coefficient and high momentum loss on collision. The experiments have been made using an excitation frequency of 15 Hz. Again, the excitation amplitude is indicated via the reduced acceleration parameter Γ. At the beginning of the experiment, the intruding aluminium disc (1mm thick, radius R) is inserted among the assembly of small beads (spheres with a diameter 2r = 1.5mm) at 1.5 cm above the bottom of the container. The pile is prepared to match as much as possible a triangular lattice (except around the intruder which induces local perturbations in the array). In order to provide a quasi 2D free motion of the intruder between the two frontal windows of the cell, three equidistant steel 1.5 mm diameter beads are pinned in the outer ring of the circular disc. The intruder is painted in black with a white dot decorating its centre. Here we limit our description to the direct visualisation of the trajectories of the small beads and of the larger disc. Note that a recent extension of the present work is devoted to the experimental analysis of the dynamic of the ascent of the intruder among the particulate [10]. Prior to vibration, we dispersed test beads among the assembly of corrugated aluminium beads. These test beads are about 1% in number, they are made of steel with the same diameter 1.5mm, and look more shiny compared to the aluminium ones. The stroboscopic flashing light is locked on the vibration such that positions are recorded for the same height of the container. Using an image processing device, we selectively record the test beads and the intruder positions. As in the preceding section, accumulation of such pictures bring up computer posed photographs (CPP) as reported in Fig.4B and 4C where we observe that exciting the cell at an acceleration larger than the gravitational acceleration induces internal motions in the pile. We recall that in the absence of intruder, the convection rolls spring up at the upper lateral corners of the granulate and when the excitation acceleration Γ is increased they expand in size towards the centre of the cell [8]. Compared to our earlier observations, the introduction of a larger disc among an assembly of monodisperse beads produces several new and different characteristic features. We observe that, depending on the excitation amplitude, two different regimes are clearly distinguished:

i) At relatively large accelerations , i.e. when $\Gamma \geq 1.5$, and in agreement with recently reported experiments [7] and simulations [19], we observe a global convective motions of all the particles in the cell. The CPP of such an experiment at $\Phi = 9$, is displayed on Fig 5B. Note that the trajectories of the small beads evidence convection rolls taking place *below* the intruding disc and carrying the bigger particle at the same speed. In agreement with [7], we observed that the probability of being convected downwards at one of the lateral walls, when the surface is reached, strongly depends on the diameter ratio.

ii) Fig.5C is the CPP of an experiment performed at a lower acceleration $\Gamma = 1.25$ and for a diameter ratio $\Phi = 13$. Without intruder, the convection rolls would be confined in the upper left and right corners of the cell. Here, two important features should be noticed. First, the extension of the convection rolls is much larger than without intruder. The roll localisation starts just *above* the intruder and the roll size is progressively reduced when the bigger particle rises. A second characteristic feature is that the upward ascent of the intruder is accompanied by several *horizonta*l block re-arrangements which are directly visualised via the horizontal tracks of the markers. Remark that in Fig.2C, the intruder is not carried upward in a convective motion of the surrounding particles, *since no upwards trail is present in the underlying environment.* This experiment was repeated at successive stages of the intruder ascent. It shows that once the intruder has left a lower region, no internal motion is any more evidenced. Thus, it is clear that segregation *can occur without a convection motor* and that an intruding particle creates in its surrounding, its own field of internal motion. Additional experiments reported in [10, 12] provide a rather clear observation of both on the progressive decompaction process as well as on the size segregation process.

5. Conclusion

Our experimental observations lead us to the fundamental question of the various time scales involved in the dynamics of vibrated sandpiles :

- Short time and small range fluctuations in the relative positions of particles turn out to be the leading motor of both the decompaction as well as of the size segregation processes

-Long time and long range block motions involved in the dynamics of sandpile such as the progressive ascent of size segregated particles, progressive decompaction or decompaction gradient, turn out to be tractable in terms of a continuous medium theory at least in the moderate excitation regime.

In other words, and this is a crucial problem to be addressed now, the question arises to understand *why and how*, the complex micro dynamics of positions fluctuations in the granulate, combine into a smooth continuum-like dependence of the observable quantities which we measured all along the present work.

References

1. H.M.Jaeger ad S.R.Nagel, Sciences **255**, 523, 1992.
2. P.Evesque and J.Rajchenbach, Phys.Rev.Lett. **62**, 44, 1989.
3. S.Douady and S.Fauve, J.Phys.(Paris), **50**, 699, 1989.
4. G.C.Barker,A.H.Mehta and M.J.Grimson, Phys.Rev.Lett. **70**, 2194, 1993.
5. R.Jullien, P.Meakin and A.Pavlovitch, Phys.Rev.Lett. **70**, 2195, 1993.
6. R.L.Brown and J.C.Richards in Principles of Powder Mechanics, Pergamon Press, 1970.
7. J.B.Knight, H.M.Jaeger, Phys.Rev.Lett. **70**, 3728, 1993.
8. E.Clement, J.Duran and J.Rajchenbach, Phys.Rev.Lett. **69**, 1189, 1992
9. K.Ahmad and J.J.Smalley Powder Tech., **8**, 69, 1973.

10. J.Duran, J.Rajchenbach and E.Clement, Phys.Rev.Lett. **70**, 2431, 1993
J.Duran, T.Mazozi, E.Clement and J.Rajchenbach, To be published, 1994.
11. P.Evesque, E.Szmatula and J.P.Denis, Europhys.Lett. **12**, 623, 1990
12. J.Duran, T.Mazozi, J.Rajchenbach and E.Clement, To be published.
13. R.L.Brown and J.C.Richards, Trans. Instn. Engrs. **37**, 59, 1959.
14. G.Ristow and H.J. Hermann, To be published in J.Phys. (Paris)
15. O.Reynolds, Phil.Mag, Ser.5, **20**, 465, 1885
16. H.A.Janssen, Ziets.d Vereins Deutsch Ing. **39**, 1405, 1895
17. A.Rosato, F.Prinz, K.J.Standburg and R.H.Swendsen, Powder Technol **49**,59,1986.;Phys.Rev.Lett. **58**, 1038, 1987.
18. J.A.C.Gallas, H.J.Hermann and S.Sokolowski, Phys.Rev.Lett **69**, 1371,1992. and Physica A, **189**, 437, 1992.

Granular flow: some experimental results

Daniel Bideau[1], F.X. Riguidel[1], A. Hansen[1,2], G. Ristow[1], X.l. Wu[3], K.J. Måløy[4] and M. Ammi[1]

[1] Groupe Matière Condensée et Matériaux (URA 804), Université de Rennes1, F-35042 Rennes Cedex, France
[2] Institutt for Fysikk, NTH, N-7034 Trondheim , Norway
[3] Department of Physics and Astronomy, University of Pittsburgh, Pittsburgh, Pensylvania 15260, USA
[4] Fysisk Institutt, Universitetet i Oslo, N-0316 Oslo 3, Norway

Abstract:

The relative influence the friction forces and interstitial air is analyzed under two experimental conditions. First, the motion of one ball on a rough surface built by gluing a monolayer of spheres on an inclined plane is studied as function of the inclination angle θ and the ratio Φ of the diameter of the rolling ball to glued spheres diameter. An analysis of the friction force is proposed, and an expression is given for the "mean path.length" of one grain moving on this surface. The second part of the paper enlights the strong influence of air in the working of hourglasses when the size of the grains (and thus the permeability of the sand) becomes small. The question of stability of arches appears to be a very central one.

1- Introduction

In the middle of this century, Bagnold[1] worked out a classification of granular flow in a fluid: according to the (increasing) concentration of grains, he distinguished three different regimes, called *macro-viscous*, *transitional* and *grain-inertia.* In macroviscous regime, viscosity is dominant, whereas in gain-inertia one, the (interstitial) fluid has no influence, and the main effects are due to grain-grain interaction. In the case of dry granular flow, this description seems not to be pertinent, and the problem is often to quantify the relative importance of friction and collisions between particles. In *rapid* granular flow, the collisions are dominant, and kinetic theories of granular flow have been developed.[2] If one separates the velocity of a particle in a mean and a fluctuating component, the fluctuating component can lead to a definition of the controversial concept of *granular temperature*,[3] similar to the kinetic temperature introduced at the molecular level in classical kinetic theory. In fact, this definition of granular temperature suffers from the lack of equipartition.[4]
At the opposite side, slow granular flow is more complex. Since the work of Coulomb[5] on solid friction, much work has been devoted to the question of the *angle of repose* θ_r of a sand pile, which is the angle between the -- stable -- surface of the

pile built dynamically and the horizontal: clearly, this angle does not depend only on the solid friction coefficient introduced by Coulomb, but also on the *geometry* and the *density* of the system. A generalization of the -dynamical- friction has been proposed recently by Jaeger et al,[6] to explain the angle hysteresis observed in avalanches. We present in the first part of this paper an experimental study of the motion of one particle on a rough surface in this context, where the influence of the roughness is expected to be important. But also in this case the statistics of the motion of one particle according to its size is central in the context of segregation. This phenomenon is currently observed in granular flow, and especially in surface flow:[7] The roughness of the surface "seen by the grain" depends on its size compared to that of the grains in the pile: the larger is the grain, the longer is its mean path length along the slope when moving.

Returning to Bagnold's work, in the grain-inertia regime, the influence of air flow is generally neglected in the context of dry granular flow. But there is some evidence of a behaviour dependent on the air, but different from the macroviscous regime, when the diameter d of the grain becomes small enough:[8] in this case, the permeability of the granular system (depending on d^2) becomes small enough and the pressure effects cannot be neglected. This interaction between fluid and flowing particles leads to a surprising effect: intermittency in hour glasses. The second part of this paper will be devoted to this topics, which we generalize in terms of stability of arches.

2- Influence of "friction": Motion of a ball on a rough surface

To modelise surface flow, we have chosen to study experimentally[9-11] the behaviour of a single ball of radius R rolling on a rough surface. To do this, we have built rigid glass plane P two meters long and one meter wide, forming an adjustable angle θ with the horizontal. Above this plane, we have placed a video camera moving on rails, and obtained VHS movies which can be analysed with an image analyser. By gluing homogeneously glass spheres of radius r on contact paper fixed to the plane, we can change its roughness which we characterize by the parameter Φ equal to the ratio R/r.

The obtained surface packing fraction $C \approx 0.7$ is sensibly smaller than generally measured in dense packings of disks ($C \approx 0.82$)[12] but leads to a roughness we expect to be comparable to that of the surface of a sandpile.

We have two control parameters θ and Φ, and a phase diagram can be drawn using these parameters, schematically drawn in figure 1. Three regimes are observed. In regime I, the motion is strongly decelerated. Regime II is characterized by a constant speed which is quickly reached; in this regime, the ball stops suddenly as it is trapped by a large “hole" formed by the space between the spheres forming the rough surface. This regime is very dependent on the disorder presented by the packing at the surface. In regime III, the scale of the motion is completely changed, as the particle jumps on the plane. We have concentrated on regime II.

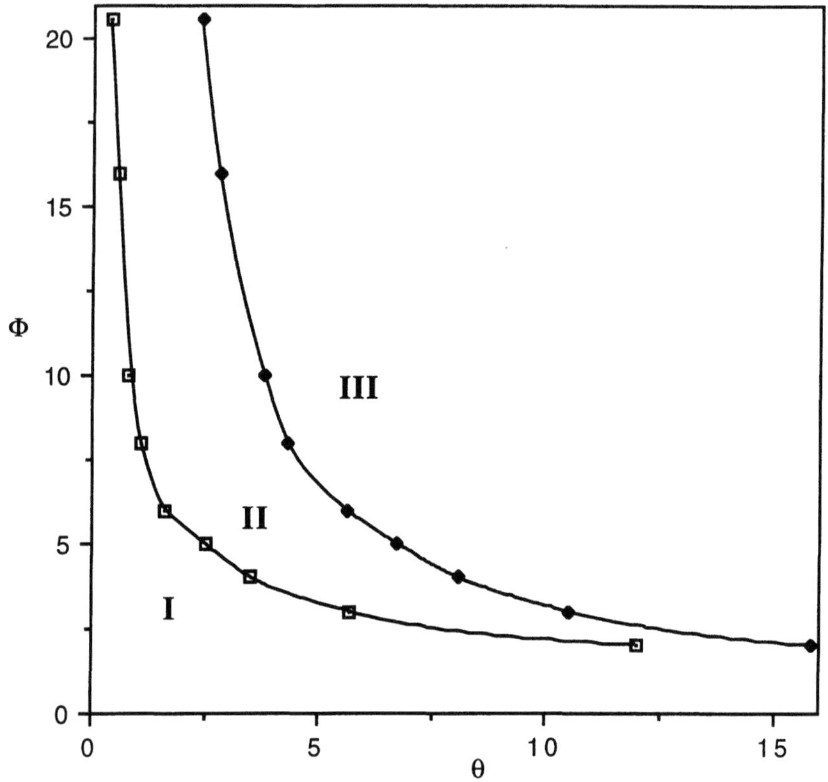

Fig. 1 Phase diagram giving the different regimes of motion of a particle moving down a rough inclined plane. The control parameters are roughnessΦ and inclination θ.

The motion of the ball results from a balance between its energy gain in the gravity field, and energy losses by collisions with the spheres forming the rough plane, and by friction on the plane.[13] Neglecting friction, we can model this motion by a Langevin-type equation[10]

$$m_* \ddot{l} + \eta(\dot{l}, \dot{\vec{x}})(\dot{l}/r) = m\, g \sin\theta \cos(\sigma(\dot{l},\dot{\vec{x}})), \qquad (1)$$

where $l(t)$ is the length of the path, $x(t)$ gives the position of the point, m_* is an effective dynamical mass which takes into account the rolling motion, m is the mass of the ball and g the gravitational constant. $\eta(l,x)$ is a fluctuating term which comes from the momentum loss in direction of motion by collisions with the spheres on the plane; it is a function of the position because the surface is disordered. Finally, $\sigma(l,x)$ is a rapidly fluctuating function reflecting the local change in direction of the motion of the rolling sphere. Starting from this equation, in which we assume that σ is small so that $\cos\sigma \approx 1$, we rewrite equation (1) in the following way:

$$m_*\ddot{\delta} + \langle\eta\rangle\frac{\dot{\delta}}{r} + \frac{\Delta\eta.v}{r} + \langle\eta\rangle\frac{v}{r} = m\ g\ \sin\theta, \tag{2}$$

where $\langle l\rangle = v$, $\dot{l} = v + \dot{\delta}$ with $\langle\dot{\delta}\rangle = 0$, and $\eta = \langle\eta\rangle + \Delta\eta$.
Averaging this equation and assuming that $\langle\eta\rangle = m\langle\gamma\rangle\Phi^a v^b$ we find relation

$$v\Phi^a = \frac{rg}{\langle\gamma\rangle v^b}\sin\theta \tag{3}$$

where $\gamma(t)$ is a random function with non zero average. This relation has been tested experimentally. When r is constant and R varies, we find a very good collapse with a = - 3\2 and b = 0, as shown in figure 2. But the agreement is not so good for the exponent corresponding to the variable r, for which the best agreement with esperiments is found for $r^{2/3}$. This weak discrepancy may be due to the fact that the mechanical characteristics, and also the geometry of the surface depend on the size of the glued beads.

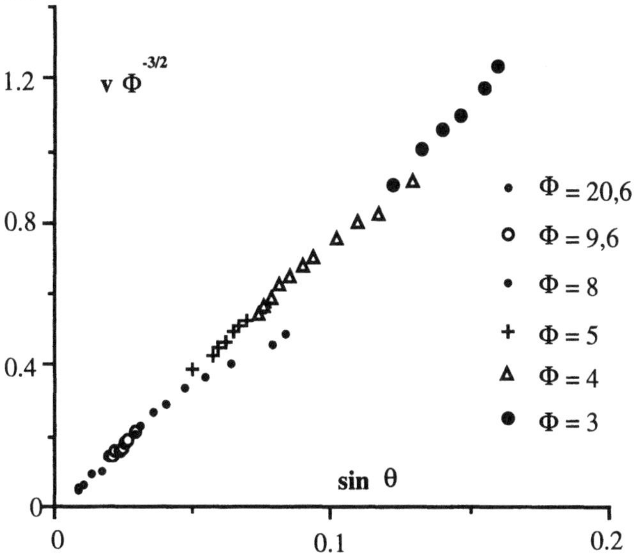

Fig. 2 Variations of $v.\Phi^{-3/2}$ versus $\sin\theta$, for different values of the particle size, on a rough surface characterized by r = 0.25 mm.

This result leads to the following expression for $\langle\eta\rangle$:

$$\langle\eta\rangle = m\ \langle\gamma\rangle\ \Phi^{-3/2} \tag{4}$$

This is an a priori surprising result: the friction force on the rolling ball is viscous! This viscous force is by nature different from that observed when a solid is moving through a fluid. We are now studying more accurately this force, and numerical simulations in a corresponding one-dimensional problem (motion on a rough line) indicate that the exponent of is strongly dependent on the deformation law at the contact between moving ball and glued balls (Hertz law for example).[11] These results

show that the most important part of the energy loss in the motion of the particle is the collision part (thus, we may neglect friction), even if the behavior of these collisions is a non-trivial one.
In terms of segregation, the important result of these experiments concern the average length L_* of the path of the moving ball, from the origin of the motion to its trapping in a hole. The shock that stops the ball is the largest it encounters along its path. Taking into account the two characteristics of this stopping shock (suddenness and scarceness), we have derived an expression for L_* of the form:[10]

$$L_* = r \exp(A\Phi^{\alpha} \sin^2 \theta) \qquad (5)$$

The best value of α in regards to our experimental results is $\alpha = 3$. This equation permits quantitative prediction on surface flow segregation, which is waited to be very strong according to the high value of α which reveals a very strong dependence of L_* on the size of the rolling "grain".

Our experiment provides a quantitative analysis of what the motion of one particle is on the surface of a "frozen" sandpile. In fact, this motion is governed by two sources of noise: (1) the "small" shocks experienced by the particle due to the roughness of the plane, which lead to a viscous friction force and (2) the "traps" that are defaults in the plane large enough to brutally stop the particle, whose contribution to segregation is essential.

3- Influence of interstitial fluid: Intermittency in the hourglass

The granular mass flow rate from a hopper or within an hourglass is generally well described by the semi empirical Beverloo's formula:[8]

$$W = C \rho_s g^{1/2} (D_0 - k\, d)^{5/2} \qquad (6)$$

Here W is the mass flow rate down the hopper (or from the upper chamber of the hourglass), ρ_s the solid density of the granular material, g the gravitational constant, D_0 and d respectively the diameters of the hopper's orifice and of the grain (supposed to be spherical), k a constant ($k \approx 1.5 - 2$). The kd term takes into account the arching effects observed near the hole when D_0 is not much larger than d. One can see that except the small kd correction term, the flux is independent on granular size.
Some deviations from this law are observed when the size of the grains becomes small, typically smaller than 500 µm.[14] It is due to a gradient in density above the "free arch" which separates the bulk in the hopper with the freely falling grains. This density gradient creates a pressure gradient in the interstitial air which increases when the grain size decreases (the permeability of a packing varies as d^2), and as a result the mass flux is reduced.
The decrease of the permeability with the grain size leads to another - and stronger - effect in granular flows in confined geometry: the flow may become intermittent. Such conditions are easy to obtain in an hourglass.[15]

Our hourglass -schematized in figure 3 - has the following geometrical characteristics: the two chambers are geometrically identical and their volume is 200 ml. The diameter of the orifice between these chambers is about 2 mm, and the slope of the shoulders is 45°. To measure the mass flow rate, the lower chamber is open to the atmosphere, but the upper chamber is closed. We have done experiments with glass beads with different uniform diameters d varying from 41 μm to 200 μm.
As shown in figure 3, the flow is characterized by an intermittency, which can be described as a succession of active (or flowing) and inactive states. Each of these states has a duration which is roughly constant for a given grain diameter. The mean values of the duration of each of these states are respectively T_a and T_i. We define the total period T as $T = T_a + T_i$. One important fact is that the observed intermittency disappears when the two chambers are opened, or when they are evacuated: clearly, the air pressure difference between the two chambers is the key to the problem.

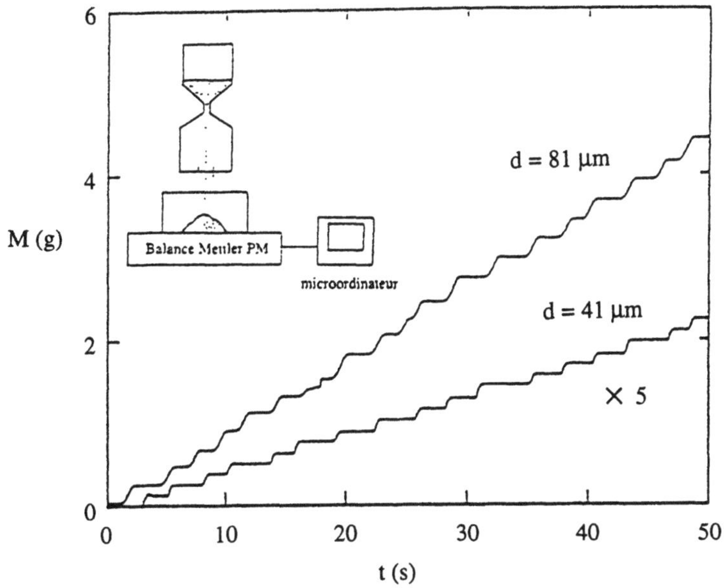

Fig. 3 Variation of the mass M flowing from the upper chamber versus time. The data for d = 41 μm have been multiplied by 5 in order to bring them to the same scale as the data for d = 81 μm. The inset shows schematically our experimental system.

The mechanism leading to this intermittency is simple. When a mass δm flows from the closed upper chamber, the free volume in it increases, and the air density decreases by $\Delta\rho_1$. This creates a drop of pressure ΔP_1 proportional to $\Delta\rho_1$. Assuming that the flux of beads is constant during the flow, the time derivative of ΔP_1 is constant. The pressure in the upper chamber decreases linearly during the active period, down to a value for which the pressure difference between the two chambers is sufficient to stabilize an arch at the vicinity of the hole. This arch blocks the flow, and the inactive period starts.
During this period, the difference of air pressure creates an air current q passing through the sand, which is related to the difference in air pressure through the Darcy equation

$$q = -\left(\frac{k\pi D_0}{2\eta}\right) \Delta P_1 \tag{7}$$

Here κ is the permeability of the packing and η the viscosity of air. Because the conic shape of the hourglass near the hole, the pressure gradient is localized to a small region which size is of the ordre of D_0, the length scale appearing in this equation. This current is proportional to the rate of change of air density $d\Delta\rho_1/dt$ and finally, we obtain the following expression

$$\frac{d\Delta P_1}{dt} = -\frac{P_2 \pi \kappa D_0}{2\eta V_1} \Delta P_1 , \tag{8}$$

where P_2 is the pressure within the lower chamber. Integrating this equation yields an exponential decrease for ΔP_1, with a pressure relaxation time of the form

$$\tau = \frac{2\eta V_1}{P_2 \pi \kappa D_0} . \tag{9}$$

According to this analysis, the pressure in the upper chamber appears to be a succession of linear decrease (active times), and exponential increase (inactive times). Two facts remain unclear :

- why an arch with linear extension larger than 20 times the diameter of the (spherical) grains remain stable during the inactive time.
- why the total period is found experimentally to be very insensitive to the size of the grains (between 41 and 160 μm).

In order to have a better understanding of this behaviour, we have performed two types of experiments.

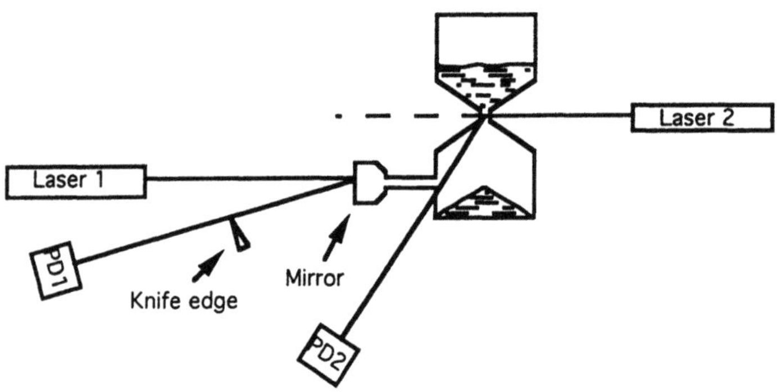

Fig. 4 Experimental setup to measure fluctuation of pressure and of position of the free arch. The hour glass is seen from above. PD1 and PD2 are photodetectors.

The first is schematised in figure 4. It consists of measuring the reflection of a laser beam at the level of the free arch just above the orifice, and, at the same time, the fluctuation of the pressure in the lower chamber. This pressure fluctuation leads to a deformation of a very thin mirror, also measured using the reflection of a laser beam. The beam is more or less cut by the knife edge (figure 4) depending on the deflection of the mirror. The diode PD1 then measures the intensity of the cut beam, and thus the pressure.

The upper curve in figure 5 gives the variations of pressure in the lower chamber, and the lower curve the light intensity reflected by the system just above the orifice. This latter measurement is recorded by photodetector PD2 in figure 4. At the scale of T, the total period, these curves confirm our analysis. The position of the free arch is clearly fluctuating: it passes from a lower level when the arch is stable to an upper level when grains flow. The surprising fact is that we observe clearly quasi regular small fluctuations of pressure during the active times, which match roughly the fluctuations in the position of the free arch. These regular fluctuations seem to be the trace of an "internal clock" in the flow, which has to be explained.

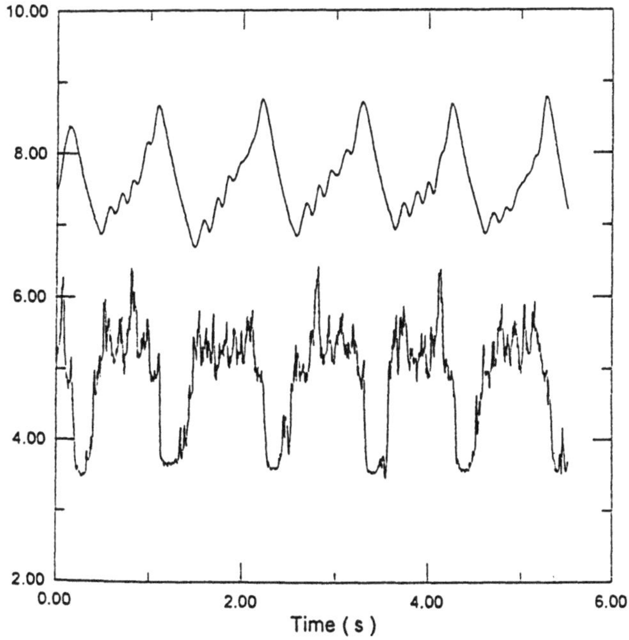

Fig.5 : The upper part gives the variations of pressure in the lower chamber, and the lower one gives the variations of light reflected just below the free arch, near the hole.

We have performed other experiments using a large centrifuge. The idea is to measure the influence of gravity on the mass flow rate (supposed to vary as $g^{1/2}$ according to the Beverloo equation) and on the stability of the arch, then on the inactive time T_i.

The mass flow rate varies with g^{α} where α is an exponent equal to 1.1 ± 0.1 (thus far from 1/2). If we suppose that the mass flow during the active time is constant with time, and varies as $g^{1/2}$, this result implies that the balance between active and

inactive time is strongly dependent on gravity, as expected. This is confirmed by the other result: as shown in figure 6, the inactive time depends strongly on the gravity, whereas the permeability is essentially independent of it.
These results seem to confirm that the stability of the arch is the key of the observed intermittency.

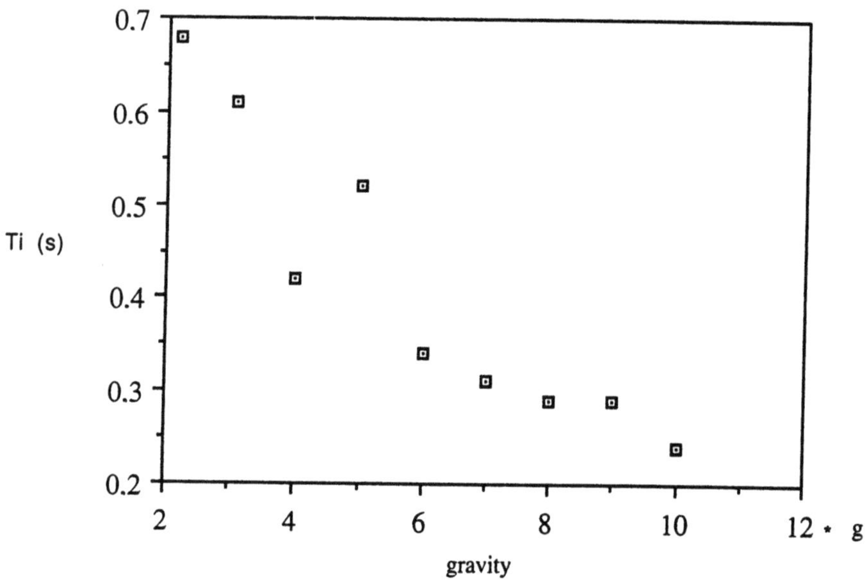

Fig. 6 : Variations of the inactive period T_i with gravity.

This question (stability of arches) is a central one not only in the academic problem of flow within an hourglass, but also in many other problems of flow of particles in a confined geometry.

Acknowledgements

We want to acknowledge support of the CNRS via the GdR "Physique des Milieux Hétérogènes Complexes".

1- R.A. Bagnold, Proc. Royal Soc. London **225**, 49 (1954)
2- J.T. Jenkins and S.B. Savage, J. Fluid Mech. **130,** 186 (1983)
3- H.J. Herrmann in "Disorder and Granular Media", D. Bideau and A. Hansen Eds, North Holland (1993)
4- E. M. Lifshitz and L. P. Pitaevskii, Statistical Physics, Course of Theoretical Physics (volume **5**), (1980) Pergamon Press, Oxford
5- C.A. Coulomb, Mémoire de Mathématiques et de Physique présentés à l'Académie des Sciences Paris, **7**, 343 (1773)

6- H.M. Jaeger, C.H. Liu, S.R. Nagel, and T.A. Witten, Europhys. Lett., **11**, 619 (1990)
7- J.C. Williams, Chem. Proc. Suppl. (1965)
R. Jullien and P. Meakin, Nature, **344,** 425 (1990)
8- R.M. Nedderman et al Chem. Eng. Sc. **11**, 1597 (1982)
9- F.X. Riguidel, R. Jullien, G.H. Ristow, A. Hansen, and D. Bideau, J. Phys. I France, **4**, 1 (1994)
10- F.X. Riguidel, A. Hansen, and D. Bideau, Europhys. Lett. submitted
11- G.H. Ristow, F.X. Riguidel, and D. Bideau, J. de Phys. I France, submitted
12- D. Bideau, A. Gervois, L. Oger, and J.P. Troadec J. Physique, **47**, 1697 (1986)
13- C.D. Jan, H.W. Shen, C.H. Ling and C.I. Chen, Eng. Mech. Proc of the Ninth Conf, 768 (1992)
14- B.J. Crewdson, A.L. Ormond, and R.M. Nedderman Powder Technology, **16**, 197 (1977)
15- X.l. Wu, K.J. Måløy, A. Hansen, M. Ammi, and D. Bideau, Phys. Rev. Lett., **71**, 1363 (1993)

Density Waves in Granular Flow

H. J. Herrmann [1 2], E. Flekkøy [2 3], K. Nagel [4], G. Peng [1] and G. Ristow [5]

[1]HLRZ, KFA Jülich, Postfach 1913, 52425 Jülich, Germany

[2]L.H.M.P., E.S.P.C.I., 10 rue Vauquelin, 75231 Paris, France

[3]54-627 M.I.T., Cambridge, MA 02139, U.S.A.

[4]ZPR, Univ. Köln, Weyertal 86, 50923 Köln, Germany

[5]FB Physik, Univ. Marburg, Renthof 6, 35037 Marburg, Germany

Abstract: Ample experimental evidence has shown the existence of spontaneous density waves in granular material flowing through pipes or hoppers. Using Molecular Dynamics Simulations we show that several types of waves exist and find that these density fluctuations follow a $1/f$ spectrum. We compare this behaviour to deterministic one-dimensional traffic models. If positions and velocities are continuous variables the model shows self-organized criticality driven by the slowest car. We also present Lattice Gas and Boltzmann Lattice Models which reproduce the experimentally observed effects. Density waves are spontaneously generated when the viscosity has a nonlinear dependence on density which characterizes granular flow.

1 Introduction

As shown already in Daniel Bideau's talk, moving dry granular media, like sand, show a rich variety of rather astonishing and scarcely understood phenomena [1,2]. Famous are the so-called "Brazil nut" segregation [3–5] and the heap formations that occur under vibrations [6–8]. A series of experiments have given evidence that under certain circumstances density patterns are generated inside the flowing medium. D. Bideau and collaborators, for instance, visualized wave-like patterns emanating from the outlet of a two dimensional wedge-shaped hopper. Also previous authors [9–12] had noted the formation of similar structures. Similarly rather erratic shock-like density waves have been observed in flow through pipes [13] and down inclined planes [14]. Another experimentally observed ubiquous phenomenon in granular media seems to be $1/f$ noise. Baxter [15] also observed power law decay in the frequency dependent forces that act on the wall of a hopper. For avalanches going down the slope of a sand pile theoretical considerations of self-organized criticality [16] led to the proposal that their size and life time distributions were power laws which was in fact only verified experimentally on very small piles [17].

We will present Molecular Dynamics simulations[13,18] showing complex density patterns with a power law spectrum. Then we present traffic jam models[19] showing similar behaviour. Finally we will reproduce the effects using a lattice gas with dissipation[20] and a Boltzmann lattice model[21].

2 Molecular Dynamics simulations

We present Molecular Dynamics (MD) simulations of inelastic particles with static and dynamic friction in two dimensional systems[13,18]. In fact, MD simulations [22] have already been applied to granular media to model segregation [5], outflow from a hopper [23,24], shear flow [25], convection cells on vibrating plates [26,27], avalanches on a sand pile [28] and others.

We consider a system of N spherical particles of equal density and with diameters d either all equal or chosen randomly from a Gaussian distribution of width w around $d_0 = 1$ mm. These particles are placed into a hopper having an opening angle θ and at the bottom an opening of diameter D. In the case $\theta = 0$ we have a pipe. When two particles i and j overlap (i.e. when their distance is smaller than the sum of their radii) three forces act on particle i: 1.) an elastic restoration force

$$\boldsymbol{f}_{el}^{(i)} = Y(|\, \boldsymbol{r}_{ij} \,| - \frac{1}{2}(d_i + d_j))\frac{\boldsymbol{r}_{ij}}{|\, \boldsymbol{r}_{ij} \,|} \ , \tag{1a}$$

where Y is the Young modulus and $\boldsymbol{r}_{ij}$ points from particle i to j; 2.) a dissipation due to the inelasticity of the collision

$$\boldsymbol{f}_{diss}^{(i)} = -\gamma m_{eff}(\boldsymbol{v}_{ij} \cdot \boldsymbol{r}_{ij})\frac{\boldsymbol{r}_{ij}}{|\, \boldsymbol{r}_{ij} \,|^2} = -\gamma m_{eff}\, v_{ij}^n \ , \tag{1b}$$

where γ is a phenomenological dissipation coefficient and $\boldsymbol{v}_{ij} = \boldsymbol{v}_i - \boldsymbol{v}_j$ the relative velocity between the particles; 3.) a shear friction force which in its simplest from can be chosen as

$$\boldsymbol{f}_{shear}^{(i)} = -\gamma_s m_{eff}(\boldsymbol{v}_{ij} \cdot \boldsymbol{t}_{ij})\frac{\boldsymbol{t}_{ij}}{|\, \boldsymbol{r}_{ij} \,|^2} = -\gamma_s m_{eff}\, v_{ij}^t \ , \tag{2a}$$

where γ_s is the shear friction coefficient and $\boldsymbol{t}_{ij} = (-r_{ij}^y, r_{ij}^x)$ is the vector $\boldsymbol{r}_{ij}$ rotated by 90°. Eq. 2a is a rather simplistic description of shear friction which is proportional to the relative velocity of the particles. In order to allow for static situations of blocking in a hopper due to arching it is important to include real static friction, i.e. which does not depend on the velocities but rather the relative angle of the surfaces [29]. When two particles start to touch each other, one puts a "virtual" spring between the contact points of the two particles. Be δs the *total* shear displacement of this spring during the contact and $k_s \delta s$ the restoring frictional force (static friction). The maximum possible value of the restoring force is then, according to Coulomb's criterion, proportional to the normal force F_n multiplied by the friction coefficient μ [5,29]. Cast into a formula this gives a friction force

$$f_{friction}^{(i)} = -\text{sign}(\delta s)\min(k_s m_{eff}\delta s, \mu F_n) \tag{2b}$$

where δs is the shear displacement integrated over the entire collision time. When particles are no longer in contact with each other the spring is removed.

When a particle collides with a wall the same forces act as if it would have encountered another particle of diameter d_0 with infinite mass at the collision point. The walls are in fact made out of small particles themselves and in order to introduce roughness on the wall these particles are chosen randomly from a distribution of two radii. The only external force acting on the system is gravity $g \approx -10\text{m/s}^2$.

In fig. 1 we see a space-time diagramm of the density inside a pipe with 600 particles and periodic boundary conditions. The particles initially have homogeneously randomly distributed initial positions and velocities (left pipe). After some time spontaneously various patterns appear in the density: On one hand one has very dark regions, nearly constant in time. Then one sees black diagonal stipes of constant velocity down the pipe. Finally there are also some lighter horizontal lines. We want to try in the following to explain these rather complex structures,

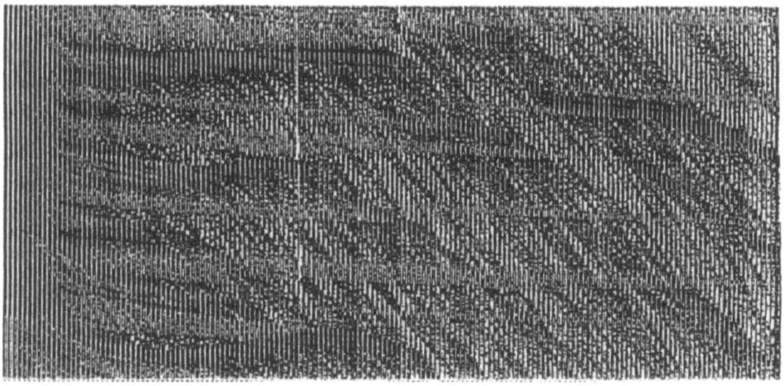

Fig. 1. Vertical pipe plotted at regular time steps next to each other. Time goes from left to right. Gravity acts from top to bottom. For more details see ref. 13.

Similar effects have also been observed in hoppers of opening angles of $\theta = 30°$ and the density at a position six particles diameters above the outlet has been measured as a function of time[18]. In fig. 2 we see the Fourier transformation of this density in a log-log plot. Clearly the data fall on a straight line over nearly two decades. The slope is about $\alpha = -1.35 \pm 0.1$ obtained by a least square fit. That means that we have a power law spectrum of the form $1/f^\alpha$.

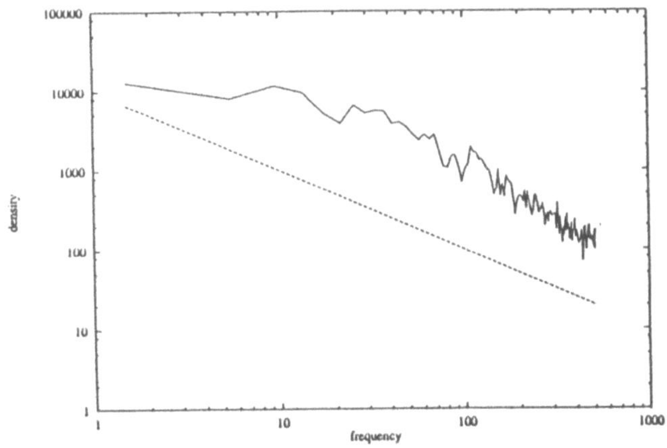

Fig. 2. Log-log plot of the spectrum of the density fluctuations in a hopper[18].

3 Traffic models

We have shown that similar to the avalanches that one observes on the surface of a sand pile also inside the bulk of granular material one has avalanche behavior which like the ones on the surface shows self-organized criticality on small scales [16]. The mechanisms that generate the patterns are similar but not identical to the original sandpile models. While the static friction similarly generates waiting times with a threshold it is not the motion of the sand itself that constitutes the avalanches but it is the group velocity of the holes between them: An individual particle can easily go from one dense region to the other by flying fast through a region of low density. There is therefore a backflow of information similar to the jamming on highways [30].

Everybody knows of the seemingly erratic motions of cars jammed on highways. One wonders whether they are due to a random behaviour of the individual drivers or if there is an intrinsic chaotic mechanism. In favour of the first hypothesis is the existence of regular kinematic waves in dissipative systems with excluded volume [31]. For this reason many traffic models include rather important statistical noise in time [30]. In favour of the second hypothesis are measurements performed on Japanese highways showing a $1/f$ spectrum in the Fourier transformed density fluctuations [32] which might stem from some self-organized criticality [16]. It is therefore interesting to see if a traffic model without noise is able to give the observed erratic behaviour and its $1/f$ spectrum.

In ref. 19 we consider a continuous one-dimensional model. The system has length L with periodic boundary conditions; and velocity v_i and position x_i of a vehicle i are continuous variables. The update rule is as follows:

If the velocity is high with respect to the gap, then the car slows down:

$$v > \Delta x - \alpha \quad \Rightarrow \quad v \to \max(0, \Delta x - 1) ; \tag{3a}$$

(the "max" is only necessary to prevent negative velocities);

else if the velocity is low with respect to the gap and slower than five, then the car accelerates:

$$v < \Delta x - \beta \ \& \ v < v_{max} \quad \Rightarrow \quad v \to v + \min(1, \gamma * \Delta x) . \tag{3b}$$

Note that this rule allows maximum speeds up to nearly six.

After the velocity has been updated for all vehicles according to the last two rules, we move all vehicles simultaneously by a distance equal to their velocities.

In our simulation[19] we used $\alpha = 0.5$, $\beta = 3.0$, and $\gamma = 0.1$. We initially placed $N = [\rho \cdot L]$ vehicles on sites 1 to N, all with velocity zero, where ρ was chosen small enough to prevent the first car hitting the first one through the periodic boundary conditions. Starting from this totally ordered initial state, the system was allowed to evolve according to the above rules, with the exception of the first vehicle, the speed of which was kept fixed at $v_{lead} = 4.99999$ after its initial acceleration. This is a simplification of the well known situation where a number of fast vehicles follow a slower one which they cannot pass.

In Fig.3a we see the time evolution of a system which had been transformed to the coordinates of the first vehicle, i.e. the positions of all cars are given relative to the first car. We see that equally spaced cars rapidly evolve into a fluctuating state (right hand side). In this new state density increases give rise to very short bursts (traffic jams) of very different sizes which redistribute the cars backwards such that in some cases they even start again in equally spaced patterns.

In order to clarify that this behaviour is an intrinsic consequence of the dynamics and not just the enhancement of numerical round-off errors, we have compared single precision with double precision calculations. The behavior of the model (i.e. the formation of the collective shocks) is robust with respect to parameter changes. More precisely, we could not find a qualitatively different behavior for changes of the parameters α, β, γ, and v_{lead} within the following range $0.1 \leq \alpha \leq 0.6$, $2.0 \leq \beta \leq 5.0$, $0.08 \leq \gamma \leq 0.12$, $4.5 \leq v_{lead} \leq 4.99999$.

We measured the distribution of times τ between consecutive "braking" events for the last vehicle (τ is the time from the end of one braking to the beginning of the next). Braking is defined as a slowing down according to the rules for the velocity update. system sizes on up to 512 processors and averaging the results. For instance for $N = 1900$ vehicles, we waited about $3 \cdot 10^5$ time steps to let the transients die out, and then measured the distribution of τ during about $1.1 \cdot 10^6$ further time steps. As seen in Fig. 3b, this distribution displays a remarkable straight line on a log-log plot, fulfilling a power law relation

$$n(\tau) \propto \tau^{-\alpha} \tag{4}$$

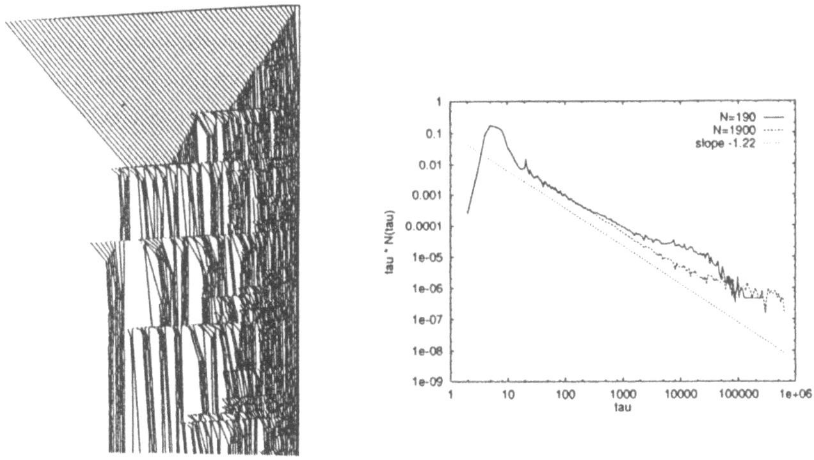

Fig. 3. (a), i.e. left: First 4200 time steps (from top to bottom) and flow of $N = 61$ cars from left to right on a lane of length $L = 1024$. (b), i.e. right: Distribution $n(\tau)$ of times τ between consecutive breaking events for $N = 190$ and $N = 1900$ cars. (From ref. 19)

with $\alpha = -2.2 \pm 0.1$. This non-trivial exponent is a strong indication for the existence of self-organized criticality[16] for this model.

Many aspects of this model are remiscent of the so-called train model for earthquake dynamics [33]. Instead of pulling at one end, the slower car may be seen as pushing against the other cars which want to move faster. This leads to a slowly increasing average density, and at some time this density locally exceeds a critical threshold. The reaction is a more or less drastic slowing down of the corresponding vehicle, which may or may not force the next vehicle to slow down as well. By this mechanism, avalanches of all sizes are generated, which may propagate through the entire traffic jam.

4 Lattice Gas and Boltzmann Models

It seems natural to describe the flow of granular media using the concepts of fluid mechanics. One must, however, consider that as opposed to classical fluids there is local dissipation of energy. Taking into account the dissipation rate in the energy balance equations it has been possible[34,35] to predict the existence of an instability: Slightly denser regions have more dissipation and therefore lower pressure which itself generates a flow that will enhance the density. So, the dissipation will be responsible for the formation of clusters of high density and these have been observed[34]. Also has it been possible to derive from this kinetic gas theory[35–37] that the viscosity increases very sharply with density.

One alternative to the direct solution of the equations of motion of fluids are the so called Lattice Gas (LG)[38] Lattice Boltzmann Models (LBM)[39]. These models are defined on a lattice with velocity vectors that can only point into few discrete directions and all have the same length. For the LG this simplification is somewhat compensated by the fact that on each site one has more real degrees of freedom (six on a triangular lattice) than in the classical numerical techniques allowing for the definition of a local shear or a local rotation.

Let us first describe the Lattice Boltzmann Model as used in ref. 20. We consider a triangular lattice, and on each site x we have six real variables $N_i(x,t)$, $i = 1,..6$, representing (counted counter clockwise) the densities of the particles going in the direction i of the lattice. (For convenience we will in the following omit the site index x and denote by N_i' the value of the particle density after collision.) One updating of the system ($t \rightarrow t+1$) is given by two steps: (1.) The collision step at which the six N_i are updated at each site through

$$N_i' = N_i + \lambda(N_i - N_i^{eq}) \tag{5}$$

and (2.) the propagation step at which each N_i is shifted to the site of the nearest neighbor in direction i. Eq. (5) produces a relaxation towards the equilibrium densities N_i^{eq} which is numerical stable provided the relaxation constant $-2 < \lambda < 0$. The value of λ sets the kinematic viscosity of the fluid. The equilibrium densities are given by

$$N_i^{eq} = \frac{\rho}{6}(1 + 2\boldsymbol{u} \cdot \boldsymbol{c}_i + 4(\boldsymbol{u} \cdot \boldsymbol{c}_i)^2 - 2\boldsymbol{u}^2) \tag{6}$$

where ρ is the mass density at site x

$$\rho = \sum_i N_i \quad , \tag{7}$$

$\boldsymbol{c}_i$ the unity vector along direction i and $\boldsymbol{u}$ the velocity vector at site x defined through the momentum density per site

$$\rho \boldsymbol{u} = \sum_i \boldsymbol{c}_i N_i \quad . \tag{8}$$

The equilibrium distribution N_i^{eq} given in Eq. (6), is chosen to give mass and momentum conservation in the collision step. The flow will be forced into the direction of the gravity $\boldsymbol{g}$, which is pointing parallel to the walls of the pipe. For that purpose an additional step is added after the collision step which is defined by $N_i'' = N_i' + \frac{1}{3}\boldsymbol{c}_i \cdot (\rho \boldsymbol{g})$. Periodic boundary conditions are imposed in the direction of gravity in which the system has a length of L_1. In the perpendicular direction one has walls separated by L_2 lattice spacings. The lattice orientation is such that one of the lattice directions is parallel to the walls. At the beginning of the simulation the average density $\bar{\rho}$ is fixed. It is an important parameter of the model which because of mass conservation stays constant in time. We initialize the system by having the same values of the N_i on each site and then let the system evolve to its steady state. In the case of the stable flows steady state is reached after 2000 or 3000 time steps. In the case of the unstable flows that develop density waves, the simulations might take up to 20000 time steps to reach steady state. At the walls we used no-slip conditions.

The relaxation parameter λ depends on the material properties including the kinematic and the bulk viscosities. Since an exact relation between λ and the material constants is not known we will lean on some approximative arguments[38] that predict a vanishing bulk viscosity. In that case one can relate λ directly to the kinematic viscosity ν through $\lambda = -\frac{1}{2}(0.25 + 2\nu)^{-1}$. We will consider that ν is a function of the local density ρ.

A salient feature of granular media is the spontaneous formation of density waves, similar to traffic jams on highways. One possibility to explain the effect that generates these waves is to assume that the viscosity depends on density. Within the kinetic gas theory of granular media[35–37] the relation $\nu \propto (\rho - \rho_c)^{1/3}$ has been derived. Since the above relation imposes a maximum density ρ_c it is rather difficult to implement it directly within the context of the LBM where the particles do not have an exclusive volume. We therefore chose a piecewise linear relation of the form $\nu = \nu_{min}$ if $\rho \leq \rho_t$ and $\nu = \nu_0 + \gamma(\rho - \overline{\rho})$ for $\rho > \rho_t$ (see fig. 4a). $\overline{\rho}$ is the average density and the threshold density ρ_t is chosen to make ν a positive continuous function of the density.

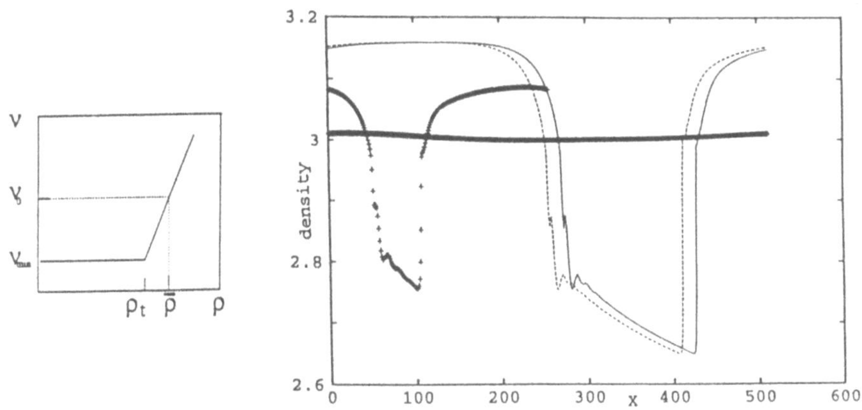

Fig. 4. (a), i.e. left: density dependence of the viscosity chosen in the simulations. (b), i.e. right: density in the center of the channel as a function of the position X along the channel for $\rho_t = 2.962$, $\bar{\rho} = 3.0$, $g = 3.33 \times 10^{-5}$ and $L_2 = 64$. The curve of crosses is for $L_1 = 256$ and 60,000 iteration steps after the initial perturbation. The other curves correspond to $L_1 = 512$ and 5000 (thick line), 60,000 (full line) and 60,025 (dashed line) iterations after the perturbation was applied. The slope $\gamma = 6.25$ and the minimum viscosity $\nu_{min} = 0.01$. (From ref. 20)

In order to generate density waves we found it necessary to introduce a small perturbation producing a 0.3% relative density difference. This perturbation was performed by introducing a small amount of momentum on one line across the pipe, keeping the mass unchanged. In fig. 4b we see that this initially very weak perturbation dramatically builds up and develops into a density wave of over 10% density contrast. For a pipe of same width but half the length, i.e. a different aspect ratio the wave has a less pronounced profile. This dependence on the aspect ratio is not to be confounded with finite size effects.

By triggering the density wave by two spatially separated perturbations, rather than just a single one, we checked that the complex shape of the waves does not reflect the detailed way in which they were initiated. We also observe that there seems to be no characteristic wavelength: Fig. 4b shows that the waves have roughly the same shape on the scale of the channel length although the amplitude depends on the system size.

Fig. 5 shows this amplitude as a function of time during 60,000 time steps. The insert shows the initial unstable phase leading to the rather drastic increase of the amplitude at the time 10,000. The first small increase in the density is due to the accelleration of the flow and can be understood from the velocity dependence in the pressure. The small jump at the time 2500 results from the perturbation. Before the instability is triggered at the time 10,000, small oscillations in the amplitude are observed. It was checked that the amplitude indeed has its steady state value at the time 60,000 by running the simulations ten times longer. The complicated relaxation towards the fully developed density wave indicates that strong non-linear effects come into play rendering a linear stability analysis meaningless. It would be interesting to understand this behaviour further.

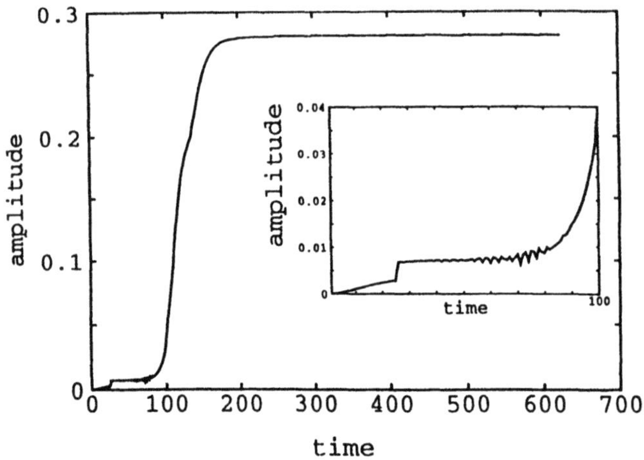

Fig. 5. The amplitude, i.e. difference between largest and smallest density, along the center of the pipe as a function of time measured in units of 100 iteration steps for $L_1 = 256$ and otherwise the same parameters as in fig. 4b. The insert is a blow-up of the behavior at early times. (From ref. 20)

A Lattice Gas model[38] can be formulated such as to include local dissipation of energy by introducing rest particles and special collision rules. For all details we refer to ref. 21. In fig. 6 we see the result for the power spectrum of the time evolution of the density in pipe flow. On top of a $1/f^\alpha$ spectrum with $\alpha = 1.33 \pm 0.02$ we see a peak which corresponds to kinematic waves. Its position depends on the length of the pipe.

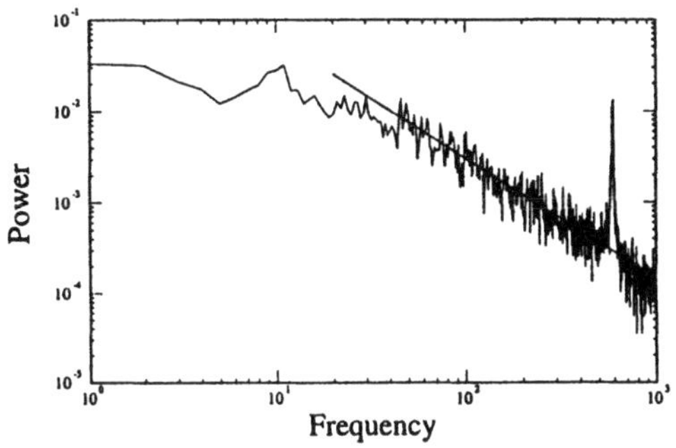

Fig. 6. Log-log plot of the Fourier transformation of the density in a pipe of length 220 and width 11. The straight line is a least square fit with slope -1.33 ± 0.02. (From ref. 21)

5 Conclusion

We have observed in various modelizations that complicated density patterns are generated. In Molecular Dynamics simulations two ingredients were found[18] essential to generate them: static friction and a large enough surface roughness of the walls. The static friction tends to align the particles, i.e. to form fronts of particles moving exactly with the same vertical velocity. These fronts are nucleated randomly at the walls and their size distribution (density contrast) goes by itself into a critical state namely a power law distribution. It therefore has the properties of self-organized criticality (SOC) [16]. Traffic jam models show[19] similar behaviour due the same effect, namely local dissipation of energy. In a fluid picture, a Lattice Boltzmann model shows[20] that the density dependence of the viscosity that one encounters in granular media generates waves of low density. A lattice gas model with local dissipation gives both, kinematic waves and SOC in the power spectrum.

Concluding, the following picture emerges: Due to the inelastic collisions between particles (also cars) an instability[34,35] tends to form clusters of high density (probably, the dark horizontal stripes in fig. 1). These clusters self-organize into a critical state giving a power law spectrum. As in classical traffic models the density dependence of the flux generates kinematic waves (probably, the dark tilted lines in fig. 1). The density dependence of the viscosity as predicted by kinetic gas theory builds up waves of low density (probably, the light horizontal lines in fig. 1). We see that simply introducing dissipation to a gas of particles produces many interesting phenomena, that occur in granular materials.

Acknowledgement

We thank Thosten Pöschel for figure 1.

References

1. *Disorder and Granular Media* eds. A. Hansen and D. Bideau (North-Holland, Amsterdam, 1992)
2. H.M. Jaeger and S.R. Nagel, Science **255**, 1523 (1992); C.S. Campbell, Annu. Rev. Fluid Mech. **22**, 57 (1990)
3. J.C. Williams, Powder Techn. **15**, 245 (1976)
4. A. Rosato, K.J. Strandburg, F. Prinz and R.H. Swendsen, Phys.Rev.Lett. **58**, 1038 (1987) and Powder Techn. **49**, 59 (1986); P. Devillard, J. Physique **51**, 369 (1990)
5. P.K. Haff and B.T. Werner, Powder Techn. **48**, 239 (1986)
6. M. Faraday, Phil. Trans. R. Soc. London **52**, 299 (1831)
7. P. Evesque and J. Rajchenbach, Phys. Rev. Lett. **62**, 44 (1989); C. R. Acad. Sci. Ser. 2, **307**, 1 (1988) and **307**, 223 (1988); C. Laroche, S. Douady and S. Fauve, J. de Physique **50**, 699 (1989); P. Evesque, J. Physique **51**, 697 (1990); J. Rajchenbach, Europhys. Lett. **16**, 149 (1991)
8. J. Walker, Sci. Am. **247**, 167 (1982); F. Dinkelacker, A. Hübler and E. Lüscher, Biol. Cybern. **56**, 51 (1987)
9. G.W. Baxter, R.P. Behringer, T. Fagert and G.A. Johnson, Phys. Rev. Lett. **62**, 2825 (1989)
10. O. Cutress and R.F. Pulfer, Powder Techn. **1**, 213 (1967)
11. A. Drescher, T.W. Cousens and P.L. Bransby, Geotechn. **28**, 27 (1978)
12. R.L. Michalowski, Powder Techn. **39**, 29 (1984); J.D. Athey, J.O. Cutress and RF. Pulfer Chem. Eng. Sci. **21**, 835 (66); P.M. Blair-Fish and P.L. Bransby, J. Eng. for Industry **95**, 17 (73); J. Lee, S.C. Cowin and J.S. Templeton, Trans. Soc. Rheol. **18**, 247 (74)
13. T. Pöschel, preprint HLRZ 89/92, J. de Physique, in press
14. D. Bideau, private communication
15. G.W. Baxter, PhD thesis
16. P. Bak, Tang and Wiesenfeld, Phys. Rev. Lett. **59**, 381 (1987)
17. G.A. Held, D.H. Solina, D.T. Keane, W.J. Horn and G. Grinstein, Phys. Rev. Lett. **65**, 1120 (1990)
18. G. Ristow and H.J. Herrmann, preprint HLRZ 2/93
19. K. Nagel and H.J. Herrmann, Physica A **199**, 254 (1993)
20. E. Flekkøy and H.J. Herrmann, Physica A **199**, 1 (1993)
21. G. Peng and H.J. Herrmann, preprint HLRZ 72/93

22. M.P. Allen and D.J. Tildesley, *Computer Simulation of Liquids*, Oxford University Press, Oxford, 1987
23. G. Ristow, J. Physique I **2**, 649 (1992) and Int. J. Mod. Phys. C **3**, 1281 (1992)
24. D.C. Hong and J.A. McLennan, Physica A **187**, 159 (1992)
25. C.S. Campbell and C.E. Brennen, J. Fluid Mech. **151**, 167 (1985); P.A. Thompson and G.S. Grest, Phys. Rev. Lett. **67**, 1751 (1991); D.M. Hanes and D.L. Inman, J. Fluid Mech. **150**, 357 (1985); O.R. Walton and R.L. Braun, J. Rheol. **30**, 949 (1986)
26. Y-h. Tagushi, Phys. Rev. Lett. **69**, 1367 (1992)
27. J.A.C. Gallas, H.J. Herrmann and S. Sokołowski, Phys. Rev. Lett., **69**, 1371 (1992)
28. J. Lee and H.J. Herrmann, J. Phys. A **26**, 373 (1993)
29. P.A. Cundall and O.D.L. Strack, Géotechnique **29**, 47 (1979)
30. K. Nagel and M. Schreckenberg, J. Physique I **2**, 2221 (1992)
31. M.J. Lighthill and G.B. Whitham, Proc. Roy. Soc. A **229**, 281 and 317 (1955); M. Leibig, preprint HLRZ 42/93
32. T. Musha and H. Higuchi, Jap. J. Appl. Phys. **15**, 1271 (1976)
33. M. de Sousa Vieira, Phys. Rew. A **46**, 6288 (1992)
34. I. Goldhirsch and G. Zanetti, Phys. Rev. Lett. **70**, 1619 (1993)
35. S. Savage, J. Fluid Mech. **241**, 109 (1992)
36. P.K. Haff, J. Fluid Mech. **134**, 401 (1983)
37. J.T. Jenkins, Arch. Rat'l. Mech. Anal., **87**, 355 (1985)
38. U. Frisch, B. Hasslacher and Y. Pomeau, Phys. Rev. Lett. **56**, 1505 (1986)
39. R. Benzi, S. Succi and M. Vergassola, Phys. Rep. **222**, 145 (1992); G.R. McNamara and G. Zanetti, Phys. Rev. Lett. **61**, 2332 (1988)

Height Fluctuations and Pressure Distribution in a Model of Random Close Packing of Mono-size Discs

M. K. Hari Menon

Tata Institute of Fundamental Research, Homi Bhabha Road,
Bombay-400 005, India.

Abstract: We study a variant of the model of random close packing of mono size discs in two dimensions studied by Meakin and Jullien. For this model we calculate the correlation function of the fluctuations of the height in the n-th layer, and show that the width varies as $n^{\frac{1}{2}}$ for large n. Assuming that the discs are frictionless we calculate the pressure distribution at the floor and show that the pressure distribution is uniform and is independent of the disorder.

1 Introduction

Packing of spheres has been studied in the literature as models for liquids, glasses and granular media, in the context of interface growth and in many other areas like biology and engineering [1,2]. In particular they are very useful for modelling many of the geometrical properties of granular media like the density, average number of contacts, radial distribution function *etc.*. One of the interesting and yet unexplained feature is the pressure distribution at the bottom of a pile of grains. Experiments have shown that the peak of the pressure distribution is not at the centre where the height of the pile is maximum, a result which is contrary to the naive expectation [3]. The reason for this is understood only qualitatively in terms of the formation of arches and the friction between the grains. A quantitative understanding is still lacking. Here we describe a model of hard disc packing in two dimensions, with the motivation of understanding the role of disorder in the observed behaviors of the pressure distribution. We have been able to calculate exactly the pressure distribution at the bottom. Recently Hong [4] and Kurt *et al.* [5] have noticed that the pressure distribution is uniform for triangular piles with equal spacing between the discs. We show that in our model this result is true even when the spacing between the discs are random.

The surface of a heap of granular material is an example of a rough surface and the growth of such surfaces has been a topic of current interest. We have studied the nature of the surface fluctuations of the disc packing in our model. We have calculated the variation of the surface fluctuations with the number of layers and the height-height correlation function for the n-th layer.

2 Description of the Model

Our model is a variant of the off-lattice ballistic model studied by Meakin and Jullien [6]. In their model mono size discs are released one by one from randomly chosen points above a horizontal line. A disc released continue to travel down until it touches another disc. Then it rolls down on this disc until it touches one more disc and if the x co-ordinate of the rolling discs lies between that of the two discs touching this it will stop rolling and get stabilised there. Otherwise it will continue to roll down till it find a stable position or touches the bottom in which case it will get fixed there. In the limit of very large height every disc in the packing generated by this algorithm will have contact only with four other discs [7]. In such a case the bulk of the packing will consists of layers of hard discs with the height of each layer fluctuating as a function of the horizontal distance. One can ask the questions how the height of the n-th layer are correlated at different places and how the width of the layer varies with n. To answer these questions the packing is built in our model by placing hard discs layer by layer. In the first layer the discs are fixed rigidly to a horizontal line over which the packing is made. A second layer of discs is then formed by placing a disc between every two adjacent discs in the first layer so that it is supported by these two discs against gravity. This process is continued and a packing of discs is formed. This is shown in figure 1. All the discs are of equal radii and there is no friction between the discs. The packing could be either between two walls (analogues to free boundary condition) or on a ring (periodic boundary condition). The distance d between the centres of the discs in the bottom layer can take any value between D and $\sqrt{3}D$, where D is the diameter of the discs. The value $d = D$ corresponds to the case when the discs touch each other. The reason for keeping d less than $\sqrt{3}d$ is that in this case every disc in the bulk of the packing will have contact with four other discs and the line joining the centres of the discs will form rhombuses. Because of this there is no arch formation by the discs. If d is allowed to take values greater than this some of the discs will have contact with more than four discs, resulting in the formation of pentagons and triangles also, which are called defects by Meakin and Jullien. In fact the formation of the rhombuses is the feature which enables us to calculate many quantities in this model exactly as explained in detail below.

Consider the triangle formed by the centres of the i-th and $i+1$-th discs in the bottom layer and the centre of the disc in the second layer between these two discs. Let θ_i and ϕ_i are the angles which the two sides joining the centres of the discs in the first and second layer make with the vertical ($\theta_i = \phi_i$). These angles are determined by the distance between the i-th and $i+1$-th discs. The formation of the rhombuses give rise to propagation of these angles with out changing their values as shown in figure 1. The set of angles $\{\theta_i\}$ propagate

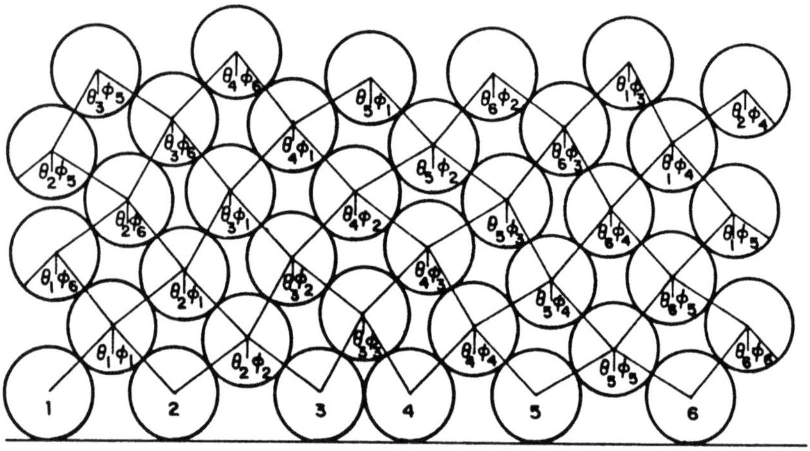

Fig. 1.. A packing of hard discs of equal size on a flat surface. The spacing between the discs in the bottom layer are random variables. Note the propagation of the angles $\{\theta_i\}$ towards left and angles $\{\phi_i\}$ towards right. We have used periodic boundary conditions.

towards the left and the set of angles $\{\phi_i\}$ propagate towards the right. If the packing had defects then whenever a propagating angle meets a defect its value will get changed. Since the distance between the discs are independent random variables, $\{\theta_i\}$ is a set of independent random variables and also $\{\phi_i\}$ is a set of independent random variables. Due to their propagation, these set of angles are enough to specify the position of all the discs in the packing.

3 Height Fluctuations of the n-th Layer

On a large scale the surface of this packing will be very rough. This is due to the propagation of the disorder from the bottom layer. We have calculated exactly the height-height correlation function and the width of these surface fluctuations. To calculate these quantities, let the surface corresponds to the n-th layer of the packing. Let us also label the discs on each layer of the packing from left to right by integers. Consider a disc labelled m on the n-th layer. One can define a wedge with this disc as apex as shown in figure 2.

The height of this disc is given by

$$h(m,n) = \sum d \cos(\theta_i) \tag{1}$$

where the summation is over all the angles covered by the base of the wedge. Since these angles are independent random variables with well defined mean and variance, by central limit theorem as $n \to \infty$, $h(m,n)$ will be a random

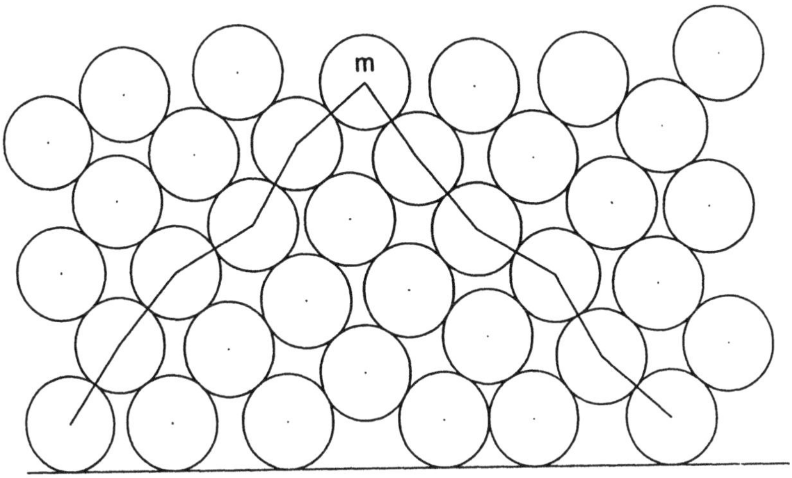

Fig. 2.. This figure shows a wedge with the disc m as its apex. The height of the disc m is determined by the angles at the bottom layer with in the base of this wedge.

variable with Gaussian distribution. Hence the width of the layer W, defined as the variance of the height fluctuations, scales with the number of layers as $W \sim (n)^{\frac{1}{2}}$. Thus the exponent β characterising the growth of the width of the interface with time is $\frac{1}{2}$ in this model. Now consider two discs labelled by m and $m+r$ (there are r discs in between them) on the surface. The correlation between their heights is given by

$$C(n,r) = \langle h(n,m+r)h(n,m)\rangle - \langle h(n,m+r)\rangle\langle h(n,m)\rangle \tag{2}$$

Here $\langle .....\rangle$ denotes the averaging over the disorder assuming that the system is self averaging. If $r > (n-1)$ then the wedges corresponding to discs m and $m+r$ will not overlap. So the heights of these two discs are independent (1). When $r \leq (n-1)$ the two wedges will overlap and there will be a correlation in the heights of these two discs. The height-height correlation function is given by

$$C(n,r) = \begin{cases} \sigma^2\ (n-1-r), & \text{if } r \leq (n-1); \\ 0, & \text{if } r > (n-1). \end{cases} \tag{3}$$

where σ is proportional to the variance of the fluctuations in the spacing between the discs. Thus the correlation length ξ is proportional to n for large n. For a fixed value of $n > r$, C(n,r) $\sim$ r and so the roughning exponent α is $\frac{1}{2}$ and the dynamical exponent z ($z = \frac{\alpha}{\beta}$) is 1 in our model. We compare our results with that of the well known models of surface growth like Hammersly-Edwards-Wilkinson (HEW) model [8,2] and the Kardar-Parisi-Zhang (KPZ) model [9]. The HEW model gives a value of $\alpha = \frac{1}{2}$ and $\beta = \frac{1}{4}$ and for KPZ model $\alpha = \frac{1}{2}$ and $\beta = \frac{1}{3}$. When the diffusion term is absent in the HEW model, then the surface growth is only due to noise. For the Gaussian noise with delta correlations in space and time this model (random rain model) gives a value of $\alpha = 0$ and $\beta = \frac{1}{2}$. For the model studied by Meakin and Jullien β is close to $\frac{1}{3}$ (0.31 ± 0.01) [6]. Though our model is similar to this model the exponent β has a larger value in our model. This is because the quantity we are looking at is the width of the layers where as Meakin and Jullien address the question of the growth of the width of the surface in their model. The crucial difference is that the surface of the packing in Meakin-Jullien model does not corresponds to a particular layer number because there the particles are added from random positions above the surface. The movement of the discs after landing on the surface to local stable positions helps to smoothen the surface and hence the width grows at a much slower rate.

4 Pressure Distribution on the Floor

Since the discs are assumed frictionless, we can calculate the forces at the bottom of the packing. The forces acting on each disc in the packing are the following. The force of gravity due to its weight $\boldsymbol{W}$, the force exerted by the two discs touching this disc from above, denoted by $\boldsymbol{F}^L$ and $\boldsymbol{F}^R$ and the reaction forces due to the two discs supporting this disc form below, denoted by $\boldsymbol{R}^L$ and $\boldsymbol{R}^R$. There are no tangential forces. For the disc i to be in equilibrium

$$\boldsymbol{R}_i^L + \boldsymbol{R}_i^R = \boldsymbol{F}_i^L + \boldsymbol{F}_i^R + \boldsymbol{W}_i \tag{4}$$

One can write down such an equation for every disc. They will be a coupled set of equations, which can be solved to give the force acting on the bottom surface by each disc in the bottom layer. But this is rather tedious if the number of discs in the packing is large.

We can also calculate the forces using the principle of virtual work. To find the vertical force exerted by the disc i on the surface imagine that this disc is displaced vertically by an amount δh. Then the angles θ_i and ϕ_{i-1} will change and all the discs in a wedge with the disc i as its apex will get displaced. This wedge is shown in figure 3. The new positions of the discs can be find out using (1) and hence the potential energy of the pile can be calculated. Then the magnitude of the force, F_i is given by

$$F_i \, \delta h = \delta(PE) \tag{5}$$

where $\delta(PE)$ is the change in the potential energy of the pile after displacing the disc i. The following results can be easily shown.

1. The displacements of all the discs to $O(\delta h)$ are independent of the angles $\{\theta_i\}$ and $\{\phi_i\}$.
2. To $O(\delta h)$ the displacement of any disc on the boundary of the wedge is $\frac{\delta h}{2}$ where as it is zero for any disc inside the wedge.

Since from (5) the force is determined by displacements of $O(\delta h)$, it is independent of all the angles and is only a function of the number of discs on the boundary of the wedge. For a packing of N layers with periodic boundary conditions, or for a triangular pile, every disc in the bottom layer will have same number of discs on the boundary of its wedge. So the vertical force exerted by every disc on the bottom surface is the same. Hence in this model the force distribution at the bottom is uniform. The fact that the pressure distribution is independent of the spacing between the discs shows that the disorder is not a relevant quantity determining the pressure distribution. It will be interesting to see to what extent the pressure distribution depends on disorder if arches are present, for example in the Visscher-Bolsterli model [10], and also to study the problem in three dimensions.

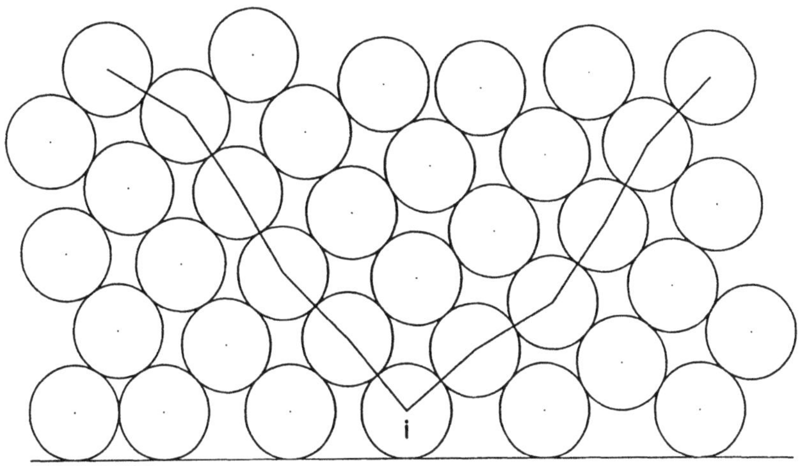

Fig. 3.. A wedge with disc i as its apex. The displacement of the disc i causes a displacement of all the discs with in and on the boundary of this wedge.

The author would like to thank Prof. B. K. Chakrabarti, for introducing him to this problem and Prof. D. Dhar for very valuable ideas and discussions and for a critical reading of the manuscript. The author also would like to thank Mr. A. A. Ali for very helpful discussions and Prof. H. J. Herrmann for bringing reference [5] to his attention.

References

1. J. D. Bernal *Proceedings of the Royal Society of London A*, **280**, 299 (1964).
2. S. F. Edwards and D. R. Wilkinson *Proceedings of the Royal Society of London A*, **381**, 17 (1982).
3. R. L. Brown and J. C. Richards *Principles of Powder Mechanics*, (Pergamon, Oxford, 1966).
4. D. C. Hong *Phys. Rev. E*, **47**, 760 (1993).
5. Kurt Liffman, Derek Y. C. Chan and D. Hughes *University of Melbourne Preprint* (1992).
6. P.Meakin and R. Jullien em J. Phys. (Paris), **48**, 1651 (1987).
7. P. Meakin and R. Jullien *Europhys. Lett.*, **14** 667 (1991).
8. J. M. Hammersley in *Proceedings of the Fifth Berkeley Symposium on Mathematical Statistics and Probability*, edited by F. Neyman (University of California, Berkeley, 1967), Vol. III.
9. M. Kardar, G. Parisi and Y. C. Zhang *Phys. Rev. Lett.* **56**, 889 (1986).
10. W. M. Visscher and M. Bolsterli *Nature*, **239**, 504 (1972).

Statics and dynamics of sandpiles: some phenomenological ideas

J.P. Bouchaud

Service de Physique de l'Etat Condensé, CEA, Orme des Merisiers, 91 191 Gif s/ Yvette CEDEX

Abstract: We propose phenomenological equations to describe how the contraints 'propagate' within a granular media. The linear part of these equations is a wave equation, where the vertical coordinate plays the rôle of time, and the horizontal coordinates the rôle of space. This means that the stress propagates along 'light' cones. The opening angle of these cones is simply the maximum angle of stability of heaps. Inclusion of non linear terms is argued to describe the 'arching' phenomenon, which has been proposed to explain the non intuitive vertical distribution pressure observed experimentally. The analogue of Péclet and Reynolds numbers can be defined, suggesting possible classifications of sandpiles. In a second part, we propose a new continuum description of the dynamics of sandpile surfaces, which involves two population of grains: immobile and rolling. We introduce a simple bilinear form for the interconversion processes of "sticking" (below the angle of repose), and "dislodgement" (for greater slopes). We find a "spinodal" angle, larger than the angle of repose, at which the surface of a tilted immobile sandpile first becomes unstable to an infinitesimal perturbation.

1 Introduction

Granular media have been the focus of intense interest in the physicist community in the last few years. Beyond their obvious practical and engineering importance [1], they represent an interesting field of classical phenomenology, since their behaviour present both solid like and liquid like features - leading to a number of fascinating properties. They have also been proposed as the paradigm of a 'new concept', self organized criticality (SOC) [2]. This proposal was followed by a series of interesting experiments, as well as theoretical and numerical work.

Here, we want to discuss phenomenologically two rather important problems (extended version of the present discussion can be found in [3,4]):

• The first one concerns the spatial distribution of stresses in different geometries (e.g. heaps or silos). One would in particular like to understand the rather unintuitive experimental result of Smid and Novosad [5], which shows that the vertical force beneath a sand heap has a double belled shape, with a local *minimum* at the center of the heap !

• Second, it would be desirable to have phenomenological equations describing the dynamics of the slope of a sandpile near criticality. We will indeed propose new equations which, in contrast to previous work [6], allow a natural accommodation of the idea of local metastability. This leads in turn to 'first order' behaviour (hysteresis) in the macroscopic dynamics of the mean slope of a sandpile, as observed experimentally, rather than 'second order' features (power-law behavior) as suggested by SOC theory.

2 The 'wave-equation' for stress propagation

It is interesting to consider the simple situation of an *ordered* bidimensional packing of spheres. Let $F_x(x,z)$, $F_z(x,z)$ denote, respectively, the x and z components of the total force acting on the north hemisphere of the grain centered at point x, z. If the angle between contacts is θ, and if the tangential force T is *everywhere* equal to $N.\tan\phi$, where N is the normal force, and ϕ the solid friction angle, mechanical equilibrium requires that:

$$\frac{\partial F_z}{\partial z} = -\frac{\tan(\theta+\phi)}{2\sin\theta}\frac{\partial F_x}{\partial x} + \frac{Mg}{2\sin\theta R} \qquad (1-a)$$

and

$$\frac{\partial F_x}{\partial z} = -\frac{1}{2\sin\theta\tan(\theta+\phi)}\frac{\partial F_z}{\partial x} \qquad (1-b)$$

where M and R are the mass and radius of the grains; we have furthermore taken the continuum limit, which assumes that the forces vary slowly on the scale of R. Note that for $\phi = 0$, $\theta = \frac{\pi}{6}$, Eqs. (1) and (2) are precisely the continuous version of the equations considered in refs. [7,8].

Now, in a real heap, the packing is somewhat disordered (local variations of θ and ϕ, of the shape and size of the grains...), and the tremedous difficulty lies in the correct description of solid friction, which may or may not be 'mobilized'. A way to by-pass this difficulty is to argue, in the spirit of phenomenological approaches based on symmetry arguments [3], that similar equations should hold *at a coarse-grained level*, but with 'renormalized' phenomenological parameters. Several other terms must in fact be included to account for important physical effects, such as arching, dispersion and dilatancy. The simplest phenomenological equations thus read [3]:

$$\frac{\partial F_z}{\partial z} = -c_1\nabla\cdot\boldsymbol{F} + \kappa_z\nabla^2F_z - \lambda_z\nabla[\boldsymbol{F}F_z] + \rho(x,z) \qquad (2-a)$$

and

$$\frac{\partial \boldsymbol{F}}{\partial z} = -c_2\nabla F_z + \kappa_\perp\nabla^2\boldsymbol{F} - \lambda_\perp[\boldsymbol{F}\cdot\nabla]\boldsymbol{F} \qquad (2-b)$$

where $\boldsymbol{F}$ is the force field in the plane perpendicular to z, ρ is the mass per unit length, while c_1, c_2 are phenomenological parameters describing in an average way the geometry of contacts, the effect of a non zero solid friction, etc...Finally, κ are 'diffusion' constants and the non-linear terms λ arise from arching effects [3].

In the simple case where $\lambda, \kappa \equiv 0$ and for a uniform $\rho \equiv \rho_0$, we find that both $\boldsymbol{F}$ and F_z obeys the *wave equation*:

$$\frac{\partial^2 F_z}{\partial z^2} = c_0^2 \frac{\partial^2 F_z}{\partial x^2} \tag{3}$$

with a 'velocity of light' $c_0 \equiv \sqrt{c_1 c_2}$. This has immediately two interesting consequences. First of all, a perturbation localized on a site x, z only affects points along the 'light cone' originating from x, z. Thus, a localized force on the surface of a powder layer should create a circular 'echo' of stress at the bottom of the layer, which should in fact allow to measure c_0. The second direct consequence of Eq. (4) is that only heaps with opening angle greater than $\tan^{-1} c_0$ can be stable: since the boundary conditions at the surface of the pile are $\boldsymbol{F} = F_z = 0$, the 'light cones' originating from any point within the pile cannot intersect its surface.

One can in fact solve completely the problem in two dimensions (or in three dimensions for a cylindrically symmetric problem with a purely radial $\boldsymbol{F}$). In the case of a heap with opening angle α such that $c \equiv \tan(\frac{\alpha}{2}) \geq c_0$, one finds that the vertical stress is given by:

$$F_z(x, z) = \begin{cases} \frac{\rho_0 c z}{(c + c_0)} & \text{for } x \leq c_0 z \\ \frac{\rho_0 c}{(c^2 - c_0^2)}(cz - x) & \text{for } x \geq c_0 z \end{cases} \tag{4}$$

Interestingly, the analysis of the same equations in the silo geometry reproduce the following well known fact: after growing linearly with the height, the total weight bearing on the bottom plate of the silo saturates when the height becomes of the order of magnitude of the width of the silo. Combined with the fact that stresses propagates along cones, the presence of partly 'absorbing' boundary conditions screens out the weight of far away grains !

The presence of a non zero κ smooths out the sharp features of the above solution, Eq. (4): this is indeed confirmed by the numerical simulations of ref. [7]. κ could be estimated by measuring how much the 'light cone' is smeared in a 'echo' experiment. Note that κ has the dimension of a length ℓ which is such that the Péclet number for a given sandpile of height h, $Pe_h = \frac{c_0^2 h}{\kappa}$, is of order 1 when $h \simeq \ell$; smaller sandpiles correspond to a regime where the 'light cone' completely looses its meaning.

One can look perturbatively how the non linear 'arching' terms λ influence the above force profile. To lowest order, we find that the correction to the normal force (as given by Eq. 4) which depends on x reads:

$$\Delta F_z(x, z) = \frac{(\lambda_z - \frac{3}{2}\lambda_\perp)\rho_0^2 \sqrt{\frac{c_2}{c_1}}}{16 c_0^2} \, x^2 \tag{5}$$

(for $c = c_0$). This shows that, as first suggested by Edwards and Oakeshott [9], a 'hole' may indeed appear in the stress distribution around $x = 0$ when $\lambda_z > \frac{3}{2}\lambda_\perp$, and furthermore that the relative amplitude of the correction is of the order of $Re \equiv \frac{\lambda\rho_0 h}{c_0}$, i.e. it grows with the height of the pile, exactly as the experiments suggest ! (see ref. [5]). *Re* plays the role of a 'Reynolds number', which becomes large in the limit of large piles. The presence of randomness in e.g. the weight of the grains would give rise, in the limit $Re \gg 1$, to *turbulent* (but static !) fluctuations in the stress distribution with $\frac{\Delta F_z}{F_z} \simeq \frac{\Delta\rho}{\rho}(\frac{R}{h})^\chi$, where χ is a new, non trivial exponent induced by the non linearities ($\chi = 1/2$ for $\lambda \equiv 0$).

3 Dynamics of the slope: metastability and spinodal angle

We have proposed [4] a new continuum treatment for sandpile surface dynamics, involving two coupled "hydrodynamic" variables [10]: the usual height variable $h(x,t)$ (the local density of *immobile* grains) and a *rolling* grain density $\mathcal{R}(x,t)$. We shall for simplicity work in 1+1 dimensions and consider sandpiles whose height $h(x,t)$ depends on a single spatial variable x. We first propose that the rolling grains are governed by a convective diffusion equation of the following form:

$$\dot{\mathcal{R}}(x,t) = -\nabla(v\mathcal{R}) + \nabla(\mathcal{D}\nabla\mathcal{R}) + \Gamma(\mathcal{R},h) \tag{6}$$

v is their drift velocity downhill along x; and $\mathcal{D}$ is a dispersion constant. The term Γ accounts for the conversion of immobile grains into rolling grains, and *vice versa*. We shall construct Γ with the help of the following physical considerations (discussed fully in Ref. [4]):

(a) We assume that an immobile grain cannot spontaneously start rolling unless it is dislodged by an already rolling grain.

(b) The local slope $-\nabla h$ of the sandpile must exceed a critical value S_c (which we associate with the angle of repose) for the dislodging process to be effective. For convenience - unless explicitely stated - we subtract off a background slope, so that S_c is zero, and the gradient of h is everywhere small. By convention, we consider piles sloping down to the right; thus $\nabla h > 0$ corresponds to a surface *less steep* than critical.

(c) If the local slope is less than critical (*i.e.*, $\nabla h > 0$) then rolling grains will tend to stick to the surface, thus being converted into immobile grains. This occurs independently for each rolling grain, and hence at a rate proportional to $\mathcal{R}$ itself.

(d) If $\nabla h = S_c$, but $\nabla^2 h \neq 0$, then conversion acts to reduce the local curvature of the surface (filling in hollows and eroding bumps) again at a rate proportional to $\mathcal{R}$.

The simplest form of Γ exhibiting the above four properties is the following [10]:

$$\Gamma(h,\mathcal{R}) = -\mathcal{R}[\gamma\nabla h + \kappa\nabla^2 h] \tag{7}$$

with $\gamma, \kappa > 0$. This expression, though nonlinear, depends linearly on each of $\mathcal{R}$ and h, a choice which offers great advantages in analytical work. Having chosen

Γ, the equation for the height $h(x,t)$ of the sandpile (in suitable units), which we define to include only immobile grains, follows by conservation as: $\dot{h} = -\Gamma$. Since Γ is linear in $\mathcal{R}$, in the absence of rolling grains *the surface is 'frozen' in a metastable state, and cannot rearrange.* A static system which is perturbed will typically evolve for a short time but then immobilize in a new state as the rolling grains come to rest. This metastability is a characteristic feature of powders, and a key ingredient of the present approach, which was missing from the previous descriptions [6]

We have studied these equations both analytically and numerically for various situations [4]. For example, if one starts with a single small bump on an otherwise flat surface (at angle S_c), and a constant rolling grain density $\mathcal{R}_0$, the bump propagates *uphill* with velocity $v_h = \gamma \mathcal{R}_0$. (It also undergoes spreading due to the dispersion term.) The uphill convection of surface features is a generic property of the model; it is related to the well known fact that traffic jams propagate in the opposite direction to the flow of cars.

We now consider a sandpile prepared in a metastable state, with $\nabla h(x,0) = -S_0 < 0$, and $\mathcal{R} = 0$. (This can be done by simply tilting the base of a sandpile at its angle of repose.) As previously discussed, since no rolling grains are present, this state does not evolve with time. Imagine, though, that we now perturb the state by creating a small number of rolling grains at the lower edge of the pile. (We could dislodge a few grains so as to spill over the lip of the supporting plate.) Interestingly, the subsequent evolution depends critically on the initial slope S_0.

Suppose at $t = 0$ a small pulse of rolling grains, $\mathcal{R}(x',0) = \Delta\, \delta(x' - x)$, is created at x, then two opposite effects will come into play. Firstly, the rolling grains convect downhill. This effect can be isolated in Eq.(6) by excluding the interconversion term Γ; we find that the density of rolling grains at x after time t decays as $\mathcal{R}(x,t) \propto \Delta \exp[-v^2 t/4\mathcal{D}]$, ignoring a weak pre-exponential factor. Secondly, the rolling grains near x increase in number by dislodging immobile ones. From Eq.(6), one finds that for $v = \mathcal{D} = 0$, $\mathcal{R}(x,t)$ would grow exponentially: $\mathcal{R} \propto \Delta \exp(\gamma S_0 t)$. Combining these two competing effects, we find that if $S_0 > S_d \simeq v^2/\gamma\mathcal{D}$, rolling grains are generated by dislodgement faster than they are convected downhill – this leads to a catastrophic avalanche. If on the other hand $S_0 < S_d$, the small perturbation is swiftly convected away, leaving the uphill profile in a new frozen state. We therefore predict the existence of a sharp 'spinodal' angle, *different from the static angle of repose*, beyond which the uphill profile is unstable to small perturbations.

We have confirmed this prediction by numerically solving our equations [4]. We find that is a clear division between initial states leading to complete relaxation of the sandpile surface, and those giving either partial or negligible relaxation. With complete relaxation ($S_0 > S_d$), the final state is at the angle of repose whereas when $S_0 < S_d$ a metastable configuration remains after the pile has come to rest. The numerical value of $S_d \simeq 20v^2/\mathcal{D}\gamma$ is somewhat larger than our naive estimate; this is not surprising since, in estimating the dislodgement contribution to the growth rate, we ignored the fact that the mean slope itself decays with time.

The parameter S_d characterizes the macroscopic hysteresis, and is usually called the Bagnold angle [1]. In our approach, this relates to physical parameters describing the dynamics of the profile: γ, v and, crucially, $\mathcal{D}$. Without dispersion ($\mathcal{D} = 0$), we would have no limit to metastability, because $\mathcal{R}$ perturbations would be immediately removed by convection and could not influence anything happening uphill from where they started. In practice of course, if a static sandpile were tilted through a large enough angle, some grains could cease to be supported by those below, and start rolling. The angle at which this occurs is, in fact, a more conventional interpretation of the Bagnold angle [1]. In our picture, such a local mechanical instability is not necessary for the global structure to be unstable. Nor is it sufficient, since for $S_0 < S_d$ the surface does not relax completely even if some isolated grains do start rolling. Therefore our 'spinodal' interpretation of the Bagnold angle is quite distinct from the usual, mechanical one. Our hysteresis mechanism is also quite different from that of Jaeger et al. [11], which was based on a nonmonotonic friction-velocity curve for an individual rolling grain.

For $S_0 > S_d$, we expect the initial profile to relax to $\nabla h = 0$ after a certain 'flushing time' $T(L)$. Here L is the spatial extent of the sandpile. A rough estimate for $T(L)$ can be made by again neglecting the time dependence of the mean slope in the evolution equation for $\mathcal{R}$, treating this as a constant, S_0. The result [4] depends on a certain "Péclet number" Pe $= vL/\mathcal{D}$; for large enough values of this quantity, and for initial slopes S_0 such that $\delta = S_0 - S_d < S_d$, one obtains $T(L) \simeq \frac{vL}{\gamma \mathcal{D} \delta}$, whereas for $\delta > S_d$ one has $T(L) \simeq \frac{L}{\sqrt{\gamma \mathcal{D} \delta}}$. This result has been confirmed numerically. (The behavior of $T(L)$ in the small Pe regime is given in Ref.[4].)

The above arguments can be extended to interpret the 'rotating drum' experiments reported in Ref.[12], in which a cylinder, partially filled with grains, is rotated around its (horizontal) axis with an angular frequency ω. If ω is very small, one expects the slope S of the pile to build up gradually, according to $S(t) = \omega t$, until the spinodal angle S_d is reached. (We presume the mechanics of the drum is imperfect enough to trigger the sandpile instability.)

However, since it takes a finite time $T(L, \omega)$ for the slope to relax after S_d is reached, the *apparent* angle S_a reached before collapse of the surface is slightly increased: $S_a \simeq S_d + \omega T(L, \omega)$. The slope after relaxation will also be slightly increased to $S_c + \omega T(L, \omega)$, since by the time an avalanche has flushed through the system, the upper edge will have moved further upward. The value of $T(L, \omega)$ depends on the relative value of the different parameters $v, \gamma, \mathcal{D}, L$ and ω, and is detailed in [4]. The above picture is valid provided the time between avalanches is large compared to the duration of the avalanches themselves. In the opposite limit, avalanches overlap strongly: this corresponds to a continuous flow regime, as described in Ref.[12-b]. The characteristic angular frequency at which the crossover from isolated avalanches to continuous flow is therefore defined as: $\frac{S_d}{\omega^*} \simeq T(L, \omega^*)$, which is in qualitative agreement with the experiments.

This scenario of regular avalanches between two limiting slopes, reminiscent of a first order phase transition [13], emerges naturally from our model and concurs qualitatively with the experimental findings of Ref.[12].

In summary, we have outlined a new continuum model for the dynamics of sandpile surfaces. The model involves two "hydrodynamic" variables: a height and a rolling grain density, whose interconversion kernel we approximate by a local bilinear form. Although the model is oversimplified in several respects, it leads naturally to the phenomenon of macroscopic hysteresis. In contrast to earlier, mechanical interpretations, the Bagnold hysteresis angle emerges from a "spinodal" stability criterion. We look forward to more detailed experimental work to test this idea.

Aknowledgments. I wish to express my gratitude towards Drs. M. E. Cates and Ravi Prakash and Sir Sam Edwards, with whom part 3 of this work was developed and from whom I have learnt a lot on these and related matters. All the errors and misjudgements are however mine.

References

1. R. A. Bagnold, 'The Physics of Blown Sand and Desert Dunes', (Methuen, London 1941), Proc. Roy. Soc. A 225, 49 (1954); R. L. Brown, J. C. Richard, 'Principles of Powder Mechanics' (Pergamon, New York, 1966).
2. P. Bak, C. Tang, K. Wiesenfeld, Phys. Rev. Lett. 59, 381 (1987), Phys. Rev. A 38, 364 (1988).
3. J.P. Bouchaud, 'Stress distribution in granular media and non linear wave equations', to appear in J. Physique II (France).
4. J.P. Bouchaud, M.E. Cates, J. R. Prakash, and S.F. Edwards, submitted to Phys. Rev. Lett., J.P. Bouchaud, M.E. Cates, J. R. Prakash, and S.F. Edwards, submitted to J. Physique II.
5. J. Smid, J. Novosad, I. Chem. E. Symp. N0 63, p. DV/v/1 (1981)
6. T. Hwa and M. Kardar, Phys. Rev. A 45, 7002 (1992)
7. K. Liffman, D. Y. Chan, B. D. Hughes, Powder Technology 72, 255 (1992)
8. D.C. Hong, Phys. Rev. E 47, 760 (1993), J. Huntley, Phys. Rev. E 48, 4099 (1993)
9. S. F. Edwards, R.B.S. Oakeshott, Physica D 38, 88 (1989)
10. Similar ideas. although with notable differences, were recently put forward in A. Mehta, G. Barker, Phys. Rev. Lett. 67, 394 (1991) and to be published in Rep. Prog. Phys., A. Mehta, in 'Granular Matter', A. Mehta Edt, Springer (1994).
11. H.M. Jaeger, C. H. Liu, S. R. Nagel, T. A. Witten, Europhys. Lett. 11, 619 (1990)
12. a- H.M. Jaeger, C. H. Liu, S. R. Nagel, Phys. Rev. Lett. 62, 40 (1989), b - J. Rajchenbach, Phys. Rev. Lett. 65, 2221 (1990)
13. S. R. Nagel, Rev. Mod. Phys. 64, 321 (1992)

Some Physical Properties of the Burridge-Knopoff Model

Jean-Pierre Vilotte [1], Jean Schmittbuhl [1],

Stéphane Roux [2]

[1]Département T.A.O., Laboratoire de Géologie, URA 1316
École Normale Supérieure, 24 Rue Lhomond, 75231 - Paris cedex 05
[2]Laboratoire de Physique et Mécanique des Milieux Hétérogènes
École Supérieure de Physique et Chimie Industrielles de Paris
2 Rue Vauquelin, 75231 - Paris cedex 05

Abstract: The dynamics of the discrete Burridge-Knopoff model, initially presented for earthquake rupture, is analyzed here as a function of the main physical parameters. This discrete array of massive spring-connected rigid blocks, subject to a velocity weakening friction, exhibits two distinct regimes of solution, depending on the ratio between the characteristic loading and propagative time scales. The first regime is often described as "chaotic", while the second is characterized by propagative solitary waves. The transition between these two regimes is numerically studied, and shown to be the transition between two distinct macrospic frictional regimes : a chaotic stick-slip and a continuous creep. ...

1 Introduction

Earthquakes and large fault networks are surface evidences of the lithospheric active deformation processes. The lithosphere is the outer shell of the earth, with a typical thickness of the order of hundred kilometers. It is divided on a small number of mobile plates,about 12, with relative velocities of the order of a few centimenters per year. The main seismic activity is localized along plate boundaries: mid-oceanic ridges, where plates are diverging; major transform faults, where plates are sliding; subduction zones, where one plate is plunging beneath an other one; collision zones , where one plate is indenting another one. The rheological behavior of the lithosphere is rather complex due to the wide range of temperature, from surface temperature to $\simeq 1300^0$C, and pressure, from atmospheric to some Giga Pascals. At short time scale, the behavior is weakly dissipative and mostly elastic allowing shear and longitudinal wave propagation. At larger time scales, the behavior depends on the pressure and temperature : at low pressure and temperature, it is elastic-brittle, while at high pressure and temperature, it is mainly temperature activated creep.

Table 1. Time scales for earthquake dynamics

Time Scale		Dynamic process
0.1 - 2 s	....	Period od destructive waves
1 - 100 s	....	Rupture time
1 h - 1 y	....	Precursors
1 - 10^2 y	....	Geodesic deformations
10^2 - 10^3 y	....	Seismic Cycle
10^3 - 10^6 y	....	Growth of fault networks
10^6 - 10^8 y	....	Plate Tectonics

Time and space scales are quite important in geophysics. Seismic rupture involves slip patch of the order of 1 to 30 meters over surfaces of 1 to 10^5 Km^2. Time scales of rupture and fault growth are quite distinct, Table 1. The seismic rupture can be considered as a frictional instability along pre-existing fault planes, and fault networks as quenched heterogeneities at the rupture time scale.

Numerous observations, high frequency radiation at the seismic source or global seismicity, has emphasized the spatio-temporal complexity of earthquakes, which is too often reduced to the Gutenberg-Richter (GR) distribution. This power law distribution relates the earthquake frequency to the magnitude, i.e. a measure of the radiated energy, and has been recently considered as the signature of a critical phenomena, [1 – 3].

In the last 15 years, severall hypotheses have been discussed in the literature about the physical origin of earthquake complexity and the radiation of seismic waves from complex rupture events, see for exemple [4 – 6]. Two distinct views have been traditionally put forward: one that consider complexity as the result of quenched heterogeneities, at the scale of the fault network (segments, branching, or curvature) or at the scale of the fault surfaces (asperities, roughness), which control the initiation, arrest and radiation of seismic rupture [7 – 10] and are responsible of strength variations; one inherited from the field of the physics of non linear processes [11 – 13] that consider complexity as the result of the non linear rupture dynamics which continuously tend to produce a critical heterogeneous state of stresses. The latter has been quite revitalized recently both by the rapidly improving precision of the high frequency radiation data [14 – 15], and by the important new perspectives which have been gained from numerous discrete and continuous models based on non linear dynamical friction laws,[11 – 13, 16 – 18]. Observations at the seismic source level do favour strong coseismic slip heterogeneities [19], strong variation of the stress drops for small earthquakes, a flattening effect on the GR distribution [20 – 22]. Stress drop distribution along the fault plane is generally assumed now to depend on the radial wave number, and the power spectrum of the slip and slip rate during an seismic event is a power law of respectively the wave number and the frequency.

Recent studies have focussed on the dynamics of discrete fault models and have brought new perspectives and questions on the subject. For the most part, the attention has been focussed on the complex behavior of discrete arrays of massive spring-connected rigid blocks subject to a velocity-weakening friction law,the original prototype of those models being the Burridge-Knopof (BK) model. The simplest versions of these are based on the original cellular-automata model of Bak and Tang [2], in which an array of identical cells or massive spring connected blocks is loaded uniformly or stochastically. A schedding of load to the four nearest neighbors is precribed whenever a cell (or a block) fails. The state after failure is then quasi-statically determined. The results obtained so far show complex failure sequences which emphasized similarities with the GR distribution law, regarded here as the signature of the self-organized criticality [23 – 25]. Models, which solves the dynamics between slip events, as the original BK model have been revitalized by Carlson and Langer (CL) [12, 26 – 30]. In this model of massive spring-connected blocks, a velocity weakenenig friction law is generally assumed. Depending on the ratios of model parameters, the BK models give rise to complex failure sequences suggestive of the GR distribution,[12, 26 – 27]. However, this regime is not unique and a transition to solitary wave solutions has been recently shown [28, 29].

In this paper we focussed on the dynamics of the BK model, and provide an overview of the different regime of solutions. In Section II, the BK model and the non dimensional parameters are introduced. In Section III, the different regimes of solution are analyzed. In conclusion, we emphasize the limitations of this model in term of seismic rupture and stress out some perspectives on the use of BK models for frictional sliding.

2 The Burridge-Knopoff Model

2.1 Physical analysis

The discrete one-dimensional Burridge-Knopoff consists of a chain of spring-connected rigid blocks lying on a rough support. Each block is connected to its immediate neighbours by horizontal springs and is uniformly loaded through a vertical spring. Frictional contacts take place between the blocks and the support, Fig. 1.

Each block j has a mass m^j, and is connected along the chain to its immediate neighbors, $j-1$, and $j+1$, by horizontal springs of rigidities ${k_h}^{j,j-1}$ and ${k_h}^{j,j+1}$ respectively. A continuous loading, is imposed through a vertical spring of rigidity ${k_v}^j$, by the motion at constant rate V of the rough support. The dynamic of a single block is the result of two distinct phases. A static phase during which, the block remains stuck on the moving support and is statically loaded

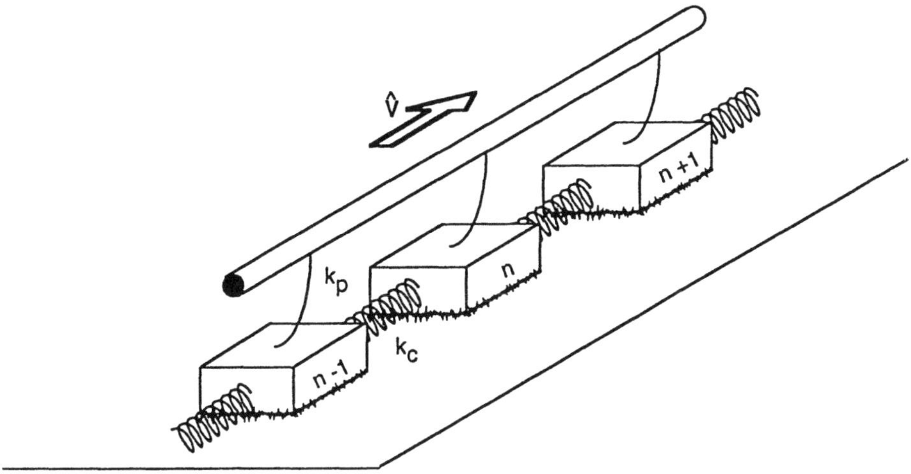

Fig. 1. One dimensional BK model

until the absolute value of the reaction force reaches a given dynamic threshold, ${F_{cr}}^j$ at which the block starts to slip relatively to the moving support. This dynamic phase is dissipative: the dissipation depends on the exact structure of the assumed dynamic friction. As a result of this dissipation, the relative velocity of the block decreases and finally the block switches back to the static phase. The dynamic frictions is assumed here to be the sum of a classical Coulomb's law and a smooth velocity weakening function, simply parametrized by a characteristic weakening velocity V_r, for which the dynamic friction is reduced by a factor two.

A reference frame X is here attached to the rigid rod. The displacement, U^j of each block j is therefore defined with reference to the fixed origin of the vertical spring k_v^j attached to it. The positive direction is defined as the direction of the imposed motion. The position of each block is randomly initialized, inducing an initial random elastic stored energy, both in the horizontal and vertical springs.

The equation of motion, at each time T, is simply:

$$m^j \ddot{U^j} = \mathcal{F}^j(T) + {\mathcal{F}_f}^j \ , \tag{1}$$

where $\mathcal{F}^j$ is the net resultant force on the block j:

$$\mathcal{F}^j(t) = {k_h}^{j,j+1}\left(U^{j+1} - U^j\right) + {k_h}^{j,j-1}\left(U^{j-1} - U^j\right) - {k_v}^j U^j + {k_v}^j VT \tag{2}$$

The friction force $\mathcal{F}_f^j$, is considered here as the sum of two terms, $\mathcal{F}_f^j = \mathcal{F_C}^j + \mathcal{F_D}^j$, where $\mathcal{F_C}$ is a classical Coulomb's law, and $\mathcal{F_D}$ a smooth velocity weakening function.

The Coulomb's friction law $\mathcal{F_C}$ is classically given by :

$$\mathcal{F_C} \in \big[-F_{cr}{}^j, F_{cr}{}^j \big] \tag{3a}$$

$$|\dot{U}^j| > 0 \Rightarrow \mathcal{F_C} = -sgn(\dot{U}^j) F_{cr}{}^j \tag{3b}$$

During the static phase, with no relative block velocity : $\dot{U}^j = 0$, and $\mathcal{F}^j < F_{cr}{}^j$. The net force on the block j, $\mathcal{F}^j$, increases due to a uniform loading through the vertical spring, k_v^j, induced by the motion of the lower support, and the possible dynamics of its two nearest neighbors. The static friction is therefore simply defined during this phase as the reaction force :

$$\mathcal{F_C} = -\mathcal{F}_j = -\Big[k_h{}^{j,j+1} (U^{j+1} - U^j) + k_h{}^{j,j-1} (U^{j-1} - U^j) - k_v{}^j U^j + k_v{}^j V(t) \Big] \tag{4a}$$

$$\mathcal{F_C} \in \big[-F_{cr}{}^j, F_{cr}^j \big] \tag{4b}$$

The Coulomb's law can be written :

$$\mathcal{F_C} = \big[-\mathcal{F}^j + sgn(\dot{U}^j) F_{cr}{}^j \big] \delta(\dot{U}^j) \Theta(F_{cr}{}^j - |\mathcal{F}^j|) - sgn(\dot{U}^j) F_{cr}{}^j \tag{5}$$

The velocity weakening function $\mathcal{F_D}$ can be written as a function of a characteristic weakening rate V_r, and the relative block velocity $\dot{U}_j$:

$$\mathcal{F_D}(\dot{U}^j, V_r) = \Big[sgn(\dot{U}^j) F_{cr}{}^j \Big(1 - \mathcal{G}\big(\frac{|\dot{U}^j|}{V_r}\big) \Big) \Big] \big(1 - \delta(\dot{U}^j) \Theta(F_{cr}{}^j - |\mathcal{F}^j|)\big) \tag{6}$$

where $\delta(.)$, and $\Theta(.)$ are respectively the Dirac and Heaviside functions. The function $\mathcal{G}$ is choosed here a smooth velocity weakening function, Fig. 2 :

$$\mathcal{G}(x) = \frac{1}{1+x} \tag{7}$$

Such a velocity weakening friction law is emperical, eventhough velocity weakening friction laws are often advocated in the geophysical litterature, there is still no physical models or experiments which could allow us to constrain the exact form of the dynamic friction for speeds relevant to the seismic rupture. For low speeds, severall experiments [30 – 32] have suggested rate and state dependent friction laws: In those direct shear friction tests an abrupt change of the driving velocity leads to gradual evolution of the frictional stress, and this transition occurs over a characteristic slip distance which does not depend on the velocity. Extrapolation of those friction laws to higher slip velocities remains however hypothetical. Slip and velocity weakening friction laws have also been advocated but still lack of experimental supports [18, 20].

It is worth noting at present the lack of an internal length in this model, which is therefore intrinsically discrete. Furtheremore, regarding the expected regularity of the solution, it can be noted that if the position and velocity of any

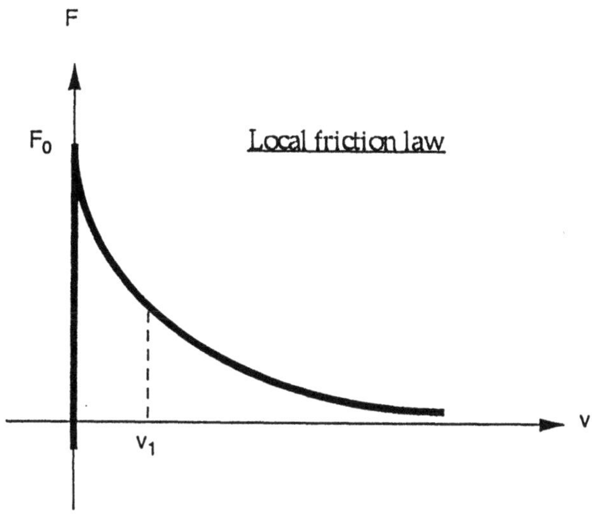

Fig. 2. Dynamic friction law

block are continuous in time, the acceleration exhibits finite discontinuities for $\dot{U} = 0$, and the jumps in acceleration are bounded. The discrete discontinuities of the acceleration are fundemental in the sense that they control the dynamical complexity of this otherwise simple model. This feature is a direct consequence of the non-smoothness of the friction law.

In the following, we consider homegenous models, for which the only heterogeneity is the small initial noise in the block positions:

$$\begin{array}{lllll} m^j & = & m & & \\ {k_v}^j & = & k_v & & \\ {k_h}^{j,j-1} & = & {k_h}^{j,j+1} & = & k_h \\ {F_{cr}}^j & = & F_{cr} & & \end{array} \tag{8}$$

2.2 Dimensionless analysis

We introduce the characteristic time, $T_* = {\omega_v}^{-1} = \sqrt{m/k_v}$, the characteristic length, $U_* = F_{cr}/k_v$, which is the length needed to bring one single block to the dynamic threshold, and therefore the characteristic velocity $V_0 = U_*/T_*$.

The system is now fully described by three dimensionless parameters: a characteristic loading speed, $\nu = V/V_0$, a characteristic rate sensitivity of the dynamic friction, which controls the dissipation, $\nu_r = V_r/V_0$, and a characteristic elastic parameter $l = \sqrt{k_h/k_v}$ which can be interpreted as the sound speed for the model. The friction is normalized using the dynamic threshold: F_{cr}. The dimensionless governing equations can now be rewritten as :

$$\ddot{u^j} = f^j(t) + f_f \tag{9}$$

where the dimensionless resultant force f^j is :

$$f_j(t) = l^2\left(u^{j+1} - 2u^j + u^{j-1}\right) - u^j + \nu t \tag{10}$$

and the normalized friction f_f :

$$\begin{aligned} f_f &= f_C + f_D \\ f_C &= \left[-f^j + sgn(\dot{u^j})\right] \ \delta(\dot{u^j})\Theta(1-|f^j|) - sgn(\dot{u^j}) \\ f_D &= \left[sgn(\dot{u^j})\left(1 - g(\tfrac{|\dot{u^j}|}{\nu_r})\right)\right] \ \{1 - \delta(\dot{u^j})\Theta(1-|f^j|)\} \end{aligned} \tag{11}$$

There is no intrinsic length in this discrete model,and the system size is defined as the number of massive blocks N along the chain. It is worth noting that due to the specific vertical coupling of the BK model, Eq. 10 is no longer a classical massive wave equation but a Klein-Gordon type of equation. Two time scales can be defined in the BK model: a dynamic one, defined as the time needed to bring one single block up to the dynamic threshold; and a dynamic one, defined as the time needed for a wave travelling at the sound speed to propagate along the chain. This allow to define the model parameter : $\theta = N\nu/l$ which is the ratio between these two time scales. The parameter $Theta$ can also be identified to the internal impedance of the system, characterizing the source of mechanical energy. Characteristic geophysical time scales,c.f. Table 1, imply that the loading time, identified as the recurrence time of an earthquake, is by severall order of magnitude greater than the dynamic time scale, identified as the rupture time. Geophysical applications would require typical a very low internal impedance,$\Theta << 1$. On the other hand, for frictional experiments these two time scales are of the same order, i.e. $\Theta \cong 1$. This lead us to distinguish two rather distinct problems: a first one for which $\Theta << 1$, involving two distinct time scales, the loading and the dynamic one, numerically decoupled; a second one for which $\Theta \simeq 1$, and the two time scales can no further be separated, numerically both loading and dynamic pahses have to be solved simultaneously.

3 Numerical Solutions

All the results presented here are obtained for one dimensional models with periodic boundary conditions. The positions of the block are initialized by imposing a small spatial perturbation (white noise of maximum amplitude 0.001). Time integration of the equation of motion is performed using a Runge-Kutta scheme, and special care has to be taken due to the dynamic friction yield. Each numerical simulation are started from random block position (all the blocks being at rest), and after discarding 1000 initial transients. The remaining of the paper, deals with the study of the system dynamics, for a given friction law, as a function of the internal impedance Θ. Specific influence of the characteristic weakening rate $\nu - r$ of the dynamic friction, which characterize the dissipative part of the dynamics is discussed in [33]. Unless otherwise stated, the velocity weakening friction law is the same in all cases and correspond to a characteristic weakening rate $\nu_r = 0.2$.

Depending on the parameter Θ, the system exhibits two rather distinct solutions: a so called "chaotic" solution, extensively studied by Carlson and Langer [11 – 12, 16], and a propagative solution of solitary wave form [28 – 29]. This is well ilustrated in Fig. 3, in which the speed of each block is plotted according to its position in space, and time. A sharp transition can be observed between these two type of solutions. For $\Theta \simeq 0.4$, unstable propagative pulses can be observed along the chain, before major slip events, the size of which is of the order of the system size. These unstable slip pulses are highly variable both in space and time. In contrary, for $\Theta \simeq 1.$, the solution is well organized and has a solitary wave form. This wave is very stable and propagates through the chain, at constant speed and constant wavelength.

Such a transition can also be observed in terms of spatially averaged variables. The spatially averaged velocity $< \dot{u}(t) >$, is plotted versus time in Fig. 4. The transition between the so-called chaotic and propagative solutions appears now as the transition between a chaotic stick-slip motion and a continuous creep. Such a transition is effectively observed in shear friction tests, mainly as a function of the apparatus stiffness. It seems therefore that, for a given friction law, the internal impedance Θ is a good order parameter for the dynamics of the system.

3.1 Separated loading and dynamic time scales: $\Theta << 1$

The loading and dynamic time scales are well separated and the dynamical phase is integrated separately until all the massive blocks relative velocities tend to zero. Then the system is loaded up to the position where one block of the chain reaches the dynamic threshold.

The so called chaotic solution is pictured in Fig. 5, in term of a quake map where slip events are plotted versus time and position along the chain. The black pixels correspond to the actual sliding blocks. The solution appears clearly composed of two types of events: a large number of erratic spatially localized quakes; and almost periodic broad events spatially extended all over the chain.

In term of the spatially averaged displacement $< u(t) >$, see Fig. 6, the solution is well characterized by an overall erratic stick-slip motion. The small quake events are plotted here overimposed on the time-averaged displacement curve. As expected, they do appear as discontinuous short time events compared to the overall motion. Large events correspond to the stick-slip jumps. These small events appear rather correlated in time and all distributed during the global loading of the system. They do prepare large events during which the system is globally unloaded.

The spatially averaged velocity, Fig. 7, exhibits typical a stick-slip behavior with short slip events, and the solution describes in the phase space $< u >_s - < v >_s$ as a loosy limit cycle, see Fig. 8.

It is quite clear from Fig. 5 and Fig. 6, that both types of events have rather different time behavior. Large events have a regular dynamics with a well defined statiscal recurrence time, Fig. 9.

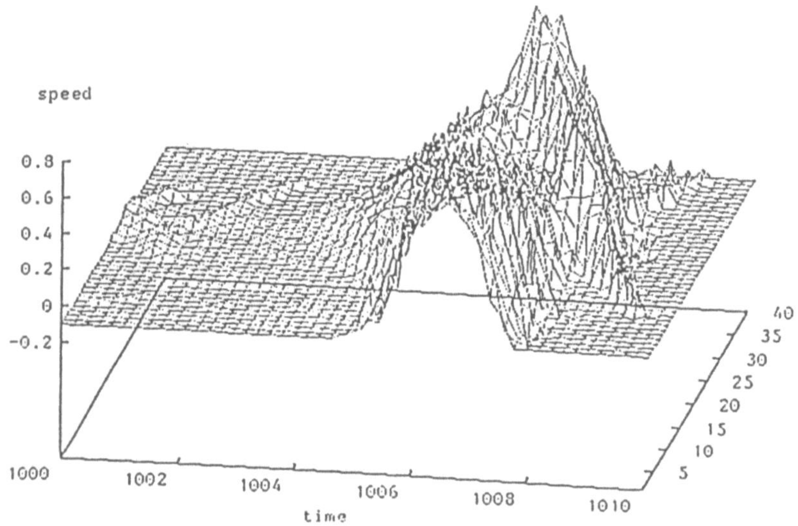

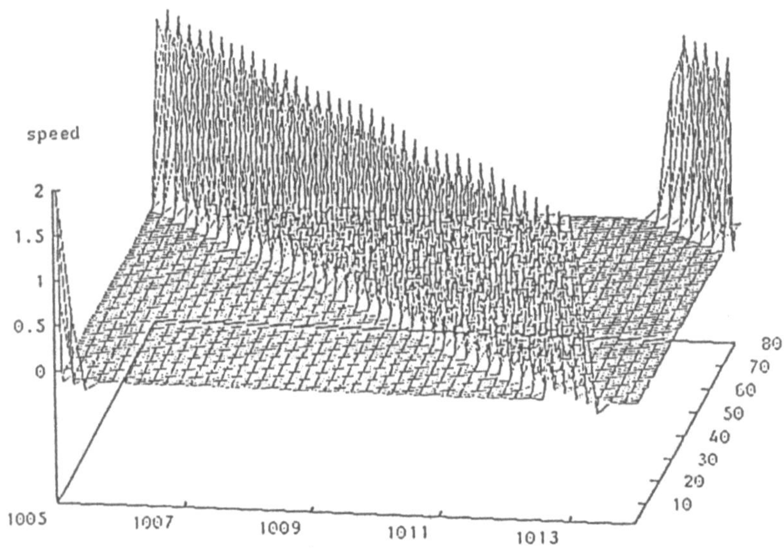

Fig. 3. "Chaotic" and solitary wave solutions for two different values of Θ : $\Theta = 0.4$, and $\Theta = 0.9$. Block velocities are plotted in the space-time diagram. Parameters of the system are $l = 10$, $\nu = 0.1$, and $\nu_r = 0.2$

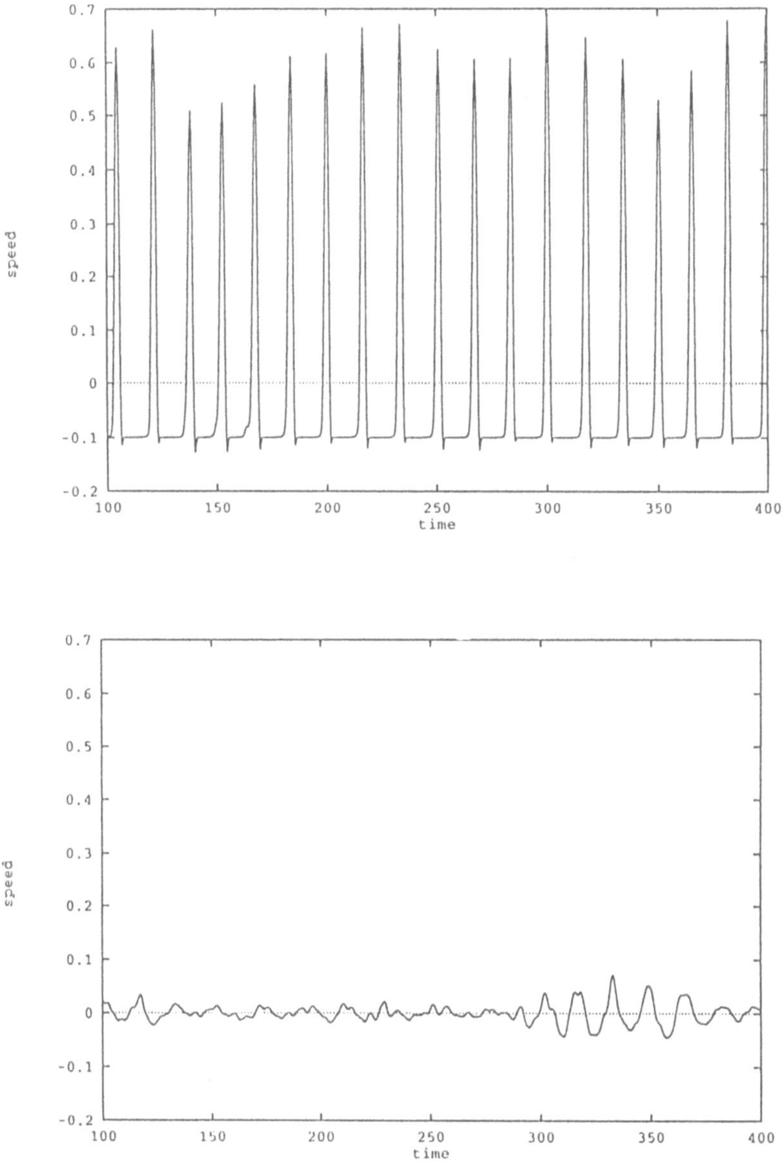

Fig. 4. Stick-slip and creep solution. Spatially averaged velocity versus time for the same systems than in Fig. 3

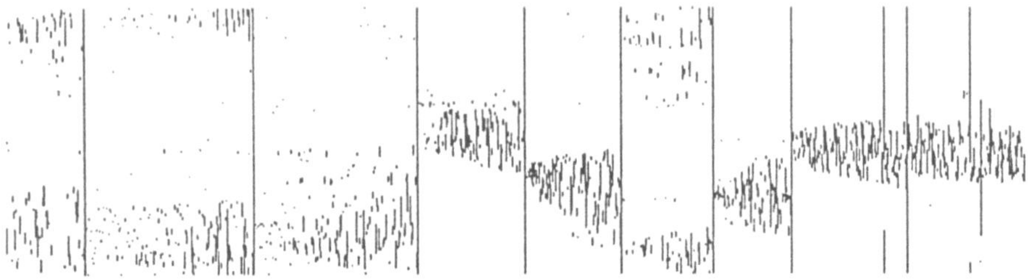

Fig. 5. Quake map : The black pixels represent the blocks which are actually sliding and a space-time diagram. Parameters of the system are $l = 10$, $\nu_r = 0.2$.

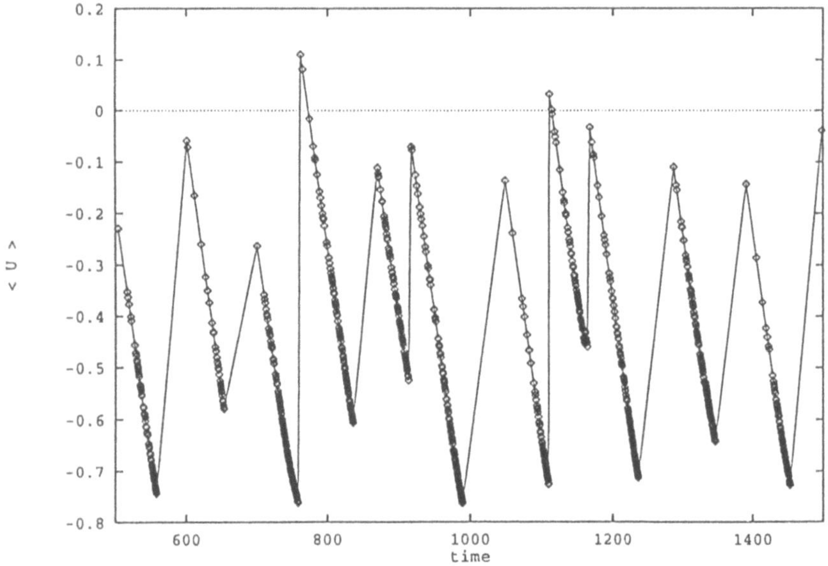

Fig. 6. Spatially averaged displacement $< u >$ versus time for the same system than in Fig. 5. The open circles represent the slip events

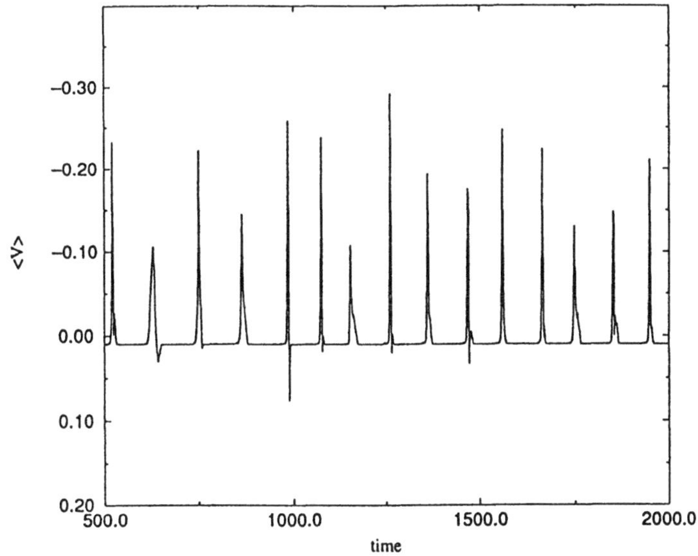

Fig. 7. Spatially averaged velocity $< v >$ versus time, same system as Fig. 5

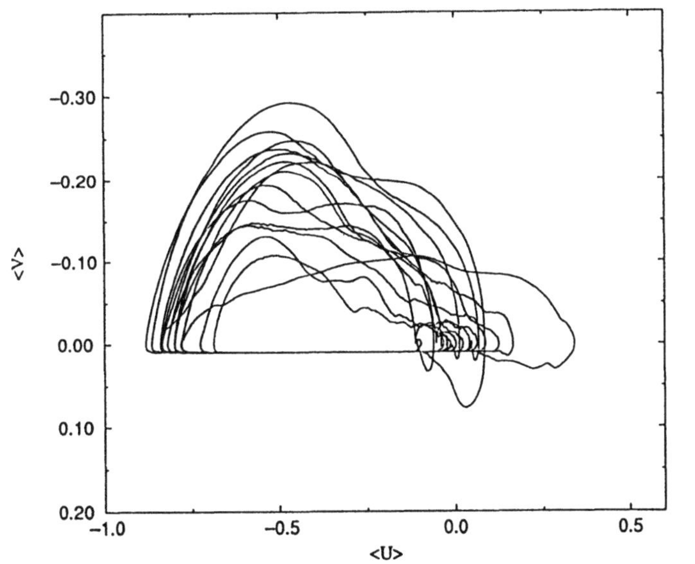

Fig. 8. Trajectories in the phase space : $< v > - < u >$, same system as Fig. 5

A similar rate of production of small events can be observed between each major one, and a statiscally averaged rate of production can defined between two major events, see Fig. 10, with $n_q \propto t^3$, where n_q is the number of quakes. Large events are therefore prepared by the succession of small time correlated events which modify the chain heterogeneity.

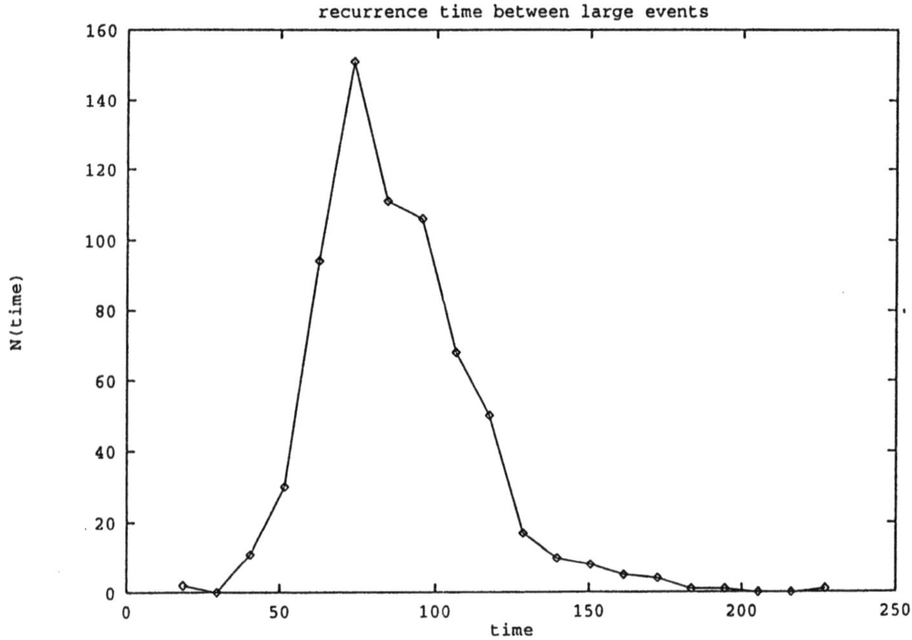

Fig. 9. Histogram of the number of large events as a function of time

In Fig. 11, the cummulative number of events is plotted versus the size of the events, i.e. the number of blocks which participate to this event, for a system size of $N = 250$, and two different values of l.

These distributions emphasize some similarity with the GR distribution for small events. They exhibit an exponential behavior for small events, $N(L) \propto \exp(L/l)$ where $N(L)$ is the number of events of size L, and l is the elastic system parameter. Such a distribution is defined up to an upper critical size L^*,which depends of the parameter l. Above this critical size, events appear unstable and reach the system size.

This dynamical behavior can also be pictured in Fig. 12, where each size of events, the mean kinetic energy is plotted versus the mean dissipative energy; and in Fig. 13, where the mean slip is plotted versus the event size. For small events, both the kinetic and dissipative energy scales with the event size L. For large events, the dissipative energy remains proportional to the system size while now the kinetic energy scales as the square of the event size. This emphasize the idea of a critical event size. For small events, i.e. smaller than the critical size L^*, the GR distribution is well defined and the mean slip, during an event, scales with the event size, which implies that the magnitude scales as the L^3. Large events exhibit a clear departure from this law. Such a behavior suggest characteristic event size which depends on the parameter l.

No clear physical explanation is available for such a critical size L^*. However a hand argument may help to understand that dynamics. For each block, a static yield offset : $f^j - 1$ can be defined, and the heterogeneity along the chain characterized as the spatial fluctuation of this yield offset. After large events, the numerically observed fluctuations of this yield offset have large amplitudes and wavelengths. The succession of small quakes tend to reduce progressively both the amplitude and the wavelength of these fluctuations. The chain, just before a major event, tend to a critical state and is characterized by the existence of zones with small amplitude - high frequency fluctuations of the yield offset. At that stage, when a local event is nucleated, it can no more be stopped and will develop all along the chain. The crititical size L^* can be associated with a critical size of these patches.

3.2 Coupled loading and dynamic time scales : $\Theta \cong 1$

Loading and propagative time scales are in this regime of the same order. The loading and dynamic phases are now solved simultaneously. The characteristic solution is here solitary waves travelling along the BK chain at constant speed and constant wavelength, c.f. Fig. 3. Such solutions appears for $\Theta \cong 0.8$, and are in sharp contrast with the so called chaotic behavior described previously.

In Fig. 14 and 15, the averaged velocity $< v(t) >$ and displacement $< u(t) >$ are respectively plotted versus time. The solitary wave solution appears globally as a creep regime of continuous sliding. In the phase plane $< u > - < v >$, Fig. 16, the dynamics tends now to be reduced to a fixed point.

It can be shown analytically [33], that a complex selection mechanism is at play, and that these solutions are controlled by the detailed structure of the velocity weakening of the friction law and the model parameter Θ. More precisely the exact velocity dependence of the friction law will control the dynamic and stopping discontinuities of the acceleration, the latter being the crucial feature for the observed dynamical complexity. In the cases reported here, the friction law is kept the same for the different numerical simulations, and only the internal impedance Θ has been varied.

In Fig. 17, the time-space evolution of the solitary wave front is reported for different internal impedance Θ (this has been obtained here at fixed l and ν, for different system sizes n). The wave speed, c, is found numerically to be slightly supersonic, while the spatial wavelength R is of the order of $R \simeq 8$ for $l = 10$, and is found to depend of l. Since the system have selected wavespeed and wavelength, the amount of energy dissipated by one solitary wave is fixed and is furthermore bounded from below due to the discrete nature of the BK model.

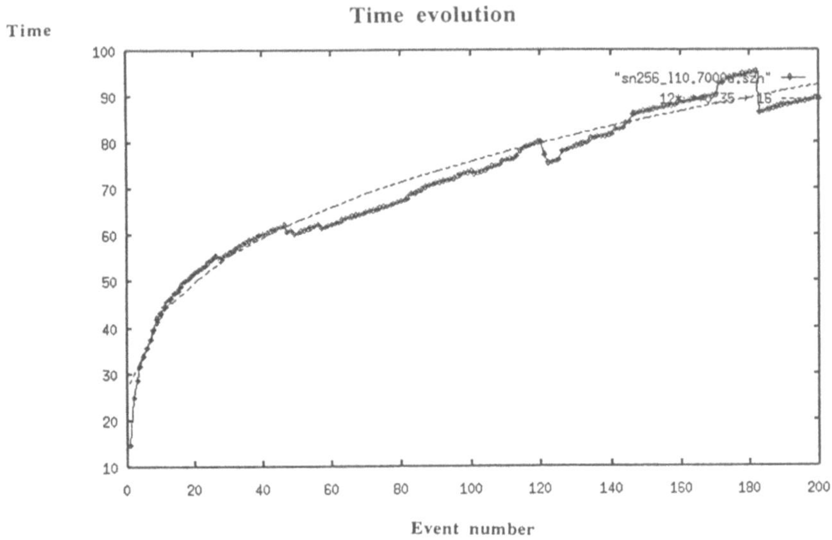

Fig. 10. **Event nunber versus time between two major events. This curve is obtained by averaging over all the major event intervals**

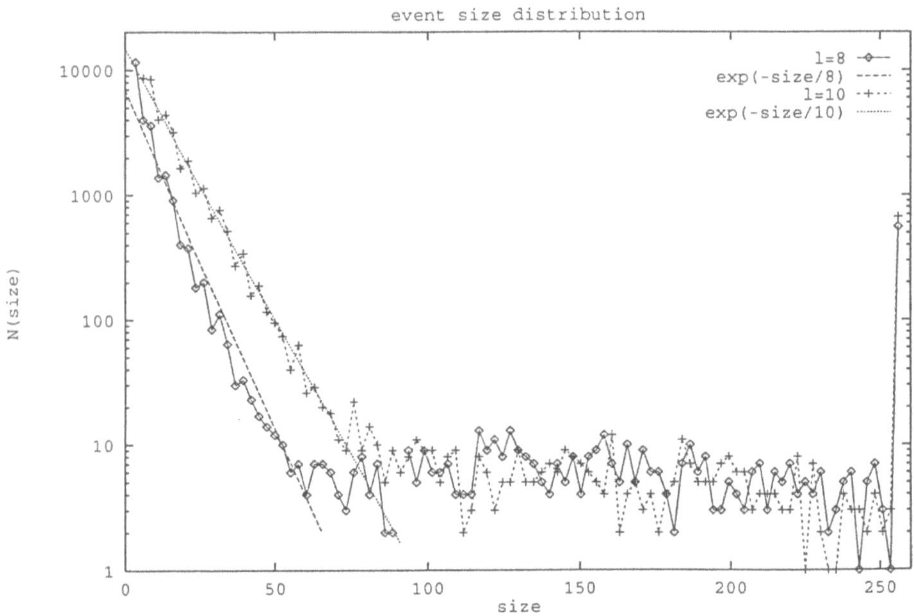

Fig. 11. **Histogram of the event sizes for two different values of l, $N = 250$, and $\nu_r = 0.2$**

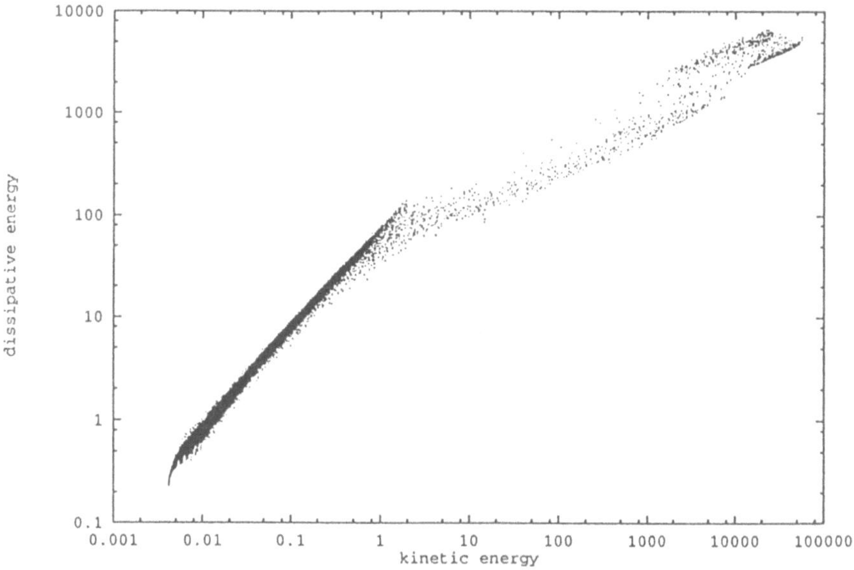

Fig. 12. Mean kinetic energy versus mean dissipative energy. This curve has been obtained by averaging over all the events of the same size.

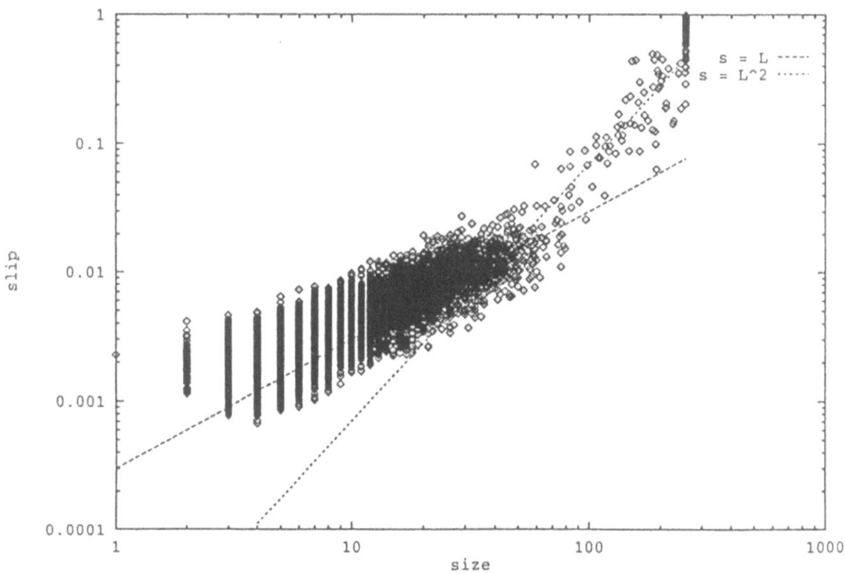

Fig. 13. Mean slip versus size of the event. This curve has been obtained by averaging over all the events of same size.

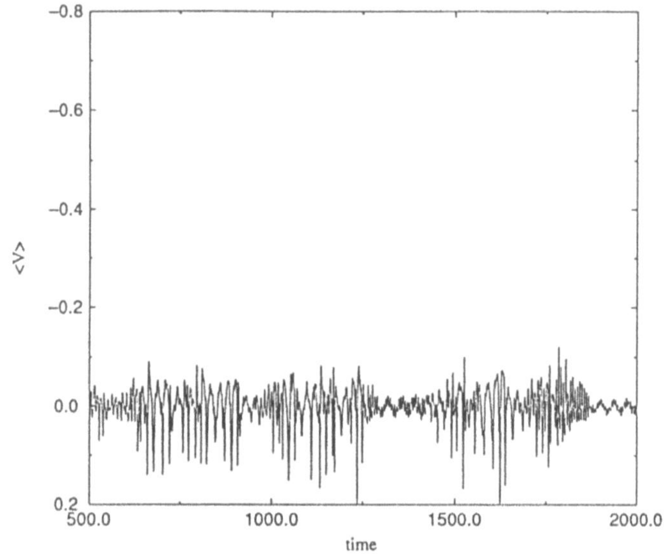

Fig. 14. Averaged velocity $< v >$ for $\Theta \cong 0.9$. The system parameters are $l = 10$, $N = 128$, $\nu = 0.078$, and $\nu_r = 0.2$.

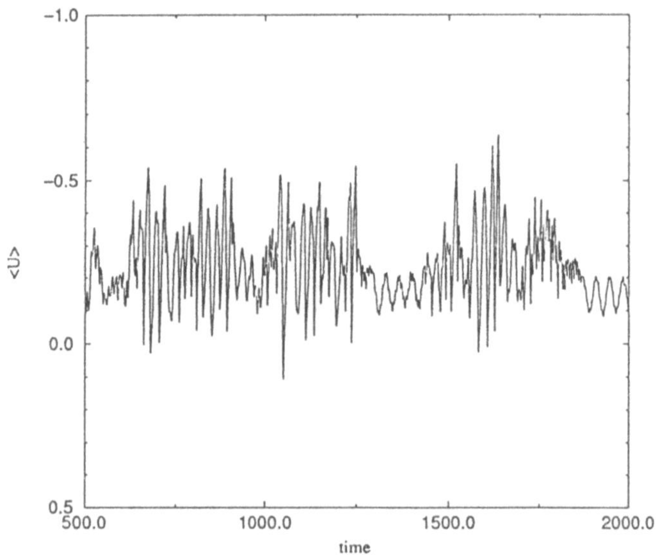

Fig. 15. Averaged displacement $< u >$ versus time, same system as Fig. 14

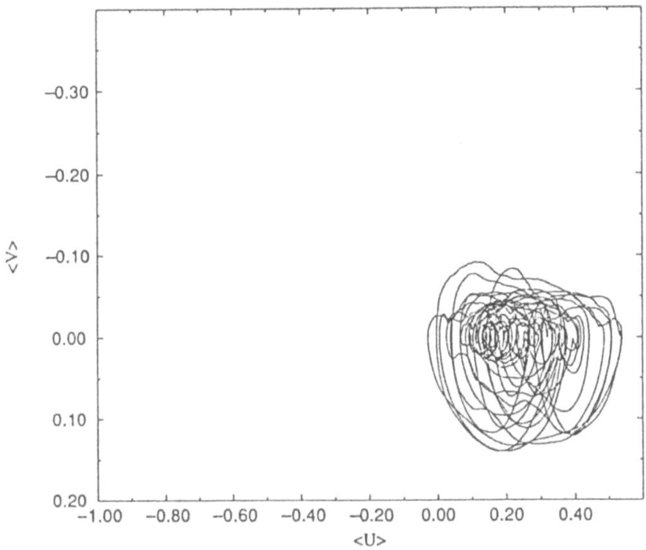

Fig. 16. Trajectories in the phase plane $< v > - < u >$, same system as Fig. 14

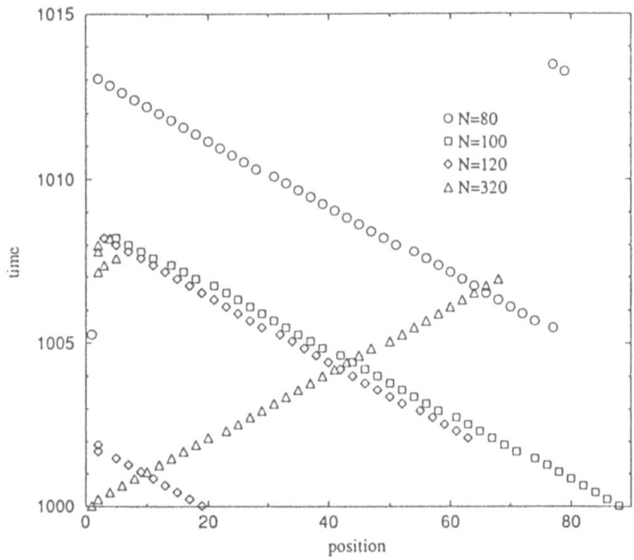

Fig. 17. Time-space evolution of the solitary wave front for different system sizes, the remaining parameters kept constant. The slope defines the wave velocity.

Increasing the internal impedance Θ, above the critical value $\Theta_c \simeq 1.$, leads to the emergence of new solitary waves. Such a process is intrinsically discrete, solution with n_s solitary waves travelling along the chain is only observed for $\Theta \simeq n_s\Theta_c$. In the intermediate regime, $\Theta \in [n_s\Theta_c, (n_s+1)\Theta_c]$, the new solitary wave is unstable until there is enough energy fed in the system. This can be observed in Fig. 18, where a second solitary wave tend to emerge from an existing stable one but can not survive.

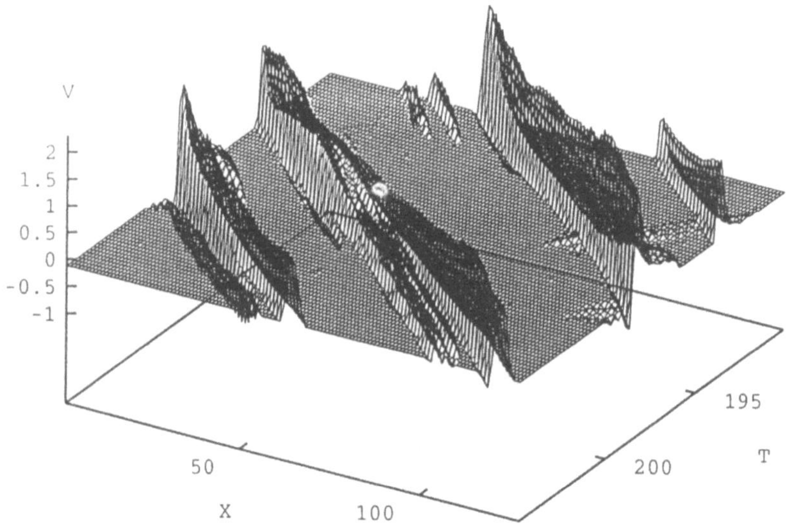

Fig. 18. Sliding velocity plotted in the space-time plane, $\Theta \cong 1.2$. The system parameters are here $l = 10$, $N = 120$, $\nu = 0.1$, and $\nu_r = 0.2$

In order to follow the transition between the so-called chaotic regime and the solitary wave one, the dynamics of the system can be characterized by the mean friction friction force Φ, i.e. averaged over time, space and a given number of independent realizations.

$$\Phi = < f_{\mathcal{F}} > = \frac{1}{K}\sum_{k=1}^{K}\frac{1}{T}\sum_{i}\frac{1}{N}\sum_{j=1}^{N} f_e{}^{j,k}(t_i)$$

where k denotes the realization number ($K = 10$), i denotes the time step number (for a fixed time step Δ), and j the block number. The force f_e is simply here:

$$f_e = \nu t - u^j$$

Such a statiscal measure is usefull in the sense that characterizing the dynamics on the basis of the space-time evolution of the local block motion is quite difficult and may depend on the size of the time window used numerically. The averaged friction can then be studied, for a fixed local friction, as a function of the system size N, of the driving velocity ν, or the sound speed li, see [28] for such study. More generally, that leads to define the averaged friction as a function of the internal impedance Θ. In Fig. 19, the variation of the mean friction force is plotted as a function of the driving velocity. This can obviously be mapped as a variation of the mean friction in term of Θ providing the change of variable $\Theta = N/l \quad \nu$, at fixed N and l.

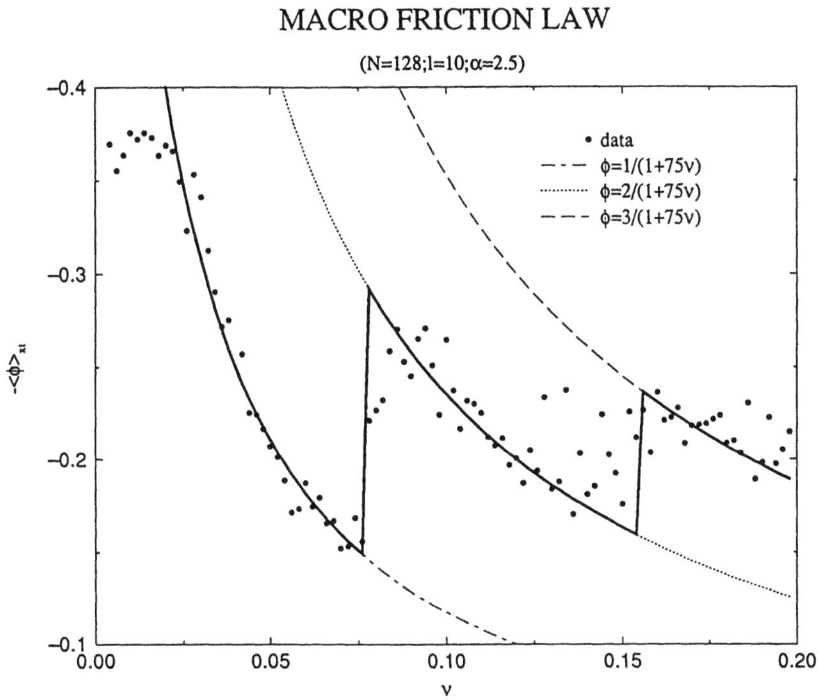

Fig. 19. Time and space averaged friction force as a function of the driving velocity $\nu = l/N \quad \Theta$

Each minimum of the curve corresponds to the emergence of a coherent solitary wave solution. The left part of the curve, before the first minimum, was shown to be transition between "chaotic" and solitary wave solutions, over a certain range of Θ, it is characterized by increasing interactions between unstable propagative pulses until the system reached a critical internal impedance from which the dissipative energy get localized in a stable solitary wave. During this transition, interactions and destabilizations of the propagative pulses induce a remanant chaotic motion and a Θ weakening mean friction force. The emergence of a new solitary wave is also preceeded by the same type of transition,i.e. the mean friction force after a sudden jump starts to decrease again when Θ is increased. The fractional part of the internal impedance is here relevant to the nature of the travelling waves. At constant l, since the wavespeed and the displacement jump across the wave front are system selected, in order to accomodate in a mean sense the prescribed sliding velocity ν, the number of travelling waves n_s has to be proportional to Θ ($n_s = E(\Theta \Delta u)$, where $E(.)$ means the integer part, and $\Delta u \cong 1$). In other words, the critical impedance $\Theta_c \simeq 1$ characterize the level of supplied mechanical energy needed to stabilize a dissipative solitary wave with system selected speed and wavelength. The fractional part of the internal impedance characterize a state in which there is not yet enough mechanical energy to sustain a new solitary wave in the system. This frustration induces a remanant chaotic motion left over by spontaneous switching among travelling wave patterns, which in turn is responsible of the mean friction weakening behavior. When l is varied for fixed ν, and N, the fractional part can also be interpreted as a mismatch between the wavelength and the system size N. This frustration also introduces remanant chaotic motion which leads to spontaneous switching among travelling wave patterns. This can also be analyzed on a section N, ν of the phase space, Fig. 20, where the minima of the mean friction force are effectively located along hyperbolic curves. This statistical behavior appears quite stable. In Fig. 21, a small level of noise is added to the the dynamic threshold along the BK chain. No significant change can be observed when compared with Fig. 21.

Interestingly enough it can be noted that each transition can be fitted with a Θ weakening function of the same functional form than the initial g function of the local friction. Indeed, such Θ weakening function can be simply written as

$$g(\Theta) = \frac{E(\Theta/\Theta_c) + 1}{1 + \Theta/\gamma^*}$$

where $E(.)$ is the integer part. As a function of the driving velocity we have:

$$g(\nu) = \frac{E(\nu/\nu_c)}{1 + \nu/\nu_r{}^*} \tag{13}$$

where ν_c is the critical driving velocity,$\nu_c = l/N \quad \Theta_c$, and ν_r^* the mean characteristic weakening rate ν_r^* is much lower, $\nu_r^* \simeq 0.013$, than the initially prescribed weakening rate $\nu_r \simeq 0.2$, and therefore the mean friction is much more rate sensitive. sensitivity parameter than the initial one, $\nu_r = 2$. Such a property leads to interesting perspectives in term of renormalization which should allow to study the self-similar properties of the friction law with the BK model, see [34, 35], where that kind of analysis is done.

4 Conclusions

Dispite the simplicity of the one dimensional discrete BK model, its dynamics is quite complex and is entirely control by the non-smoothness of the assumed friction law. For a fixed friction law, this sytem exhibits two regimes of solution depending on the value of a single model parameter, the internal impedance Θ which characterize the supplied mechanical energy, or the the ratio between characteristic loading and dynamic time scales.

For $\Theta << 1$, the solution is characterized locally by unstable small propagative pulses which bring progressively the system toward a critical state where large events of the size of the system can take place. In term of spatially averaged behavior, this regime corresponds to an erratic overall stick-slip motion. This small propagatice pulses are controlled by the velocity weakening dependence of the friction law. When the friction law is independent of the velocity, this regime degenerates in a regular one characterized by recurrent large quakes. Interestingly enough this dynamical pattern is rather similar to the one observed for dynamical crack propagation model [18], while the physics of the system is rather different. This emphasize the fact, that most of this observed dynamical complexity is controlled by the stoping discontinuity which in the BK model is on the acceleration while for the crack modek it is on the velocity.

For $\Theta \geq 1$, the solution is characterized by solitary wavwes, with system selected wave speed and wavelength. The number n_s of solitary waves depend on the value of the parameter Θ, $n_s \propto \Theta$. In term of spatially averaged behavior, this regime corresponds to a continuous creep.

For geophysical applications, $\Theta << 1$, since the recurence time for an earthquake is by severall order of magnitude greater than the rupture time. $\Theta \cong 1$ is more relevant to the dynamics of direct shear frictional tests, where a transition between stick-slip and creep motion is effectively observed depending on the imposed rate of sliding or the apparatus stiffness.

The BK model bears however quite an uncertain relation to the problem of the dynamic seismic rupture. Elastic interaction in this model is simplify to nearest blocks interactions, and do not take into account non local interactions as produced by classical elasticity. This implies that stress concentration at the front of the rupture zone is poorly scaled and can lead to a rather different dynamics. Energy radiation is not also correctly taken into account in this model, or only by an ad-hoc viscous damping along the vertical leaf, which implies a large dynamic overshoot of the final slip after rupture. The vertical coupling in the BK model leads to a Klein-Gordon type of equation and not to a massive wave equation. Finally, in this model waves can only travel with supersonic speed. Such supersoniv wave propagation is also observed for local stopping phase in the crack model with a velocity weakening friction law [18].

If the BK model is probably not a very convincing discrete model for dynamic rupture, it may be of interest in the case on frictional sliding. In particular, the transition between erratic stick-slip and continuous sliding motion bears some ressemblance with the actual experiments. An interesting property here, is that

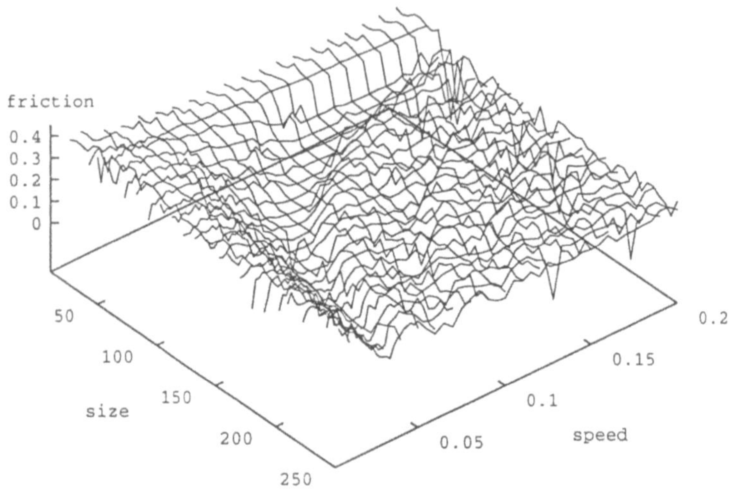

Fig. 20. Mean friction force plotted in the phase plane section N_ν

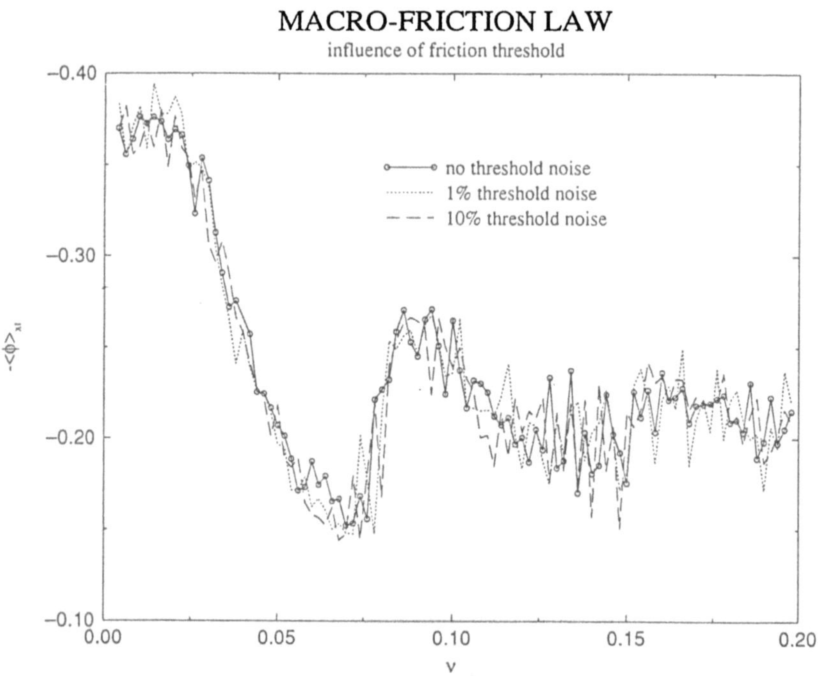

Fig. 21. Mean friction force versus driving velocity ν, when a small noise is added to the dynamic threshold along the BK chain

the averaged friction force comming out from the model has the same functional form than the prescriber local one. This open some perspectives in term of renormalization in order to precize the stability of the local functional form of the dynamic friction law.

References

1. P. Bak, C. Tang, K. Wiesenfeld: Phys. Rev. Lett. **59** 381 (1987)
2. P. Bak, C. Tang: J. Geophys. Res. **94** 15635 (1989)
3. J.B. Rundle: J. Geophys. Res. **98** 21823 (1993)
4. H. Kanamori, G. Stewart: J. Geophys. Res. **83** 3427 (1976)
5. B.V. Kostrov, S. Das: *Principles of earthquake source mechanics* (Cambridge University Press, Cambridge) (1988)
6. R. Madariaga: J. Geophys. Res. **84** 2243 (1979)
7. Y.Y. Kagan: J. NonLinear Sci. **1** 1 (1991)
8. W.L. Power, T.E. Tullis, J.D. Weeks: J. Geophys. Res. **93** 15268 (1988)
9. J. Schmittbuhl, S. Gentier, S. Roux: Geophys. Res. Lett. **66** 2476 (1993)
10. P. Okubo, K. Aki: J. Geophys. Res. **92** 345 (1987)
11. J.M. Carlson, J.S. Langer: Phys. Rev. Lett. **62** 2632 (1989)
12. J.M. Carlson, J.S. Langer: Phys. Rev. A **40** 6470 (1989)
13. Z. Olami, H.J. Feder, K. Christensen: Phys. Rev. Lett. **68** 1244 (1992)
14. T.H. Heaton: Phys. Erath Planet. Sci. **64** 1 (1990)
15. J.N. Brune: Rev. Geophys. **29 supplement** 688 (1991)
16. J.S. Langer, C. Tang: Phys. Rev. Lett., **67** 1043 (1991)
17. J.R. Rice: J. Geophys. Res. submitted (1993)
18. A. Cochard, R. Madariaga: Pure Appl. Geophys. submitted (1994)
19. A.A. Gusev: Pure Appl. Geophys. **130** 635 (1989)
20. K. Aki: J. Geophys. Res. **92** 1349 (1987)
21. P.A. Rydelek, I.S. Sacks: Nature **337** 251 (1989)
22. S.E. Hough, H. Gue, A. Lerner-Larm, S. Seeber, C. Scholz: Seismological Res. Lett. **63** 24 (1992)
23. J.B. Rundle, D.A. Jackson: B.S.S.A. **87** 1363 (1977)
24. S. Brown, J.B. Rundle, C. Scholz: Geophys. Res. Lett. **18** 215 (1991)
25. K. Ito, M. Matsuzaki: J. Geophys. Res. **95** 6853 (1990)
26. J.M. Carlson: Phys. Rev. A **44** 6226 (1991)
27. H. Nakanishi: Phys. Rev. A **41** 7086 (1990); Phys. Rev. A **43** 6613 (1991).
28. J. Schmittbuhl, J.P. Vilotte, S. Roux: Europhys. Lett. **21** 375 (1993)
29. G. Vasconcelos, M. de Sousa Vieira, S.R. Nagel: preprint
30. J.H. Dietrich: J. Geophys. Res. **84** 2161 (1979)
31. T.E. Tullis, J.D. Weeks: Pure Appl. Geophys. **124** 383 (1985)
32. J.R. Rice: Pure Appl. Geophys. **121** 443 (1983)
33. J. Schmittbuhl, J.P. Vilotte, S. Roux: Phys. Rev. E submitted (1994)
34. J. Schmittbuhl, J.P. Vilotte, S. Roux: preprint
35. M. de Sousa Veira, H.J. Herrmann: preprint

Jerky Flow, Stick-Slip in Geological Materials and Earthquake Models

G. Ananthakrishna[1] and Hema Ramachandran[2]

[1] Materials Research Centre, Indian Institute of Science, Bangalore 560 012, INDIA
[2] Seismic Array Station, Bhabha Atomic Research Centre, Gauribidanur, 561 206, INDIA

Abstract. There have been numerous studies on friction in geological materials in order to understand earthquakes [1]. The results are, in many respects similar to those on the Portevin-Le Chatelier effect (PLC). The stress drops in geological materials under different loading conditions and their occurrence within a window of values of temperature and deformation rates have their analogues in the PLC. The change in friction with a change in the slipping speed is reminiscent of type A serration. Further, the velocity weakening friction law used in earthquake models is similar to the negative strain rate behaviour of the flow stress in metals exhibiting the PLC. Some time ago, we derived the above-mentioned features of the PLC using a nonlinear dynamical model consisting of three different types of dislocations [2,3]. In this article, show that the velocity weakening friction law, a crucial input to earthquake models, can be derived along similar lines. In analogy with the three types of dislocations used in the case of the PLC, the basic entities used in this model of a geological fault are potential locks, locks and gouge particles, with some transformations between them. We next consider the dynamics of the homogeneous Burridge-Knopoff model of earthquakes, which uses a simplified form of the velocity weakening friction law. Apart from mimicking the empirical Gutenberg-Richter law, it exhibits soliton-like solutions as well. We have studied this model in detail as a function of the various parameters and have determined its phase diagram.

1. Introduction

A number of instabilities of dynamic origin bear formal resemblance to each other. They exhibit what may be generically referred to as 'stick-slip behaviour'. We list below some examples from mechanics that have been studied in detail. An example which bears a great resemblence to the central theme of the paper is the peeling of a Scotch tape. When the tape is pulled at a constant velocity, the experimental strain energy released shows an unstable region as a function of the drive velocity [4]. Another phenomenon that has been known for some time, and has been studied in detail only recently is the discontinuous periodic flow of sand in an hour glass. Continuous fluid-like flow is seen only in a small domain of the ratio of the particle size to the orifice size [5]. Periodic flow is seen on either sides of the window, with discontinous flow being observed on the lower side and intermittent flow on the higher side. Although the dynamics is complex, the behaviour is formally similar to the stick-slip behaviour. (Indeed the example of the hour-glass is analogous to the plastic instability of the PLC found in creep test.) Yet another example, which will be considered in great detail is the phenomenon of the Portevin-Le Chatelier effect (PLC), also referred to as jerky flow or serrated yielding. Here again, it is known experimentally [6] that the system exhibits a negative strain rate sensitivity (SRS) of the flow stress. A number of theories which attempt to describe such 'stick-slip' phenomena use this aspect of 'negative flow rate characteristics' as a crucial input [6]. In the case of the PLC theories [7,8], the negative SRS is an input. It is only recently that it has been possible to derive the negative SRS from a dynamical model of the PLC [2,3]. Another field where the dynamics is studied with such instability characteristics introduced at a phenomenological level is the homogeneous Burridge-Knopoff model [9] for earthquakes. A simple velocity weakening friction law is assumed to operate along the region of contact of tectonic plates in relative motion. It is due to this that earthquake-like events mimicking the Gutenberg-Richter [10] law are seen. *It is the purpose of this work to examine the role of nonlinear dynamics in the origin of the stick-slip behaviour in earthquakes and its connection to the PLC.* There has, so far, been no theoretical attempt to describe the origin of the static-dynamic friction from a dynamical point of view, despite the fact that it is often an input in the studies of earthquakes [11]. We demonstrate here that the velocity weakening friction law used in the earthquake models is not merely a convenient assumption, but one that can be derived on the basis of the nonlinear interactions between the various entities in the model we propose.

The scheme of this article is as follows - We first discuss the similarities between the PLC effect and the friction experiments on geological samples. We then outline a dynamical model for the PLC effect consisting of three types of dislocations. Using this we show that apart from several experimental features, the negative strain rate sensitivity, which is the central theme of this paper, can be derived. We then construct on similar lines, a dynamical model for the case of stick-slip motion on geological faults. A strain dependent velocity weakening friction law is obtained. A simpler form of the velocity weakening friction law is then used in the Burridge-Knopoff model to show the existence of several phases including one of random motion. We obtain the phase diagram consisting of the chaotic solutions and the soliton solutions.

2. Jerky flow and friction studies on geological materials

Let us first recall the principal features of the jerky flow. Metals subjected to a constant rate of tensile deformation normally exhibit a single yield drop. However, under some metallurgical conditions, successive or repeated yield drops are seen. This is referred to as the PLC effect or the jerky flow. The phenomenon has been observed in several metals [12] such as iron with impurities (eg., C), commercial aluminium, stainless steel, brass, alloys of aluminium and magnesium, etc. Repeated yield drops occur only within certain windows

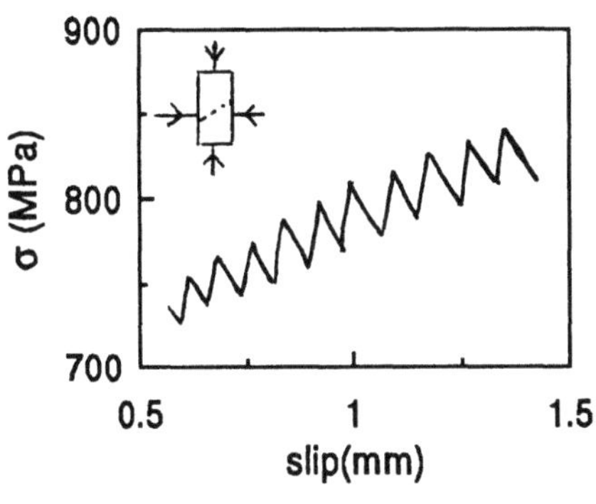

Fig. 1

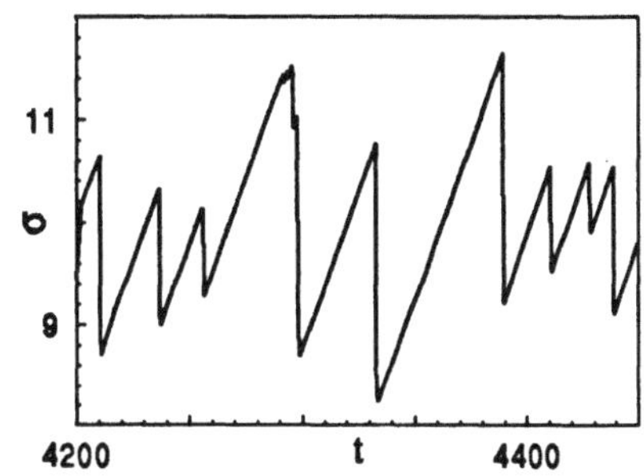

Fig. 2

Fig. 1. : Force vs displacement for Westerley granite. The loading configuration shown in inset. Fig. 2. : The stress-strain curve for $Cu_{10}Al_{90}$ alloy, deformed at a strain rate of $3.3 \times 10^{-6} s^{-1}$ at a temperature of 620K.

of strain rates and temperature. Also, the phenomenon has a critical strain beyond which alone the instablity manifests. The occurrence of repeated yield drops is also sensitive to the concentration of the solute atoms. There are several types of serrations designated A, B,

C and D, each having its own signature. Each of these yield drops is accompanied by the formation and propagation of dislocation bands. As we shall see, several of these aspects of the PLC effect are similar to those exhibited by geological samples.

Figure 1 shows a typical force vs diplacement in experiments on samples of Westerly granite, with the loading configuration given schematically in the inset [1]. These plots bear a striking resemblance to the stress-strain curve in a constant strain rate test in metals. One such plot [13], is shown for a Cu-Al alloy in Fig. 2. The dependence of the load drops on the slip rate and the temperature in rock samples is similar to that in jerky flow (Fig.3). No stress drops are seen at low temperatures for fixed but slow slip rates $\sim 10^{-5} mm/s$. However, finite stress drops are seen which grow in amplitude as a function of temperature upto intermediate values of temperature, thereafter decreasing in amplitude for any further increase in temperature. The load drops are absent once again at higher temperatures. Further stress drops occur only for intermediate ranges of slip rates. Thus, as in the case of jerky flow, the repeated instability occurs only within a window of parameters.

The load drops in the case of the Westerly granite bear a formal resemblance to the stress drops seen in type A serrations in metals, where for a metal with a smooth stress-strain curve, a sudden increase in the strain rate leads to an yield drop. This is shown schematically in Fig.4. This is accompanied by a limiting stress value which is larger for a larger strain rate [12]. This is exactly what is exhibited by rock samples as well[1]. For a velocity increase, a transient stress peak appears before reaching a final steady state value larger than that for a lower slip rate. If the slip rate is reduced, there is a sharp stress drop accompanied by an increase in the stress

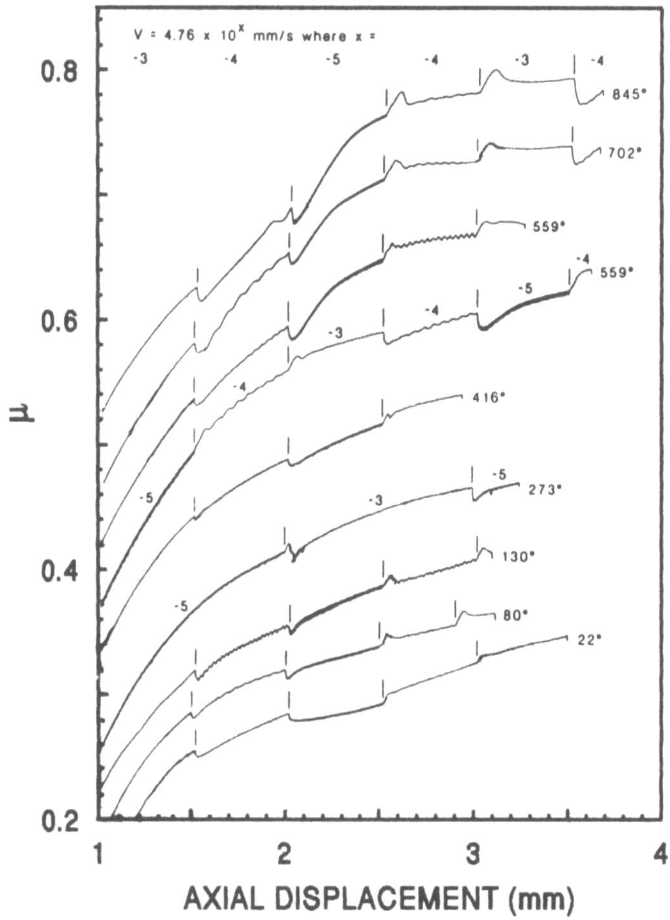

Fig. 3. : The dependence of load drops and slip rates on temperature shown for granite.

level to a final value. These similarities between the PLC effect and the results on rock friction studies are intriguing. Though the underlying mechanisms in the two cases are quite different, these results suggest that they are formally similar and may be regarded as belonging to the same generic class. These statements will become transparent when we outline the model for the stick-slip behaviour along the fault on lines similar to that of the PLC effect. (See section 7 also.)

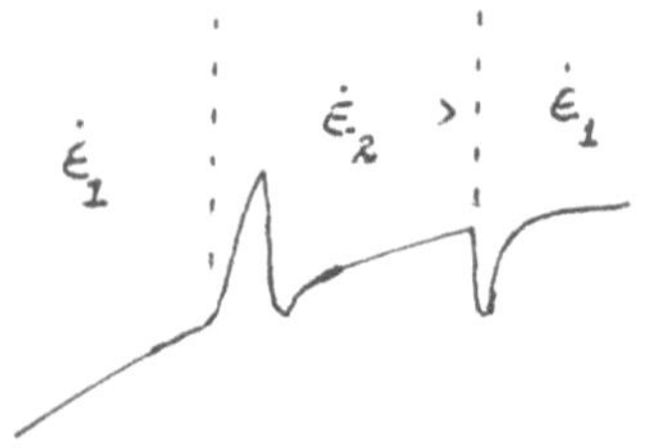

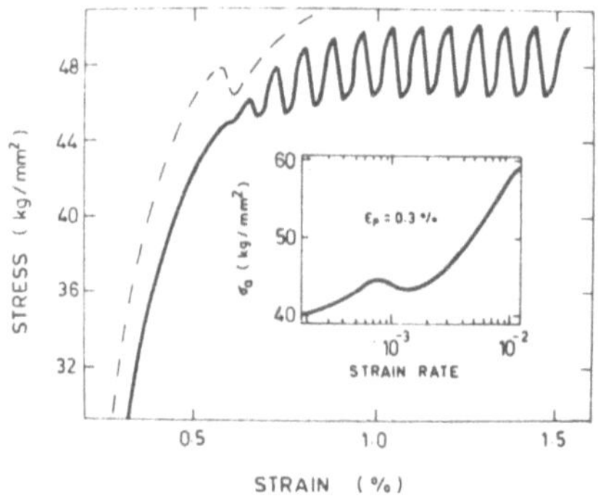

Fig. 4 Fig. 5

Fig. 4. : Schematic representation of the stress-strain curve for type A serrations under change in strain rates. Fig. 5. : The theoretically calculated yield drops. Note the derived negative strain rate dependence of the yield stress shown in the inset.

3. A dynamical model for the PLC

The PLC effect is known to have a microscopic origin arising from dynamic strain ageing which results from the interaction of dislocations with mobile point defects. This induces a negative strain rate sensitivity triggering the onset of the PLC effect. The basic idea was formulated in the work of Cottrell [14]. Most of the present theories [7,8] use the dynamic strain ageing concept in one form or another. None of these models attempt to understand the emergence of the full temporal oscillatory behaviour of the stress strain curve. In this sense all the known phenomenological models are static. Further no effort has been made to understand the connection between the interaction of the defects participating at the microscopic level and the macroscopically measured quantities such as the onset of the PLC at a critical strain, the strain rate range over which the PLC effect occurs, etc. An attempt to understand the problem in the above perspective was made by our group several years ago [2,3]. We proposed a theory involving three types of dislocations and some transformations between them. Even though the spatial inhomogeneous structure was ignored and only the temporal oscillatory state was sought to be described, the model proved to be very successful in that it could explain most of the experimentally observed features [2,3]. The basic idea could be summarized by stating that limit cycle solutions arise due to nonlinear interaction between the three different types of dislocations, suggesting a new mathematical mechanism for the PLC. Recently, the extension of the model to account for the formation and propagation of dislocation bands has also been attempted [15].

The model consists of three types of dislocations - the mobile dislocations, N_m, the immobile dislocations, N_{im}, and those with clouds of solute atoms, N_i. The rate equations for the densities of dislocations are:

$$\dot{N}_m = \theta V_m N_m - \beta N_m^2 - \beta N_m N_{im} + \gamma N_{im} - \alpha N_m \,, \tag{1}$$

$$\dot{N}_{im} = k\beta N_m^2 - \beta N_{im} N_m - \gamma N_{im} + \alpha' N_i \tag{2}$$

$$\dot{N}_i = \alpha N_m - \alpha' N_i . \tag{3}$$

The dot here refers to the time derivative. The first term in Eq.(1) is the rate of production of dislocations due to cross glide with a rate constant θ. V_m is the velocity of the mobile dislocations, which in general, depends on some power of the applied stress σ_a. The second term refers to two mobile dislocations either annihilating or immobilizing with rate constants $(1-k)\beta$ and $k\beta$ respectively ($k\beta N_m^2$

in Eq.2). The third term represents the annihilation of a mobile dislocation with an immobile one. The fourth term represents the remobilization of immobile dislocations due to stress or thermal activation (see γN_{im} in Eq. 2). The last term represents the process of solute atoms gathering around the mobile dislocations thereby slowing down such dislocations. We consider such dislocations as new entities N_i. This process results in an outgoing term in Eq.(1) and an incoming term in Eq.(3). As more and more solute atoms gather, the dislocations eventually stop. We represent this by a loss term in Eq.(3) and a gain term in Eq.(2). Thus, α refers to the concentration of the solute atoms which participate in slowing down the dislocations, and $(\alpha')^{-1}$ refers to the time constant for slowing down to occur. The above equations should be coupled to the machine equation describing the rate of change of the stress developed in the sample. This is given by

$$\dot{\sigma}_a = \kappa[\dot{\epsilon} - bN_m V_m] , \tag{4}$$

where κ is the effective stiffness, b is the Burgers vector and $\dot{\epsilon}$ is the applied strain rate. Using a power law dependence for $V_m = V_0(\sigma_a/\sigma_0)^m$, Eq.(1-4) can be cast into a dimensionless form by using the scaled variables:

$$\begin{aligned} x &= N_m(\beta/\gamma),\ y = N_{im}(\beta/\theta V_0),\ z = N_i(\beta\alpha'/\gamma\alpha), \\ \phi &= (\sigma_a/\sigma_0),\ a = (\alpha/\theta V_0),\ b_0 = (\gamma/\theta V_0),\ c = (\alpha'/\theta V_0), \qquad (5) \\ d &= (\kappa\gamma b V_0/\beta\sigma_0),\ \tau = \theta V_0 t, \text{ and } e = (\dot{\epsilon}\beta/\gamma b V_0). \end{aligned}$$

rewritten as

$$\dot{x} = \phi^m x - b_0 x^2 - ax + y - xy, \tag{6}$$

$$\dot{y} = b_0[k b_0 x^2 - xy - y + az], \tag{7}$$

$$\dot{z} = c[x - z], \tag{8}$$

$$\dot{\phi} = d[e - \phi^m x], \tag{9}$$

There is range of the parameters a, b_0, c, d, m and k for which the densities of dislocations x, y and z are oscillatory. It is obvious that every time the quantity $\phi^m x$ (or $bN_m V_m$) exceeds e there will be an yield drop. The theoretically calculated multiple yield drop curve is shown in Fig. 5. Although this model ignores the spatial dependence, surprisingly many qualitative results such as negative strain rate behaviour of the flow stress, the existence of two critical strain rates within which the phenomenon is observed, the existence of a critical strain, dependence of the amplitude of serration on the strain rate and on the strain, bounds on the concentration of solute atoms and many other features are predicted. *The most important prediction is that the negative strain rate sensitivty of the flow stress arises naturally as a consequence of bifurcation.* The negative strain rate sensitivity of the flow stress is shown in the inset of Fig. 5. The interested reader may refer to the original papers[2,3].

4. A dynamical model for the stick-slip behaviour on faults

Earthquakes are often slip events on the interface of two tectonic plates that are in relative motion caused by mantle-wide convection. We confine our attention to the interfaces of the plates. They have complicated inhomogenieties. Some recent studies indicate that the fault topology is self-affine[16]. In an idealised, one-dimensional representation of the fault (Fig. 6), we picture the two plates as rough surfaces with protrusions. As the plates move past each other, some of the protrusions of one plate may get entangled with those on the other, to form 'locks' that are immobile. Protrusions that are not locked, but are free to move without hindrance are termed 'potential locks'. As some regions are locked, the imposed drive velocity v_d produces strain. This accumulates along the region of contact, being large at some places and small at others. Each part of the fault has a velocity depending on the local shear stress $v(\sigma)$. The locked regions will get 'unlocked' and slip when the local stress exceeds a critical

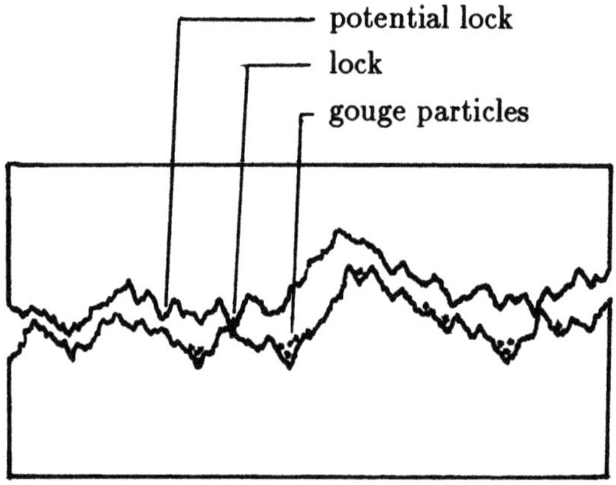

Fig. 6. : A schematic representation of the fault.

value. The 'unlocking' may occur either by elastic or plastic deformation of the protrusions, or by fracture. The last of these leads to the formation of gouge particles, which accumulate at the interface. Thus, regions that are moving at one point of time may be locked and stationary at a later time, and vice-versa. In the most simplified mean-field approach, where the spatial aspects are ignored, we consider a system with the following three entities - potential locks that are in motion (h), locks that are immobile (L) as long as the stress is below a critical value, and the gouge particles (g) that are produced due to wear and tear. The much smoother surface (s) produced due to wear and tear is not explicitly considered. Let ρ_h, ρ_L and ρ_g be the densities of the three entities respectively. The following are the simplest transformations that can be envisaged :

$$h + h \longrightarrow L \tag{10}$$

$$L \longrightarrow Ag + k_1 h + s \tag{11}$$

$$h + h + g \longrightarrow L \tag{12}$$

$$g \longrightarrow s \tag{13}$$

The first transformation refers to two potential locks meeting and forming a lock. The second represents the fact that a lock may fracture giving rise to gouge particles with a rate constant A, with the remaining fraction forming a potential lock (representing a much less dominant asperity). The third transformation denotes the fact that gouge particles can be sandwiched between the two colliding potential locks leading to the formation of locks. The last equation refers to the loss of the gouge material, which can happen in several ways. For instance, it may be lost as it falls into a deep crevice or it may get stuck to the smooth surface. (This surface, that plays the role of pool chemicals in chemical reactions is not explicitly considered in the model.) The rate equations corresponding to these are:

$$\dot{\rho}_h = -\beta v(\sigma)\rho_h^2 + k_1\psi(\sigma)\rho_L - \Lambda v(\sigma)\rho_h^2\rho_g, \tag{14}$$

$$\dot{\rho}_L = \beta v(\sigma)\rho_h^2 - \psi(\sigma)\rho_L + 2\Lambda v(\sigma)\rho_h^2\rho_g, \tag{15}$$

$$\dot{\rho}_g = \zeta\psi(\sigma)\rho_L - \Lambda v(\sigma)\rho_h^2\rho_g - \chi\rho_g. \tag{16}$$

The coefficients of each term represent the rate constants. There are two stress dependent factors $v(\sigma)$ and $\psi(\sigma)$. The first refers to the local velocity while the second represents the breaking stress at which the locks break. Both of these could have a power law dependence on the stress. In the first case, the power law for $v(\sigma)$ can arise due to plastic deformation. A possible justification for the latter is to include in some way, the self-affine nature (power law) of the fault surfaces. This is equivalent to using this information about the geometry in a passive way. The forms we use are: $v(\sigma) = v_0[(\sigma_c - \sigma)/\sigma_c]^n\Theta(\sigma_c - \sigma)$ and $\psi(\sigma) = \psi_0[(\sigma_c - \sigma)/\sigma_c]^m\Theta(\sigma_c - \sigma)$, where Θ is the usual step function. The value of m is chosen to be consistent with the roughness exponent. Lastly, these equations should be coupled to the rate of change of the shear stress developed at the interface due to the constant macroscopic drift velocity v_d. Under some reasonable assumptions this can be written as:

$$\dot{\sigma} = G[v_d - \bar{b}v(\sigma)\rho_h]\,, \tag{17}$$

where G is the shear modulus of the fault and $\bar{b}$ is a constant. The above equations can be written in a scaled form as

$$\begin{aligned}
\dot{X} &= \Theta(1-\phi)[-(1-\phi)^n X^2(1+Z) + k_1(1-\phi)^m Y] && (18)\\
\dot{Y} &= (1-\phi)^n\Theta(1-\phi)X^2(2+Z) - (1-\phi)^m\Theta(1-\phi)Y && (19)\\
\dot{Z} &= [\zeta(1-\phi)^m Y - (1-\phi)^n Z]p\Theta(1-\phi)X^2 - rZ && (20)\\
\dot{\phi} &= \Omega[\nu_d - (1-\phi)^n X], && (21)
\end{aligned}$$

where the differentiation is with respect to a variable $\tau = t\psi_0$ and the variables X, Y and Z are given by,

$$X = (\rho_h\beta v_0/\psi_0), Y = (\rho_L\beta v_0/\psi_0),\ Z = (\rho_g\Lambda/v_0),\ \text{and}\ \phi = (\sigma/\sigma_c), \tag{22}$$

and the parameters p, r and Ω are defined by

$$p = (\psi_0\Lambda/\beta^2 v_0), r = (\chi/\psi_0), \Omega = (G\bar{b}/\sigma_0\psi_0), \text{and}\ \nu_d = (v_d\beta/\psi_0\bar{b}). \tag{23}$$

These equations have a domain of instability for a range of the values of the parameters p, r and Ω. It is straightforward to carry out the analysis on lines similar to the earlier model. We list the principal result; the details will be reported elsewhere. For a fixed value of the strain, we find that there is a window of drive velocity for which the velocity-weakening law is seen. This also means that there is a minimum drive velocity v_{d_c} below which there is no instability. For such values, we should find only a creep of the faults. Although this model emphasizes the origin of the velocity-weakening law, it is desirable to improve the model and make it more realistic. For instance, one criticism could be that as the sliding progresses, the dynamics should be controlled by the layer of gouge particles. However, while this may be true in the case of laboratory experiments, over the geological time scales, where the mean time between the large earthquakes is $\sim$ 100 years, cementing would probably take place. This is represented by Eq.(13). In any case, attempts should also be made to make contact with several physical processes which may be operative on geological faults, for instance, the effect of pore pressure on the velocity-weakening law. This is in progress.

5. The homogeneous Burridge-Knopoff model

In order to understand stick-slip motion, especially with reference to earthquakes, Burridge and Knopoff introduced a simple spring-block model several years ago [9]. The homogeneous version of this model is completely deterministic, and gives rise to earthquake-like events

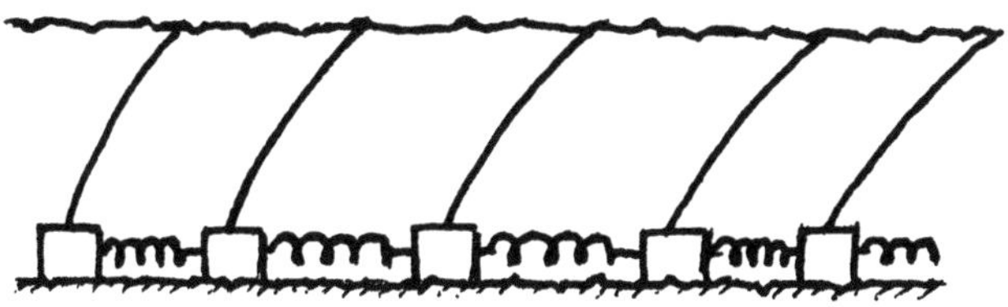

Fig. 7. : The one dimensional Burridge-Knopoff model.

entirely due to its dynamics. This model has, of late, attracted considerable attention due to the increased interest in the physics of dynamical systems. As has been mentioned earlier, the stick-slip behaviour arises because a velocity-weakening friction has been assumed. From the physical point of view, as well as from the model proposed in the earlier section, a negative slope in friction is to be expected only for a window of parameters, implying that earthquake-like events occur only for certain range of physical quantities. There should also be a lower critical value v_{c1}, below which creep should occur, and above which instability should occur. However, the functional form of friction used in the Burridge-Knopoff model has a negative slope from zero drive-velocity. Despite the simplification, the model exhibits rich dynamics.

The Burridge-Knopoff model, applicable to plates grazing past each other, like the San Andreas fault, concentrates on the region of contact of the two plates. The edge of one of the plates is represented by a chain of blocks. Each block of mass m is connected to its two neighbours by identical connecting springs of stiffness k_c and to the bulk of the plate by means of a pulling spring of stiffness k_p (Fig. 7). The second plate is represented as a rough surface on which this chain rests, and its presence is included only in terms of the frictional force it presents to the former. The constant drive velocity causes the springs k_p to pull the blocks with a velocity v with respect to the rough surface. A velocity-weakening frictional force $F(\dot{X}_j) = F_o\phi(\dot{X}_j/v_f)$ with $-F_o < F(0) < F_o$, is assumed. The equation of motion is -

$$m\frac{d^2X_j}{dt^2} = k_c[X_{j+1} - 2X_j + X_{j-1}] - k_pX_j - F_0\phi[(\dot{X}_j + v)/v_f] \quad (24)$$

where X_j denotes the displacement of the jth block, v_f is a characteristic velocity at which the frictional force falls to half the maximum value, and the over-dot on X_j refers to the time derivative. The form of $\phi(y)$ is $sgn(y)/(1 + |y|)$. In the notation of Ref 11, Eq.(24) reads

$$\frac{d^2U_j}{d\tau^2} = l^2[U_{j+1} - 2U_j + U_{j-1}] - U_j - \phi[2\alpha(\dot{U}_j + \nu)] \quad (25)$$

where U_j is the dimensionless displacement of the jth block and τ

the time-like variable. The scaled parameters l $(= \sqrt{k_c/k_p})$, α and ν are measures of the rigidity, friction and interplate velocity.

This model has been extensively studied [11], and many features of its dynamics are well known. Identical initial conditions for all the blocks gives rise to coherent alternate stick and slip motion. The blocks stick as long as the restoring force due to the pulling springs in insufficient to overcome the frictional force. Once the critical strain is reached, the blocks slip (in coherence, due to identical initial conditions), releasing the accumulated strain. The slip event has a time duration τ_s $(\sim \pi)$, and repeats after a duration τ_L. The dynamics is more interesting once a slight randomness is introduced in the initial conditions [11]. The blocks no longer slip in unison. Due to the intrinsic instability in the system, small perturbations grow, and slip events ranging from one-block event to those extending over the entire fault ("great-events", occurring roughly every τ_L) are seen in the steady-state. Carlson, Langer and coworkers term such motion as "chaotic solutions" and they find that these earthquake-like events of different magnitudes have statistics that mimic the empirical Gutenberg-Richter law [10]. Vasconcelos et al [17] suggest that the scaling behaviour of the events exhibits two distinct regimes. They have also studied the model in the limit of constant friction.

Another interesting motion exhibited by the system is one with a soliton-like travelling waves. This was first reported by Schmittbuhl et al [18] who have studied such solutions as a function of the parameter $\theta = \nu \cdot N$. We have studied this phase in detail [19], and found that the friction parameter α and the drive velocity ν play decisive roles in the generation and sustenance of the travelling modes. We have carried out a systematic study of the influence of each of the parameters α, ν, and l on the dynamics of the system. We have also considered the number of blocks, N, as a parameter in order to study the effect of finite size. Apart from the earthquake-like events, and soliton-like travelling waves, we find two new phases of uncorrelated motion. We have also mapped out the phase diagram, and have obtained, from simple order-of-magnitude considerations, the phase boundary between earthquake-like events and solition solutions.

5.1 Numerical results

Using the fourth-order Runge-Kutta method, Eq.(25) was solved for U_j and $\dot{U}_j$. α was varied from 0.25 to 120, ν between 0.008 and 0.1 and l between 0.5 and 15. The size N of the system was varied from 50 to 400, in steps of 50. A few systems with $N < 50$ were studied.

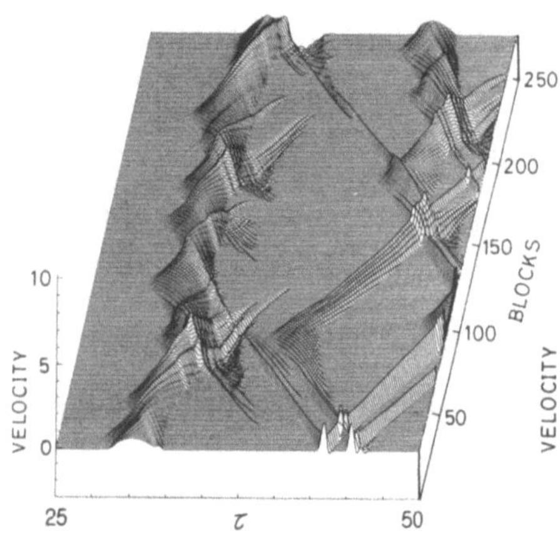

Fig. 8

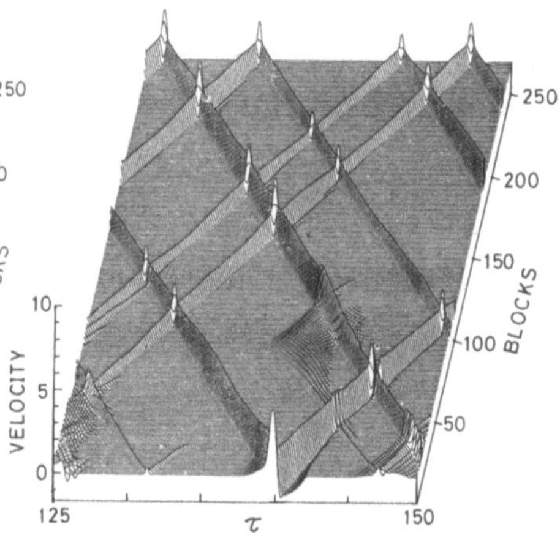

Fig. 9

Fig. 8. : The velocity-space-time profile for $N = 250, \nu = 0.05, \alpha = 4.0$ and $l = 10$, at a very early stage of simulation showing the destabilization of a great event and the emergence of travelling pulses. Fig. 9. : The velocity-space-time plot for the system in the travelling wave regime at $N = 250, \nu = 0.05, \alpha = 2.5$ and $l = 10$. Though at an early stage of computation, the soliton-like nature of the travelling pulse is evident. A fork-like remnant of a great event is also seen.

The boundaries were assumed free. The blocks had a small initial random displacement from equilibrium and zero initial velocity.

To appreciate the results better, we first describe the typical time sequence of results of the numerical simulation. The situation is best understood from the velocity-space-time profile. For a certain range of parameters, long-lived soliton-like travelling pulses are found to be the steady-state solutions. Great events are seen in the initial stages of simulation, with changes in the nature of events occurring over several loading cycles. In the first few cycles, short-lived, localised slip pulses emanate from the great events (Fig. 8), their lifetime increasing with each cycle. Simultaneously, the great events are reduced in magnitude and spatial extent, acquiring a fork-like appearance in the velocity-space-time plot (see for example, Fig. 9). Within several cycles, these travelling pulses are so long-lived that these may be considered to be travelling waves. Thereafter, the fork-like events appear only occassionally. We find that travelling waves have unbounded lifetime, usually surviving to the end of our computation. They have a temporal width of τ_s/l. These waves behave exactly like solitons, passing through one another, without attennuation or amplification and reflecting off boundaries. At the point of intersection, the amplitude of the two waves superpose (see for example Fig. 9). There is a range of values of the parameters in between those corresponding to chaotic solutions and those corresponding to travelling waves, for which the steady-state solutions are travelling pulses with a finite lifetime. For these set of parameters, the system always exhibits travelling pulses of finite lifetime with no truly long-lived travelling wave, however long the simulation is continued. New pulses are created at fork-like events as old ones die out, maintaining a statistical constancy. We shall shortly identify this case as corresponding to the region of coexistence of phases. In all discussions that follow, we shall refer to steady-state solutions only. All results have been compiled after at least $15\tau_L$ had elapsed. Our computations usually lasted about $1500 - 2000\tau$.

We have considered a four-dimensional parameter-space spanned by the variables α, $\dot{\nu}$, l and N. Our numerical studies, carried out on a grid indicate that different phases are stable at different regions of the phase-space. We have computed the phase diagram for the model (Fig.10). Every point on the diagram represents the average

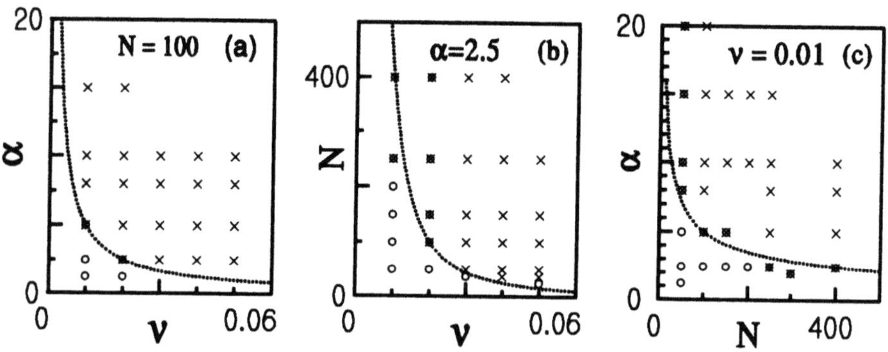

Fig. 10. : Sections of the phase diagram showing (a) the $\alpha - \nu$ plane at $N = 100$ and $l = 10$ (b) the $N - \nu$ plane at $\alpha = 250$ and $l = 10$ and (c) the $\alpha - N$ plane at $\nu = 0.01$ and $l = 10$. o denotes chaotic solutions, $\times$ travelling waves, $\otimes$ the coexisting phases. The dotted line is the curve $4N(\alpha\nu)^2 = 1$.

of five runs of $\sim 2000\tau$ each, with two or three of them starting from random initial conditions. The remaining started from 'near steady state' values, obtained after $\sim 12\tau_L$, beginning from random initial conditions. We consider a state as 'near steady state' when 4 successive frames of 50τ are statistically equivalent. We shall first describe the various sections of the phase-diagram with l held at 10. The $\alpha - \nu$ section (Fig.10a) for $N = 100$ shows two solutions - 'chaotic solutions' and travelling waves marked by o and $\times$ respectively. The dotted line demarcates the two phases. Earthquake-like events or 'chaotic solutions' are seen in a restricted region of phase space, while travelling waves are seen for a very wide range. Transition from one phase to another can be accomplished as a function of α, ν or N. Consider a representative case of $N = 100$, $\alpha = 2.5$, $\nu = 0.01$ and $l = 10$, for which great events are seen. As α is increased, short-lived pulses emanate from the great events. These have a finite steady state lifetime which increases as the transition is approached (ie. as α is increased in this case). Simultaneously, great events degenerate into fork-like events. We regard the region of phase space that exhibit both 'chaotic events' (mainly remnants of great events) and travelling pulses of a finite lifetime as the region of coexistence, and have marked these by $\otimes$ in the phase diagram. A further increase in α ($\alpha_c \geq 6$ in this case) results in travelling waves with unbounded lifetime. The transition from 'chaotic' motion to travelling waves can also be effected by increasing ν ($\nu_c \simeq 0.02$ for $\alpha = 2.5$). The range of α for which travelling waves are seen increases with increasing ν, simultaneously reducing the region of great events. We note that for $\nu > 0.04$, great events are no longer seen for any value of $\alpha > 1$.

The $N - \nu$ section (Fig. 10b) for $\alpha = 2.5$ and $l = 10$ appears quite similar to $\alpha - \nu$ section, except that the region of coexistence is much larger. It is clear from the figure that for low values of relative interplate velocity, ν, travelling waves can be supported only in very large systems.

The section of the phase diagram in the $\alpha - N$ plane is given in Fig.10c for $\nu = 0.01$ and $l = 10$. As before, the region of 'chaotic' solutions and travelling waves are separated by the dotted curve. In small systems, ($N < 50$) great events are seen as expected for small α, no travelling waves are seen as α is increased. For $\alpha > 8$ and $N = 50$, the great events trigger off several pulses simultaneously over the entire fault, all of which die out very rapidly (Fig. 11). This recurs once every τ_L. As α is increased, the number of such pulses also increases. We believe that the proliferation and rapid decay of the travelling pulses arises because of conflicting demands on energy input and dissipation for small systems. The travelling pulses that are created cannot be sustained as they dissipate too much energy. Such events are also regarded as coexistence points.

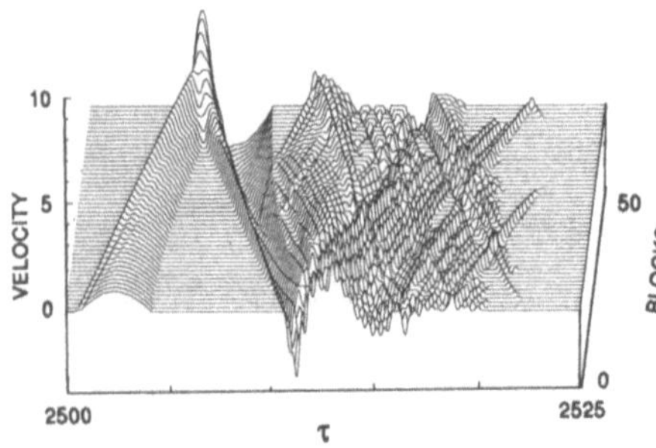

Fig. 11

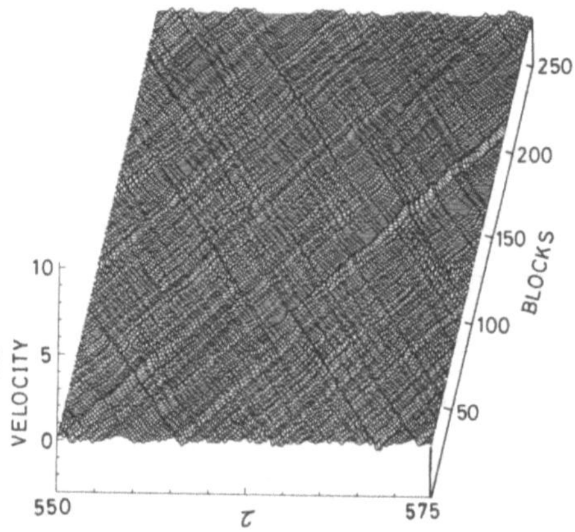

Fig. 12

Fig. 11. : The velocity-space-time plot for $N = 50, \nu = 0.01, \alpha = 20$ and $l = 10$.

Fig. 12. : The velocity-space-time plot for $N = 250, \nu = 0.05, \alpha = 120$ and $l = 10$.

We now obtain an expression for the phase boundary based on total energy considerations. For transparency of arguments, we adopt the notation of Eq.(24). The total energy (TE) of the system is the sum of the potential energy (PE) in the springs, the kinetic energy (KE) of the blocks and the energy dissipated (D) due to friction. Consider chaotic solutions. We note that the small events with smaller velocities (and hence larger friction) dissipate more energy than the great events which have large velocities. The small events have velocities in the range $10^{-3}v_f$ to $0.1v_f$. However, due to fact that events of all sizes occur (with variable correlations in space and time), this does not provide an efficient mechanism for energy dissipation. We also note that the nett effect of all events (including the great events) is to maintain an average speed of v. Thus, the average KE of N blocks is $Nmv^2/2$. The time-averaged total energy (over a full cycle equal to the time between two great events), is $TE_C = PE_C + mv^2N/2 + D_C$. The subscript C refers to the chaotic events. For the case of the system supporting a single travelling wave, similar contributions exist, with one major difference. The movement of all blocks is accomplished in a sequential way, by a travelling slip wave, with $\sim n \ll N$ blocks slipping at a time with a slip velocity $\sim v_f$ which is large. (Note that the friction is appreciably reduced for velocities $\sim v_f$). In contrast, the distribution of instantaneous velocities of the remaining $N-n$ blocks is confined to much smaller velocities ($\sim 10^{-3}v_f$) which are less than the velocities during chaotic events. This is borne out by our simulations as well. The dissipation is therefore larger in the case of travelling waves. Thus, the time averaged total energy $TE_{TW} = PE_{TW} + mv_f^2 f(n)/2 + D_{TW}$, where TW denotes a travelling wave, $f(n)$ the area under the travelling wave, PE_{TW} the potential energy, and D_{TW} the dissipation due to the $N-n$ blocks. $f(n)$ can be determined only if the functional form of the travelling wave is known Dissipation due to the n blocks slipping is negligible on account of the large slip velocities. It is not easy to estimate the PE terms, however, they are likely to be comparable. Since the energy input to the system is a constant, we have $mv^2N/2 + D_C \simeq mv_f^2 f(n)/2 + D_{TW}$. Travelling waves will be favoured when $D_{TW} > D_C$, or $N(v/v_f)^2 > f(n)$. Since $2\alpha\nu = v/v_f$, the transition will occur when $N(2\alpha\nu)^2 = \psi(l)$. (Since n is found to depend on l alone, we use $f(n) = \psi(l)$). From our simulations, we have estimated $\psi(l)$ to be unity for $l = 10$. Thus $\Lambda = N[2\alpha\nu]^2$, can be regarded as a more appropriate order parameter for $l = 10$, than $\theta = N\nu$ suggested by Schmittbuhl et al [8]. The dotted curve in Figs. 10a-c, separating the two phases, in fact, corresponds to $\Lambda(N, \nu, \alpha, l = 10) = 1$.

We find that the number of travelling waves n_{TW} increases with each of the parameters α, ν and N [19]. Changes as a function of α are most interesting. While n_{TW} increases from 5 to 10 when α increases from 2.5 to 10 ($N = 250$, $\nu = 0.05$), a further increase in α gives rise to several smaller amplitude travelling waves in between the long-lived travelling waves along the fault. The lifetimes of these

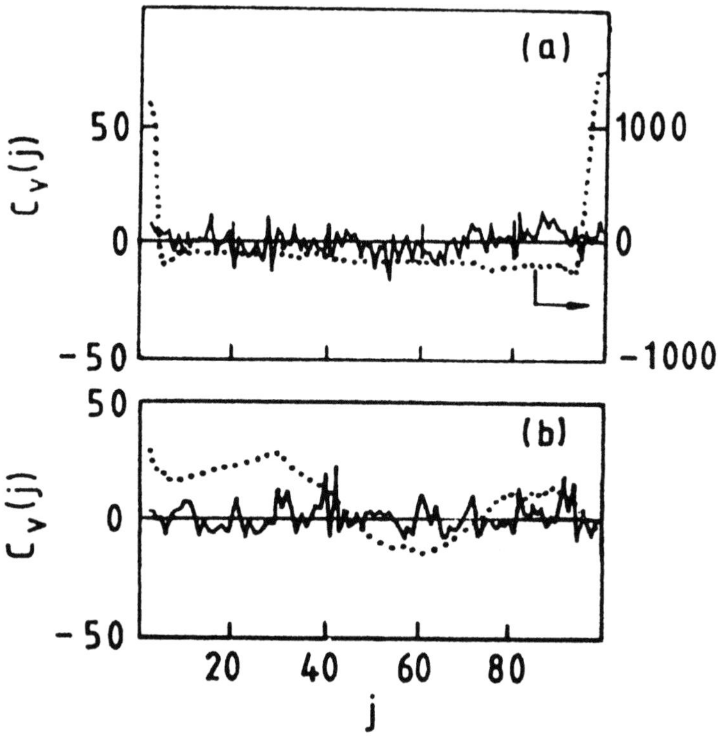

Fig. 13. : $C_v(j)$ as a function of j for $N = 250$ and $\nu = 0.05$. (a) The bold curve corresponds to $\alpha = 120$ and $l = 10$. The dotted curve corresponds to $\alpha = 2.5$ and $l = 10$. (b) The bold curve corresponds to $\alpha = 2.5$ and $l = 0.5$. The dotted curve corresponds to $\alpha = 0.25$ and $l = 10$.

small amplitude waves decrease as α is increased beyond 20. A balance exists between the number of travelling waves created and those decaying. Finally, for very large values of α (eg. 120), n_{TW} is so large that individual travelling waves can no longer be discerned (Fig. 12). This causes a loss of correlation in the velocity as can be seen from the plots of the velocity correlation function $C_v(j) = [< \dot{U}_i.\dot{U}_{i+j} > - < \dot{U}_i >^2]/ < \dot{U}_i >^2$ for $\alpha = 120$ (Fig.13a). Also shown is the case $\alpha = 2.5$ (dotted curve). While the former lacks correlation (bold curve), the latter shows a definite correlation over two different length scales - one corresponding to the width of the travelling waves and the other corresponding to the separation between the waves along the fault. The rapid fluctuations seen in Fig. 13a are smoothened on time-averaging.

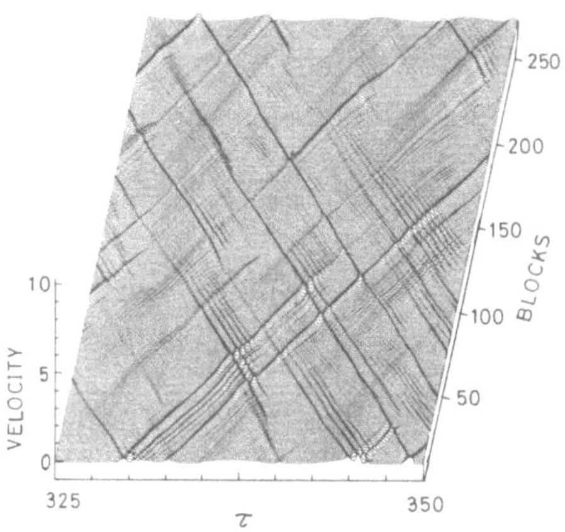

Fig. 14

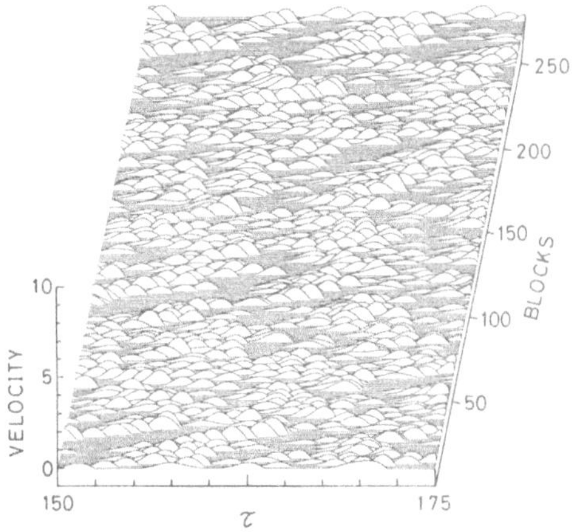

Fig. 15

Fig. 14. : The velocity-space-time plot for $N = 250, \alpha = 0.5, \nu = 0.05$ and $l = 10$. The steady-state is attained quite rapidly. Fig. 15. : The velocity-space-time plot for $N = 250, \alpha = 1.0, \nu = 0.05$ and $l = 1$. The steady-state is attained quite rapidly.

A decrease in α from 2.5 through 1 causes the velocity-space-time surface to look very wrinkled. All blocks are in continuous motion now [17], and the situation is significantly different from the case when $\alpha > 1$. Numerous propagating solutions of small magnitude and short lifetime are seen (Fig. 14), having a much smoother variation in velocity as compared to the large a limit. This is reflected in the velocity-velocity correlation function for $\alpha = 0.25$ (dotted curve in Fig. 13b).

There is yet another class of motion - that of random uncorrelated slipping of blocks, brought about by reducing l. A large value of l implies that the connecting springs are quite stiff, and so the blocks move in a strongly correlated fashion. A very small value of l, on the other hand, represents very weak connecting srings permitting blocks to slip nearly independently (see Fig.15 for $l = 1$). For $l < 1$, the motion of the blocks becomes random. Fig. (13b) shows the complete lack of correlation for $l = 0.5$ (bold curve). This is not significantly different from $\alpha = 120$ case, though the velocity-space time profiles (Figs. 12 and 15) look quite different in the two cases.

Simulations for $l = 15, 10, 8, 5$ and 2 (for various values of α and ν) show that the velocity of propagation of the travelling wave solutions depends linearly on l and is independent of both α and ν. Thus the travelling waves reported here are not the same as those considered by Langer and Tang [20] since the velocity of the latter depends on both α and l.

6. Discussion and conclusions

The central theme of this paper is to analyse the origin of stick-slip behaviour from the point of view of dynamical systems. The dynamics of systems exhibiting stick-slip behaviour is normally handled by an assumed negative flow-rate characteristics at a phenomenological level. We have shown that this negative flow-rate characteristic is a consequence of nonlinear interaction between participating entities. For example, in the case of the PLC effect, these correspond to the mobile dislocations, the immobile dislocations and those with clouds of solute atoms. In an analogous way, in the case of stick-slip behaviour of faults, the corresponding entities are the potential locks, the locks and the gouge particles. Just as in the PLC effect, mobile dislocations on interaction form immobile dislocation locks, in the case of the fault model, mobile potential locks transform to immobile locks on contact. The cloud of solute atoms gathering around dislocations has a correspondence in the gouge particles accumulating at the fault. The mechanism operating is a cyclic tranformation between these three entities. From the mathematical point of view, the negative flow-rate characteristic arises as a natural consequence of Hopf bifurcation as a function of the drive parameters (the applied strain rate or the relative velocity of the tectonic plates). The model has been prompted by the broad similarities between the PLC effect and the rock-friction studies. However, it should be mentioned that this model is much too simplistic a representation of the fault dynamics. The model can be improved by including various mechanisms operating at the faults. Any correspondence with real life situation, however remote, needs an estimation of the values of the various parameters in the model. This is currently under investigation.

As an exercise in the study of the dynamics of earthquakes, the well-known Burridge-Knopoff model has been investigated as a function of all the parameters entering the picture. We find that although soliton-like solutions are seen for a wide range of values of the parameters, and in particular, in the thermodynamic limit, we believe, that for the drift velocities of the tectonic plates only earthquake-like events are likely. This remark has been made possible by the fact that the full phase diagram, involving the chaotic solutions and the soliton solutions has been obtained.

References

[1] D. A. Lockner, R. Summer and J. D. Byerlee, Pure and Appl. Geophys., **124**, 445 (1986); D. A. Lockner and J. D. Byerlee, US Geological Survey, Rep. No. 86-417 (1986).

[2] G. Ananthakrishna, and D. Sahoo. , J.Phys., **D14**, 2081 (1981); M. C. Valsakumar and G. Ananthakrishna, J. Phys., **D16**, 1055 (1983).

[3] G. Ananthakrishna and M. C. Valsakumar, J. Phys.,**D15**, L171 (1982); G. Ananthakrishna, Bull. Mater. Sci., **6**, 665 (1984) ; G. Ananthakrishna and M. C. Valsakumar, Phys. Lett., **A95**, 69 (1983) ; G. Ananthakrishna, *Directions in Chaos* Ed. Hao Bai-Lin,(Singapore, World Scientific 1990), p 133; G. Ananthakrishna in *Non Linear Phenomena in Materials Science* , Ed. L. P. Kubin and G. Martin, (Switzerland, Trans Tech, 1988); G. Ananthakrishna in *Non Linear Phenomena in Materials Science II*, Ed. G. Martin and L. P. Kubin, (Switzerland, Trans Tech, 1992).

[4] D. Maugis and M. Barquins in *Adhesion 12* , Ed. K.W. Allen (Elsevier 1989).

[5] X -l. Wu, K. J. Maloy, A. Hansen, M. Ammi and D. Bideau, Phys. Rev. Lett., **71**, 1363 (1993).

[6] S. R. Bodner and A. Rosen, J. Mech. Phys. Solids, **15**, 63 (1967); P. Penning, Acta Metall., **20**, 1169 (1972).

[7] Y. Estrin and L. P. Kubin, J. Mech. Behaviour of Mater., **2**, 255 (1989) ; L. P. Kubin and Y. Estrin, Acta Metall., **38**, 697 (1990) and references contained therein.

[8] E. Aifantis in 'Nonlinear Phenomena in Materials Science ' (1988) and the references therein ; V. Jean-Claude and C. Fressengeas, C. R. Acad. Sci., Paris, **315**, Series II, p869 (1993).

[9] R. Burridge and L. Knopoff, Bull.Seismol. Soc. Am., **57**, 341 (1967).

[10] B. Gutenberg and C. F. Richter, Ann. Geophys., **9**, 1 (1956).

[11] J. M. Carlson and J. S. Langer, Phys. Rev. Lett., **62**, 2632 (1989). J. M. Carlson and J. S. Langer, Phys. Rev., **A40**, 6470 (1989). B. E. Shaw, J. M. Carlson and J. S. Langer, J. Geophys. Res., **97**, 479 (1992); J. M. Carlson, Phys. Rev., **A44**, 6226 (1991); J. M. Carlson, J Geophys. Res., **B96**, 4255 (1991).

[12] B. J. Brindley and P. J. Worthington, Metall. Rev., **145**, 101 (1970).

[13] C. Engelke, J. Plessing and H. Neuhauser, private communication.

[14] A. H. Cottrell, Phil. Mag., **44**, 829, (1953).

[15] G. Ananthakrishna, Scripta Metall., **29**, 1183 (1993).

[16] C. Sammis, G. King and R. Biegel, Pageophys. **125**, 776 (1987); W. L. Power, T. E. Tullis, S. R. Brown, G. N. Boitnott and C. H. Scholz, Geophys. Res. Lett., **14**, 29 (1987).

[17] G. L. Vasconcelos, M. de Sousa Vieira and S. R. Nagel, Physica, **A191**, 69 (1992); M. de Sousa Vieira, G. L. Vasconcelos and S. R. Nagel, Phys. Rev., **E47**, 2221 (1993).

[18] J. Schmittbuhl, J. P. Vilotte and S. Roux, Europhys. Lett., **21**, 375 (1993).

[19] Hema Ramachandran, G. Ananthakrishna and R.N. Bharthur, Communicated for publication; Hema Ramachandran and G. Ananthakrishna, communicated for publication.

[20] J. S. Langer and C. Tang, Phys. Rev. Lett., **67** 1043 (1991).

Dynamics and Structure of Displacement Fronts in Two–Dimensional Porous Media

Knut Jørgen Måløy, Liv Furuberg, Einar Hinrichsen,
Jens Feder and Torstein Jøssang

Department of Physics, University of Oslo,
Box 1048, Blindern, 0316 Oslo 3, Norway

Abstract: The dynamics and structure of drainage in 2D porous media are discussed. We restrict the discussion to the extreme cases of slow injection rates, where capillary forces govern the dispacement, and fast injection rates, where viscous forces dominate the process.

At low constant injection rates the dynamics is investigated by measuring the pressure fluctuations in the displaced fluid. These fluctuations exhibit sudden jumps that identify bursts where the front proceeds abruptly. A modified invasion percolation algorithm that includes a capacitive interface volume models the experiment. In general the pressure step distribution has a power law behaviour with crossover to exponential behaviour.

In the constant pressure, fast flow, viscous fingering experiments, we discuss the displacement structures in terms of tree-like fractals. The different branches can be characterized by a unique branch order in a way that exhibit scaling. This permits a self-similar characterization of the structure analogous to the procedure usual used to describe deterministic fractals. The dynamics of the viscous fingers are also discussed, and compared with a modified DLA simulation.

Keywords: Fluctuation phenomena, Brownian motion, Fractals, Porous media

1 Introduction

The large number of different geometrical structures observed in fluid-fluid displacement in porous media [1,2] reflects the rich variety of physics which can be investigated in these experiments. The patterns range from compact to ramified and fractal [3]. In general the displacement fronts are sensitive to the local quenched randomness of the porous medium [1,4,5,6] wettability, [1,7] injection rates and the physical parameters of the fluids as well as imposed global fields [8,9].

In this paper we will restrict the discussion to drainage, where a non-wetting fluid is injected into a porous medium saturated with a wetting fluid. The systems considered are two-dimensional or quasi two-dimensional. Furthermore we will only consider the two extreme regimes of slow and fast injection rates. Typical structures are shown in Fig. 1. The dynamics of these two displacement regimes are very different. In the slow regime there are periods of stable front

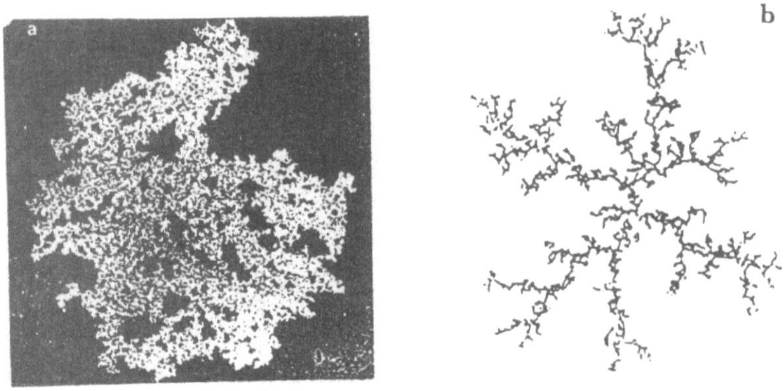

Figure 1: (a) Invasion percolation of air (white) displacing glycerol (black) in a two-dimensional model. The fractal box counting dimension of this structure was found to $D_b = 1.8$. (b) Viscous fingering pattern formed by displacing glycerol from a 2D porous medium consisting of a single layer of 1mm glass spheres. Air was injected near the center of the model with a pressure of $20mbar$ with respect to the edge of the cell. The fractal cluster dimension $D_c = 1.62$

configurations followed by abrupt bursts where a portion of the front proceeds quickly. In the high velocity, viscous fingering regime (VF) the dynamics is controlled by the large pressure gradients in the far tips of the structure with an effective screening of the interior part. In the external active zone the fingers spreads continuously outwards leaving behind a frozen ramified structure as seen in Fig. 1. Simulations and experiments [4,10,11] has previously been performed to examine the *dynamics* in the high velocity, viscous fingering (VF) regime. At low injection rates, in the capillary regime, a few [11,12,13] experiments on front dynamics have been presented.

In section 3 the structure and dynamics of slow drainage are described. The experimentally measured pressure curve (see Fig. 2) is compared with one obtained from a modified invasion percolation algorithm [12]. This algorithm includes a *capacitive interface volume* that give rise to pressure variations due to adjustments of the local radii of curvature along the front. The pressure curves can be analyzed using the height distributions of the characteristic vertical jumps and the distribution of build up times between the pressure jumps. Using parameters estimated from the experiments, the simulated distributions are in good agreement with the experimental ones (see Fig. 4). More general simulations show that the exponential pressure jump distributions observed are crossover effects from power-law distributions (Fig. 5) because of limited capacity of the front to take up a fluid volume.

Section 4 describes the structure and the dynamics of fast viscous fingering in a similar experimental system. In this regime the fingering structure grows mainly at the tips, and the internal region is screened by the outher fingers. Here, the fingering structure is analyzed like tree-like fractals with a hierarcical branching structure [15]. By introducing a branching ratio and a length ratio between subsequent branching orders, the structures can be analyzed in a way similar to the procedure usually used to analyse deterministic fractals. We also compare the dynamics of the experimental fingers with a modified DLA simulation in a rectangular geometry. This DLA simulation [10,42] will in addition to describing the structure give a time dependence in a form suitable for comparison with experiments.

2 Experimental Setup

The porous model used consisted of 1-mm glass spheres sandwiched between two contact papers. This produced a two dimensional, random packing of spheres [4,10] with porosity $\phi \simeq 0.7$ and average pore volume 2 mm^3.

A rectangular model of dimension 120 × 175 mm was made for the pressure measurements at slow injection rates. In these experiments air displaced water that was withdrawn from one end of the model with a syringe pump at a low, constant flow rate of $Q = 0.35$ cm^3/hour or $Q = 0.048$ pores/s. The model was open for invasion of air from the opposite end. The pressure fluctuations were measured with a pressure sensor of own construction [14], having a resolution better than 0.10 mm water and designed for measurements of small pressures. An experiment consisted of pressure measurements taken with time intervals of 1 s and totally about 70000 observations.

In the fast viscous fingering experiment we used circular models of radius $r = 50cm$ and rectangular models of dimension 50 × 80 cm. Air at a constant pressure was introduced into the porous layer filled with glycerine through a 1-mm hole at the center of the plexiglass disc or from one of the short sides in the case of rectangular models. The pressure of the injected air was measured during the experiment and the pressure fluctuations at the injection point was less than 0.5 %, indicating a constant pressure experiment. The transparent model was illuminated with a uniform lightning from below and pictures were taken with a camera from above. The images were analyzed with a Data Copy scanning digitizing camera interfaced to an Apollo DN3000 workstation that gave a resolution of 2000 × 2000 pixels.

3 Structure and Dynamics of Slow Drainage

If the pumping rate is sufficiently slow the viscous pressure drops in the fluids are negligible and the capillary pressure (defined as the pressure difference between the two fluid phases) is considered to be constant along the fluid-fluid interface. The dynamics and the structure of drainage at slow injection rates is therefore controlled by the varying capillary pressure thresholds of the pore throats along the interface of the invaded cluster. In the idealized case the front will proceed at the most favorable position (invade the widest pore throat in this case). This process called invasion percolation was first studied theoretically by

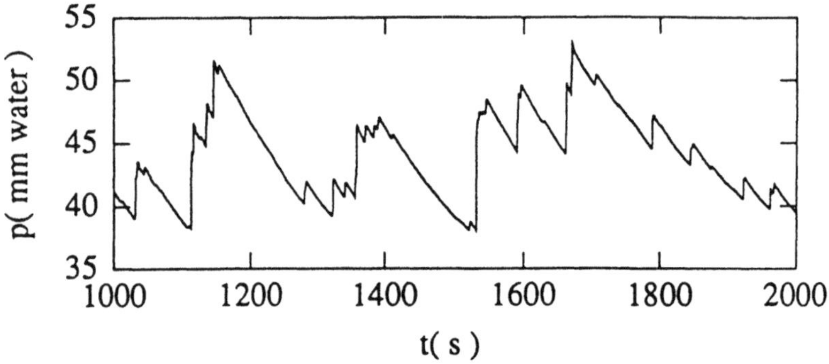

Figure 2: The pressure in the water during drainage as a function of time.

Wilkinson and Willemsen [16]. They found that the number of sites belonging to the invaded cluster scaled with the size L of the lattice as $M \sim AL^D$, with $D = 1.89$ in a two dimensional system. In the case where the displaced fluid is incompressible, trapping of the displaced fluid will create a less dense structure with a fractal dimension $D = 1.83$ [16].

The first experimental verification of invasion percolation with trapping was done by Lenormand and Zarcone [17] who experimentally investigated the invasion of air into glycerol in a micro-model of $2.5 \cdot 10^5$ ducts. They found a fractal dimension $D = 1.82 \pm 0.01$, consistent with the numerical simulations of Wilkinson and Willemsen.

Early observations [18,19,20] showed that constant rate drainage at very low injection rates takes place in a discontinuous way on the pore level. The interface between the fluids deforms continuously for a while, being pinned to the grains of the porous medium along the front. Occasionally the interface becomes unstable, the fluid in a number of pores is spontaneously displaced in a burst, and a new interface is established. These bursts are accompanied by sudden jumps in the pressure in the displaced fluid, so called 'Haines Jumps' [18] (see Fig. 2).

The dynamics of the invasion percolation model was studied by Furuberg et. al and Roux and Guyon [21,22]. A burst was defined geometrically as the volume of pores invaded through one pore on the interface before invasion stops in this area and takes place trough a pore elsewhere on the interface. Roux and Guyon [22] introduced a hierarcical type of bursts. For these bursts they found a power-law distribution of the burst sizes, and related the exponent to percolation exponents [23]. Gouyet [24,25] simulated drainage in a gravitational field, and measured the distribution of invaded areas.

3.1 Experiments

The water pressure curve measured during water extraction, as explained in the previous section, is shown in Fig. 2. The invasion process in our experiments clearly consists of two stages. In the first stage the water interface is in capillary equilibrium and the interface is stable. The constant extraction of water does not immediately displace water from full pores but the invading air is stored in the throats at the water/air interface resulting in fine readjustments of the interface position. The capillary pressure is gradually increased at a rate that depends on the extraction rate, the length of the interface, and the detailed pore geometry. When the capillary pressure has grown so large that it overcomes the capillary pressure threshold at the easiest of the interface throats, the front becomes unstable. Then a number of pores are invaded during a very short time interval and a sharp jump is observed in the pressure measurements (see Fig. 2). The water from the pores invaded by air is redistributed to the interface, quickly reducing the capacitive air volume and the capillary pressure. A burst is geometrically defined to be the total number of pores invaded after the system becomes unstable until it again becomes stable.

The model we use is small, the fluids have small viscosities, and our pumping rate is very slow. This means that viscous pressure gradients are quickly relaxed through the system compared to the pumping rate. If we used a larger system, more viscous fluids, or a larger pumping rate, viscous effects would appear. The ratio between the viscous and the capillary pressure over a distance L is given by the capillary number $Ca^* = \frac{\mu v_q L}{\gamma a}$. Here γ is the interface tension, μ the viscosity, L the system size, and v_q the superficial velocity. In our experiments this number is found to be $Ca^* = 10^{-6}$ which imply that the viscous effects can be neglected.

3.2 Simulations

This process is simulated by a modified invasion percolation model. The algorithm leads to invaded structures that are geometrically identical to those generated by invasion percolation with trapping [16]. Pressure and time is however defined in a physical way and we obtain pressure curves corresponding to that of Fig. 2.

A square lattice of size $L_1 \times L_2$ that consists of nodes (pores) connected by bonds (throats) represents the porous medium. Bonds are assigned random numbers p_i drawn from a uniform distribution on [0,1] representing capillary pressure thresholds in the throats. Nodes are assigned a *volume* v_i drawn uniformly from $[\frac{1}{2}, \frac{3}{2}]$. Initially the lattice is filled by "water". "Air" will invade from one of the short sides of the lattice at a constant rate of one average node volume per time unit ($q = 1$), corresponding to extraction of "water" at the opposite side at the same rate. The initial capillary pressure is $p_{cap} = 0$.

The bond with the smallest random number $p_{i,min}$ (smallest capillary pressure threshold) along the initial front is identified. In our model we then calculate the time Δt needed to build up the capillary pressure to this value, given the extraction rate q. In the buildup phase the capillary pressure increases linearly with extracted volume $\Delta p_{cap} = (\Delta t q)/(n_f \kappa)$. The rate depends on n_f, the number of bonds along the front, and κ, giving the strength of the linear dependence of the capillary pressure on the average capacitive volume stored per bond. In experiments the capillary pressure buildup is not linearly dependent on the extracted water volume (see figure 2), but the linear relation captures the essential dynamics.

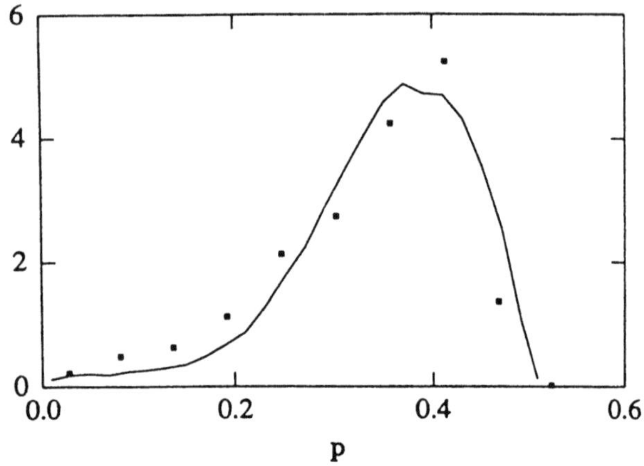

Figure 3: Normalized distributions of pressure where bursts start. Dots are experimental data points and solid line is simulation.

At the calculated time $t = (p_{i,min} - 0)(n_f\kappa)/q$ the bond with $p_{i,min}$ and the connected node is invaded and a burst starts. In our model the capillary pressure immediately decreases by $\Delta p_{cap} = v_i/(n_f\kappa)$ because the air volume that invades the node is removed from the air volume stored along the front. Displacement from the widest throat continues until the capillary pressure is too small to invade any of the available throats. According to the experiment invasion of a burst happens so quickly that the time is assumed constant during a burst. We assume that the system is in capillary equilibrium immediately after the burst. During the burst there has been a sharp jump Δp_{cap} in the capillary pressure

$$\Delta p_{cap} = \sum_i (v_i/(n_f\kappa)) \, , \tag{1}$$

where the sum is over all nodes invaded in the burst. After a burst is finished, there is a new buildup period, where the capillary pressure is built up to the level of the smallest available capillary pressure threshold again.

3.3 Results and Discussion

The linear decrease in capillary pressure during a burst introduces a characteristic pressure jump $\Delta\hat{p}$ and a characteristic number of sites invaded in a burst $\hat{s}$. We expect a scaling form for the burst size distribution

$$n(s) = s^{-\hat{\tau}} f(s/\hat{s}) \, . \tag{2}$$

The characteristic burst size is obtained by using the one to one correspondence between the uniformly distributed capillary pressures and the fraction of occupied sites f in normal percolation. Considering bursts starting at the critical pressure, p_c [23], we expect the crossover length to be visible when the correlation length ξ, corresponding to the pressure drop $\Delta p_{cap}^{-\nu} \sim (s/(n_f\kappa))^{-\nu} \sim \xi \sim s^{1/D}$ ($\nu = 1.34$ is the correlation exponent for percolation [23] and $D = 1.83$), is equal to the characteristic length of the bursts. This gives the scaling of a typical largest cluster size:

$$\hat{s} = (\kappa n_f)^{\nu D/(1+\nu D)}. \tag{3}$$

To deduce the exponent $\hat{\tau}$ that governs the burst size distribution for invasion percolation, we follow the argument of Roux and Guyon [22], based on the percolation cluster size distribution [23]. In the limit where $\kappa n_f \to \infty$, the pressure does not decrease during a burst and all bursts start at p_c. The burst size distribution is a power-law, limited only by the system size. The number of s-clusters per site at p_c in percolation is $n_s \sim s^{-\tau}$ [23]. Since the probability to hit a s burst in the invasion percolation process is proportional to the number of sites at the perimeter of the burst and not the total number of sites in the burst, the exponent $(1-\tau)$ must be changed to $(D_e/D)-\tau$, where D_e is the fractal dimension of the external perimeter, measured to be [8] $D_e = 1.39 \pm 0.02$. The fractal dimension of invasion percolation with trapping D is different from that of percolation $D' = 1.89$. The relation between the distribution of cluster sizes s' in invasion percolation with trapping and and cluster sizes s in percolation can be found by assuming that the linear sizes of the clusters are considered to be equal in the two cases $s'^{1/D'} = s^{1/D}$. The distributions then are related by $n(s') = n(s)ds/ds'$. τ then has to be substituted by [22] $D/D'(\tau-1)+1$ for invasion percolation cluster distribution. Using the relation between the burst size s and the pressure jump Δp_{cap}: $\Delta p_{cap} \simeq s/(n_f\kappa)$ we find the pressure jump distribution

$$n(\Delta p_{cap}) = \Delta p_{cap}^{-\hat{\tau}} f(\Delta p_{cap}/\Delta\hat{p})(n_f\kappa)^{(1-\hat{\tau})}\,, \tag{4}$$

with $\hat{\tau} = -D_e/D' + D/D'(\tau-1)+1 = 1.35$ and $\Delta\hat{p} \sim (\kappa n_f)^{-0.29}$. From simulations [12] with different values of κ. we found that the crossover function $f(\Delta p/\Delta\hat{p})$, is constant for small values of Δp_{cap} but crosses over to an exponential decay for $\Delta p_{cap} \gg \Delta\hat{p}$. The best fit was obtained for $\hat{\tau} = 1.30 \pm 0.05$ and $\Delta\hat{p} \sim (\kappa n_f)^{-a}$, $a = 0.3 \pm 0.05$ in good agreement with theoretical predictions.

The pressure differences measured in the experiment cannot be immediately compared to pressure differences in the simulations, since the capillary pressure threshold distribution of the porous medium is unknown. In simulations there are bursts starting at pressures in the range $[0-0.55]$, while the corresponding experimental pressures are approximately in the range of $[0-16]$ mm water. If we assume a uniform capillary threshold distribution in the porous medium, we estimate the relation between the experimentally measured pressures p and the corresponding simulation pressures p_{cap}: $p_{cap} = 0.034p/(\text{mm water})$. The experimental and numerical distributions of pressures where bursts start are shown in Fig. 3, with transformed experimental pressure values. The consistence between the distributions is good and this indicates that the experimental capillary pressure threshold distribution is almost uniform in the range of capillary pressures where bursts most often start. Since this is the part of the throat size distribution that is important for the dynamics, it is possible to compare the experiment to a simulation with a uniform throat size distribution.

We have estimated the experimental value of κ from the slope of that part of the experimental pressure/time curve where the capillary pressure builds up and the interface is in capillary equilibrium. We know that the water–air front is initially flat, giving $n_f \simeq 100$ for the first build up regions. The extraction rate gives $0.048\delta t$ pores invaded during a time interval δt. Using the linear relation between simulated and experimental pressures, we obtain $\kappa \simeq 0.2$.

We performed 100 simulations where we used this value of κ, a lattice of $L_1 = 100$ and $L_2 = 200$ corresponding to the experimental model, and stopped the simulations after 750 bursts which is the typical total number of bursts in an experiment. Fig. 4 shows the exponentially decreasing cumulative pressure jump distributions for three different experiments together with the simulations. Pressures $P = \Delta p_{cap} / < \Delta p_{cap} >$, are divided by $< \Delta p_{cap} >$ which is the mean value of the pressure jumps Δp_{cap} larger than the experimental resolution of 0.1mm water.

Also the cumulative distributions of Δt, the time between subsequent bursts, are shown in Fig. 4 for both experiments and simulations, using the dimensionless time $T = \Delta t / < \Delta t >$. $< \Delta t >$ is the mean value of the build up time, averaged over all times larger than the time it takes to build up the pressure corresponding to a capacitive volume of one typical pore volume. For times shorter than this the build up time distribution is not expected to be correct because it will strongly depend on the distribution of the detailed pore volume.

It is possible to analyze the pressure jumps in a different way which extend the scaling regime. Let us choose a range in pressures corresponding to a band of width ϵ centered around p_1 close to the percolation treshold. We can now draw a line paralel to the t axis from the first unstable pressure in figure 2 within this band. Then consider the sum of pressure jumps $\Delta p_b = \sum \Delta p_{cap}$ within the time interval before the line crosses the capillary pressure curve again. This sum Δp_b will correspond to a pressure jump in the limit when $\kappa n_f \to \infty$. In the band chosen, all pressures where bursts start are treated in this way, giving a distribution of Δp_b. This distribution is expected to be governed by the exponent $\hat{\tau}$ when p_1 is close to the percolation treshold and ϵ is small. To increase the statistics we have however choosen a finite interval $\epsilon = 0.05$ and $p_1 = 0.495$ in the simulations and a similar pressure interval estimated for the experiments. From the simulations we observe a slight increase in the exponent by increasing ϵ. Fig. 5 shows the cumulative distribution of Δp_b together with the cumulative distribution of Δp_{cap} for the experiments. The distribution of Δp_b is consistent with a power law behaviour with an exponent equal to 0.45, for both experiments and simulations with an uncertainty ± 0.1 for the experiments and ± 0.05 for the simulations. The reason why we get a slightly higher exponent than that estimated for equation 4 ($\hat{\tau} - 1 = 0.35$), is that crossover effects are introduced when ϵ is not infinitely narrow around p_c. The simulations are averaged over 50 different realizations on a system of similar size as the experiment.

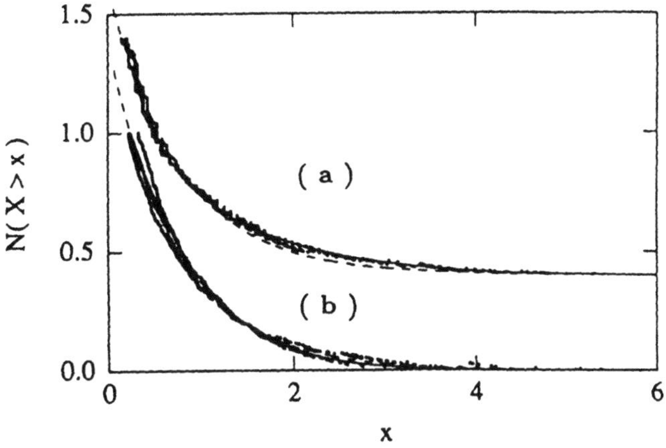

Figure 4: (a) Cumulative distribution of time between subsequent bursts, $X = T$. Distributions in (a) are vertically shifted by 0.4. (b) Cumulative distribution of pressure jumps, $X = P$. Data from three different experiments are shown as points, simulations as a solid line and the exponential function as a stippled The best experimental fits to the function $C\exp(-\beta X)$ are $\beta = 1.27$ in (a) and $\beta = 1.31$ in (b).

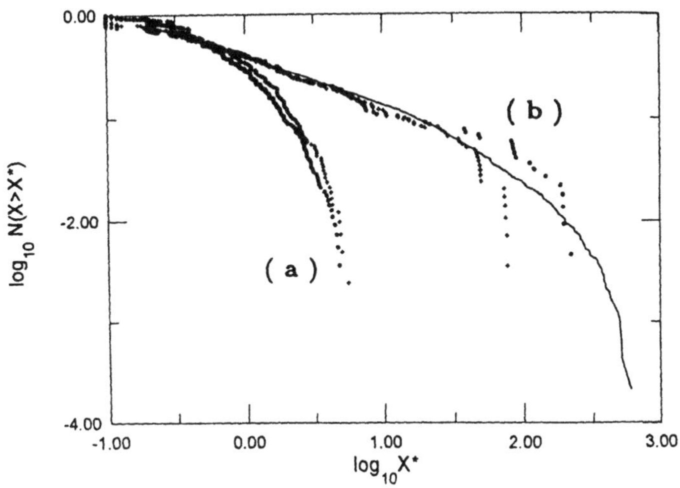

Figure 5: Cumulative pressure jump distribution $N(X > X^*)$. (a): The data points shows two experiments ($X = \Delta p_{cap}$). (b): The same experiments as in (a) and the simulation shown as solid line ($X = \Delta p_b$).

4 Branching Structure and Dynamics of VF and DLA

When a low-viscosity non wetting fluid is injected into a high-viscosity wetting fluid in a porous medium tree-like fractal structures is observed with a fractal dimension $D = 1.62$ [4]. These structures as seen in Fig. 1 (b) resembles the tree-like structures observed in the diffusion-limited-aggregation (DLA) process first invented by Witten and Sander [26]. The similarity between the fast fluid-fluid flow in a porous media and the DLA process was first discussed by Paterson [27] and a more detailed analyses has been given by Kadanoff [28]. A large variety of other processes such as dissolution of porous materials [29,30], dielectric breakdown [31], random dendritic growth [32,33] and electrodeposition [34,35] has also been shown to lead to structures similar to the DLA structure.

Many of the patterns in nature are tree-like structures with no or very few loops like the viscous fingering cluster shown in Fig. 1 (b). It is therefore of interest to classify these structures with a procedure that emphasize on the branching structure itself. We have used the concept of branch orders [15], first introduced by Horton [36] to characterize river systems, to analyze the viscous fingering and the DLA structures.

A related procedure has been used by Vannimenus [37,38] to study tree-structures using 'Strahler Numbers' [39] to analyze DLA and other structures by using a ramification matrix. The Sthraler classification scheme of branch orders is related to the procedure by Horton, and the two approaches should be considered as complementary.

4.1 Definition of Branching Structure

For tree-like structures like DLA and viscous fingers it is possible to assign branch orders in the following way. Each branch defines a continuous line, starting at a tip, and ending on another branch of lower order if it is not the 'trunk' (zeroth order branch). More than two branches may meet at a single point. The highest order branches are those which have no side branches. The next to highest order branches have side branches of the highest order, and so on. If branches of the same order n meet, the longest branch is relabeled and assigned order $n-1$. The minimal side branch order of a given point on a branch is the lowest side branch order found when going from the tip to this point. A unique assignment of branch orders is obtained if the minimal side branch order does not decrease in steps larger than one when going from the tip to the root point.

The branching ratio r_N is defined as the ratio between the number of two subsequent orders. In the same way one can define the length ratio r_L which is the ratio between the length of two subsequent orders.

Figure 6: A recursively defined tree structure. Thick line: zeroth order branch, stipled line: first order branch and thin line: second order branch. The last figure is the result after seven iterations, se text for a full explanation.

4.2 A Simple Deterministic Tree-Structure

The hierarcical ordering of self-similar fractal trees is best illustrated by discussing a deterministic algorithm as defined in Fig. 6. The *initiator* shown in (a) is a line segment. The *generator* (b) replaces this line segment by $\mathcal{N} = 5$ new line segments each scaled down by the ratio $r = 1/3$. In (c) the generator is applied again giving the second generation *prefractal*. After an infinite number of applications of this algorithm one arrives at a fractal set. This set is self-similar and has a similarity-dimension [3,2] $D_s = \log(\mathcal{N})/\log(1/r) = 1.46\ldots$ equal to the fractal dimension of the set. Each piece of the fractal is a scaled-down version of the whole set, as is required for self-similar fractals.

Table 1. Fractal dimension for DLA and VF.

		branch order ratios	
$r_N = N_n/N_{n-1}$	number	5.2 ± 0.2	4.8 ± 0.5
$r_L = L_n/L_{n-1}$	length	0.35 ± 0.01	0.34 ± 0.04
$D_s = \dfrac{\ln(r_N)}{\ln(1/r_L)}$	self-sim.	1.6 ± 0.1	1.5 ± 0.1
$N(\delta) \sim \delta^{-D_b}$	box	$1.62 \pm 0.02^{\dagger}$	1.51 ± 0.06
$N(\delta) \sim \delta^{-D_b}$	box‡	1.67 ± 0.03	—
$M(R_g) \sim R_g^{D_g}$	growth	1.710 ± 0.005	—
$M(r) \sim r^{D_c}$	cluster	$1.69 \pm 0.01^{\dagger}$	$1.62 \pm 0.05^{\dagger}$

† Estimated by scaling different cluster sizes onto the same curve.

‡ Box-counting only points with $r < R_g$.

The branches of this structure may be characterized as follows: In (a) is shown the trunk of a simple tree-like structure, which we identify as a branch of order 0 and length 1. In (b) we have added two side branches of order 1 and length $1/3$. In (c) the structure contains one branch of order 0, 2 branches of order 1, and 10 new branches of order 2. These new branches have a length of $1/9$. At iteration n of this process, we will have a branching prefractal with branches of order $0, 1 \ldots n$. If we continue this The number N_n and length L_n of branches of order n are given by

$$L_n = L_m r_L^{n-m} \text{ and } N_n = N_m r_N^{n-m}, \tag{5}$$

where in this example $r_L = 1/3$, $r_N = 5$, and $m = 1$ is a lower cut-off for the validity of the scaling relations. In general the similarity-dimension is given by $D_s = \ln(r_N)/\ln(1/r_L)$ for tree structures with fixed r_N and r_L.

4.3 Branching Structure, Selfsimilarity and Box Counting

This result is also obtained by a 'box-counting' procedure. Let the structure be covered by small 'boxes' having shapes chosen for optimal coverage but with the same length $\delta = L_n$ and width $\sim 2L_{n-1} \simeq 2r_L\delta$. Using (5) we find that the number of boxes needed to cover the fractal is given by

$$N(\delta) = \sum_{i=0}^{n} N_i L_i/\delta = A\left(\frac{\delta}{L_m}\right)^{-D} + B\left(\frac{\delta}{L_m}\right)^{-1}, \tag{6}$$

where $D = \ln(r_N)/\ln(1/r_L)$ is the fractal dimension. A and B are constants. The branches of order $n = 0$ to $n = m-1$, that do not follow the scaling-relation in (5), contribute only to the last term in (6). This method of finding the fractal dimension using optimally shaped pieces to cover the fractal is closer to the Hausdorff-Besicovitch definition than the box counting method [3]. Neglecting the corrections to scaling, i.e. the last term, in (6), we find that the number of pieces $N(\delta)$ of 'diameter' δ needed to cover the fractal is given by

$$N(\delta) \sim \left(\frac{\delta}{L_m}\right)^{-D}, \text{ with } D = D_s = \frac{\ln(r_N)}{\ln(1/r_L)}. \tag{7}$$

This result shows that the fractal dimension for scaling trees is given by the similarity dimension. If one instead covers the fractal with square boxes of side δ, one finds that the number of boxes needed to cover the fractal is $N(\delta) \sim \delta^{-D_b}$, where D_b is the box-dimension [3]. By looking at the dependence of the gyration radius of the cluster R_g one finds that the radius of gyration R_g scales with the mass of the DLA and viscous fingering clusters as $R_g^{D_g}$ [40].

4.4 Structure of Viscous Fingers and DLA

We have analyzed viscous fingering patterns obtained from experiments in a two dimensional porous medium [15,10] and also computer simulated off-lattice DLA

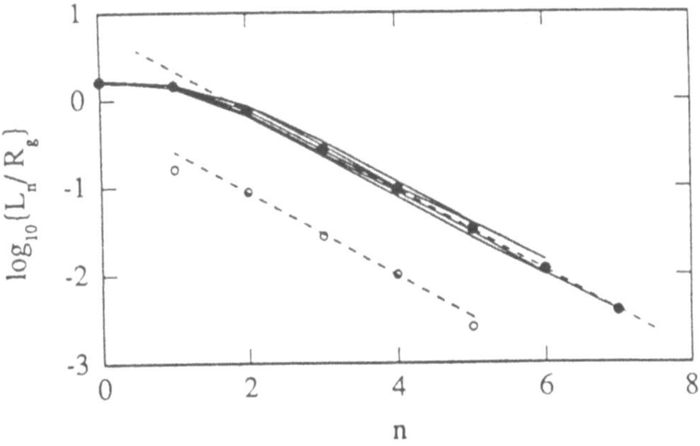

Figure 7: The scaled average length L_n/R_g plotted against branch order n. L_n is given by the tip-to-root distance in DLA (•) and the average mass in the VF experiments (∘). Lines are fits of $L_n \sim r_L^n$ to the data, with $r_L = 0.34$ for VF and 0.36 for DLA. The experimental points have been shifted down for clarity.

clusters using these ideas [15]. These structures have almost no loops, and may therefore be described as trees.

We generated 200 DLA clusters of size 10^3, 50 of size $5 \cdot 10^3$, 30 of size $2.5 \cdot 10^4$, 20 of size $5 \cdot 10^4$, and 10 of size $2.5 \cdot 10^5$. The fractal dimensions obtained for these clusters are summarized in Tab. 4.2. There has been some discussion whether or not D_b is equal to D_g [41]. The cross-over regime clearly visible for $r > R_g$ in the cumulative mass within radius r; $M(r) = M \cdot (r/R_g)^{D_c} f(r/R_g)$ [10]. may account for these discrepancies. Here M is the total mass or equivalently the total number of particles of a *given* cluster, and f is a cross-over function. D_c is called the cluster dimension. Naively box-counting the whole cluster at a given stage of growth, gives values of D_b in the range 1.59 to 1.63, depending on the range of δ used in the fit. If the box-counting is limited to the part of the cluster inside a box of size R_g, we find values of D_b in the range 1.63 to 1.69. Even though this leads to an estimate of D_b lower than D_g, they may be equal within errors. Measuring D_c of the cumulative mass $M(r)$ for $r < R_g$ where f is constant, gives $D_c = 1.69$.

The experiments performed with a circular model 40 cm in diameter. The fractal dimensions given in Tab. 4.2 were obtained by digitizing photographs of the VF structures The branch orders were, however, identified manually from the pictures.

The branching ratio r_N and length ratio r_L were obtained by fitting the number and length of branches of order n to the power laws $N_n \sim r_N^n$ and $L_n \sim r_L^n$. In the simulations, the average length of a branch was defined as either the average mass of the branch or the average tip-to-root distance of a branch. Both of these definitions gave approximately the same value for r_L. In the experiments, the average length of a branch was identified as the average mass. In Figs. 7 and 8 we show the experimental result together with results obtained

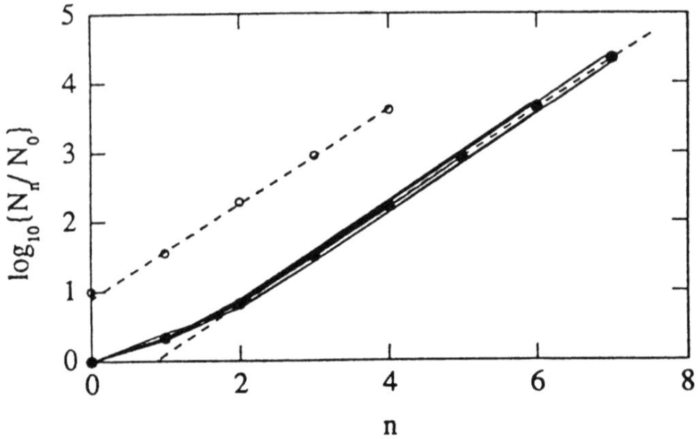

Figure 8: The number of branches N_n at a given order as a function of branch order n for VF experiments (∘), and DLA simulations (•). The lines represents fits of $N_n \sim r_N^n$ to the data, with $r_N = 4.8$ for VF and 5.2 for DLA. The experimental points have been shifted up one decade for clarity.

from simulations for all the different cluster sizes. The points shown are the average values for these different cluster sizes. For a given size, the values fall on a straight line either above or below the average curve. The size of the cluster is, however, not correlated with this deviation from the average value. This scatter is instead related to the fact that averaging is done with fixed mass and not for a fixed maximal order. Note also that the scaling of the different cluster sizes in Fig. 7, show that $R_g \sim L_n$ for all orders n, also orders $n < m$ ($\simeq 2$ for DLA) that do not follow Eq. (5). Using this result in Eq. (7) gives the identification $D_g = D_s$ since the total mass $M \sim N(\delta) \sim R_g^{D_s}$. The values of r_L and r_N do not change if the analysis is limited to the part of the cluster within R_g. The scaling relations are therefore not changed in the cross-over regime.

4.5 Dynamics of Viscous Fingers

The dynamics and structure at high injection rates is controlled by the growth of the external tips with a strong screening of the internal part of the structure. This leaves behind a frozen fractal structure as shown in Fig. 1. The dynamics of the viscous fingers has been investigated by measuring the growth of the longest finger as function of time in a rectangular model. The time dependence of the height L of the highest finger for four different experiments is shown in Fig. 9, and compared with a modified DLA simulation [10,42] first introduced by Meakin. The experiments were performed with a fixed pressure of $20 mbar$ between the two ends of the model. Similar measurements [4,10] and simulations [10,42] have been done in a circular geometry. A good agreement is found between the simulations and the experiments in both circular and rectangular geometry.

The analogy between DLA and VF is based on the equivalence between the

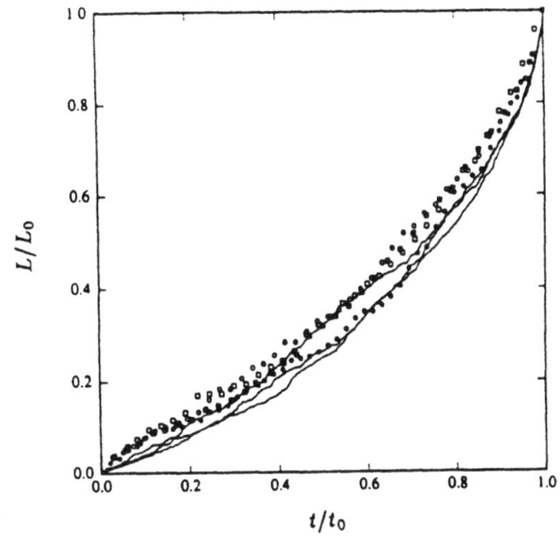

Figure 9: The time development of the longest finger in four different experiments (data points) and in 3 simulated processes (full lines). L is the height of the highest finger, and L_0 is the height of the cell. ime t is normalized with the break-through time t_0

pressure field and the probability field probed by the random walker, both obtaining Laplace's equation and subjected to the same boundary conditions [27].

In the DLA model used here the time is captured in the following way: One of the unoccupied surface sites is selected at random and the random walker is started from that site. After each time the random walker is launched from the surface site the time is incremented by $1/N_s$ where N_s is the total number of surface sites. If the random walker reach the edge of the cell, the random walker is stopped and the site from which the walker started is filled. On the other hand if the walker reach an other unoccupied surface site the random walker is stopped and an other random walker is launched with an equal probability from one of the surface sites of the cluster. In the rectangular model the particles is launched initially with an equal probability from a line.

The pressure field at a position $\mathbf{r}$ is proportional to the probability that a random walker hits that position. By starting the random walker with equal probability at each interface site we satisfy the boundary condition with a constant pressure on the interface of the cluster. By terminating the random walker when it reach the other side of the model corresponds to keeping the pressure equal to zero at that side. A detailed discussion of this simulation is found in Ref. [10,42].

5 Conclusion

In conclusion we find close agreement between simulations and experiments in the case of slow drainage. Since both the pressure jump distribution and the build up time distribution show exponential decay, we conclude that the capillary pressure in the experiments depend strongly on the capacitive interface volume and that the experiments are in the crossover regime. We find a fractal structure that arises by a dynamic process that is not characterized by power-laws, but by exponential functions. In the limit when $kn_f \to \infty$ the dynamics is governed by power law distributions.

We conclude that DLA and VF patterns are tree-like structures having fixed bifurcation and length ratios independent of branch order. From these two ratios, a fractal dimension may be defined analogous to the the self-similarity dimension of deterministic self-similar fractals such as the deterministic tree in Fig. 6. This dimension is consistent with values obtained from other methods. We have shown that DLA and VF clusters are self-similar in terms of branch order.

The dynamics of viscous fingering in a rectangular model has been been studied. The length of the longest finger as function of time fits very well to a modified DLA model which allows the time dependence of the growth process to be investigated.

Acknowledgment

We thank Amnon Aharony, Aleksandar Birovljev, Vidar Frette, Alex Hansen, Paul Meakin, Unni Oxaal, and Stèphane Roux for stimulating discussions. This work has been supported by the Research Council of Norway (NFR).

References

1. R. Lenormand, E. Touboul, and C. Zarcone, *J. Fluid Mech.* **189**, 165 (1988) R. Lenormand, C. Zarcone and A. Sarr, *J. Fluid Mech.* **135**, 337 (1983)
2. J. Feder, *Fractals Plenum, New York* , (1988)
3. B. B. Mandelbrot, *The Fractal Geometry of Nature, W. H. Freeman, San Fransisco* , (1982)
4. K. J. Måløy, J. Feder, and T. Jøssang, *Phys. Rev. Lett.* **55**, 2688 (1985)
5. J. D. Chen and D. Wilkinson, *Phys. Rev. Lett* **55**, 1892 (1985)
6. U. Oxaal, F. Boger, J. Feder, T. Jøssang, P. Meakin and A. Aharony, *Phys. Rev* **A44**, 5038 (1991)
7. M. Cieplak and M. Robbins, *Phys. Rev. Lett* **88**, 2042 (1988) M. Cieplak and M. Robbins, *Phys. Rev* **B41**, 11508 (1990) B. Koiler, H. Ji, and M. O. Robbins, *Phys. Rev* **B45**, 7762 (1992)
8. A. Birovljev, L. Furuberg, J. Feder, T. Jøssang, K. J. Måløy and A. Aharony, *Phys. Rev. Lett* **67**, 584 (1991)
9. V. Frette, J. Feder, T. Jøssang and P. Meakin, *Phys. Rev. Lett.* **68**, 3164 (1992)
10. K. J. Måløy, F. Boger, J. Feder, T. Jøssang, and P. Meakin, *Phys. Rev. A* **36**, 318 (1987)
11. U. Oxaal, M. Murat, F. Boger, A. Aharony, J. Feder, and T. Jøssang, *Nature* **329**, 32 (1987)

12. K. J. Måløy, L. Furuberg, J. Feder and T. Jøssang, *Phys. Rev. Lett* **68**, 2161 (1992)
13. D. Salin,, *these proceedings* , ()
14. K. J. Måløy, *Preprint* , (1993)
15. E. L. Hinrichsen, K. J. Måløy, J. Feder and T. Jøssang, *J. Phys. A* **22**, L271 (1989)
16. D. Wilkinson. and J. F. Willemsen, *J. Phys. A: Match. Gen.* **16**, 3365 (1983)
17. R. Lenormand C. and Zarcone, *Phys. Rev. Lett.* **54**, 2226 (1985)
18. W. B. Haines, *J. Agr. Sci.* **20**, 97 (1930)
19. J. C. Melrose, *Soc. Petr. Eng. J* **5**, 259 (1965)
20. W. B Morrow, *Ind. Eng. Chem.* **62**, 32 (1970)
21. L. Furuberg, J. Feder, A. Aharony, and T. Jøssang, *Phys. Rev. Lett* **61**, 2117 (1988)
22. S. Roux and E. Guyon, *J. Phys. A: Math. Gen.* **22**, 3693 (1989)
23. D. Stauffer, *Introduction to Percolation Theory, D. S. Taylor and Francis, London* , (1985)
24. J. F. Gouyet and Y. Boughaleb, *Phys. Rev. B* **40**, 4760 (1989)
25. J. F. Gouyet, *Physica A* **168**, 581 (1990)
26. T. A. Witten and L. M. Sander, *Phys. Rev. Lett* **47**, 1400 (1981)
27. L. Paterson, *Phys. Rev. Lett* **52**, 1621 (1984)
28. L. P. Kadanoff, *J. Sat. Phys.* **39**, 267 (1985)
29. G. Daccord, *Phys. Rev. Lett* **58**, 479 (1987)
30. G. Daccord and R. Lenormand, *Nature* **325**, 41 (1987)
31. L. Niemeyer, L. Pietronero and A. J. Wiesmann, *Phys. Rev. Lett* **52**, 1033 (1984)
32. W. T. Elam, S. A. Wolf, J. Sprague, D. V. Gubser, D. Van Vechten, G. L. Barz Jr. and P. Meakin, *Phys. Rev. Lett* **54**, 701 (1985)
33. H. Honjo, S. Ohta and M. Matshushita, *J. Phys. Soc. Japan* **55**, 2487 (1986)
34. M. Matsushita, M. Sano, Y. Hayakawa, and Y. Sawada, *Phys. Rev. Lett* **55**, 286 (1984)
35. R. M. Brady and R. C. Ball, *Nature* **309**, 225 (1984)
36. R. E. Horton, *Geol. Soc. Am. Bull.* **56**, 275 (1945)
37. J. Vannimenus, in *Universalities in Condensed Matter* R. Jullien, L. Peliti, R. Rammal, N. Boccara eds. (Springer, Berlin 1988);
38. J. Vannimenus and X. G. Viennot, *J. Stat. Phys* **54**, 1529 (1989)
39. A. N. Strahler, *Bull. Geol. Soc. America* **63**, 275 (1952)
40. P. Meakin, *Phys. Rev.* **A27**, 1495 (1983)
41. F. Argoul, A. Arneodo, G. Grasseau and H. L. Swinney, *Phys. Rev. Lett.* **61**, 2558 (1988) P. Meakin, in the proceedings from the Erice meeting on Fractals, 1988, (editor L. Pietronero, Plenum New York).
42. P. Meakin, *in Time-Dependent Effects in Disorder Materials, eds. R. Pynn and T. Riste, Plenum Press* **167**, 45 (1987)

Heterogeneous Porous Media: Fronts and Noise

M.Chaouche[1], N.Rakotomalala[1], D.Salin[1], B.Xu[2] and Y.C.Yortsos[2]

[1]Acoustique et Optique de la Matière Condensée. Université Pierre et Marie Curie - B78, 4 Place Jussieu, Paris 75252 Cedex 05, France

[2]Department of Chemical Engineering and Petroleum Engineering Program, University of Southern California, Los Angeles, CA 90089-1211, USA.

Abstract : Capillary effects can be important in immiscible flows in heterogeneous media, particularly at low capillary numbers (Ca). We present experiments and simulations of slow drainage in 3-D porous media, either homogeneous and in the presence of buoyancy or heterogeneous and in its absence. An acoustic technique allows for an accurate study of the 3-D fronts and the cross-over region. Our results suggest that both cases can be described by invasion percolation in a gradient. Both front tails scale with the corresponding Bond numbers as $\sigma_{ft} \sim B^{-.47}$ in agreement with the theory. An analogous scaling for viscous effects is also given. The noise of these fronts are found correlated in the form of a fractional Brownian motion (fBm) of a Hurst exponent $H \approx .5$. At higher Ca, experiments performed in 3-D porous media with sharp changes in permeability, exhibit a saturation profile response closely linked to the permeability variations. This viscous response to heterogeneity provides an opportunity to investigate and determine correlated (even at all scales, i.e. fBm), permeability fields.

1 Introduction

Important in oil recovery, hydrology and chemical engineering, immiscible displacement in porous media finds as much interest in the physics of disordered media. The front between two immiscible fluids exhibits characteristic patterns depending on the displacement process[1]. Fast, viscous-dominated displacement in the limit of large viscosity ratio (viscous fingering) is modeled by Diffusion-Limited-Aggregation (DLA)[2]. Invasion percolation (IP)[3] describes slow displacements where capillary forces dominate over viscous and lead to percolation patterns[4]. In the absence of pressure gradients, the latter have the fractal properties, including the critical exponents, of ordinary percolation. For example, the displacing front has the geometry of the incipient percolation cluster, which is a self-similar fractal object of a specified fractal dimension. For these regimes to apply requires macroscopically homogeneous media in the absence of external forces.

However, this is not the case in many practical applications. For example, in a vertical displacement involving fluids of different densities, gravity competes with capillary forces by exerting hydrostatic pressure gradients. The gravity number (also called the Bond number), $B_g=\Delta\rho g_x a^2/\gamma$, measures this competition. Here $\Delta\rho$ is the difference between the density of wetting and non-wetting fluids, g_x the acceleration of gravity in the flow direction x, a the mean pore size and γ the interfacial tension. If the displacement is downdip (injection of the lighter fluid from the top), gravity acts to limit the fractal regime to a finite extent and the front becomes self-affine. Its extent is measured by the front width σ in 2-D, or the front-tail width σ_{ft}, in 3-D geometries, both of which obey a scaling law with the Bond number, as shown below. A compact, Euclidean pattern follows the fractal front. The opposite case, where the displacement is updip (injection of the lighter fluid from the bottom, $B_g< 0$), leads to gravity fingering[6]. Buoyancy effects for the stable case ($B_g>0$) have been modelled with gradient percolation[7-9] (GP) and with invasion percolation in a gradient[5] (IPG), which are modified versions of ordinary and invasion percolation, respectively.

Another external force that may also act in a similar fashion arises from heterogeneity. Recent work[10] on immiscible displacement in heterogeneous porous media has indicated that spatial permeability variations also act in ways similar to gravity, with the additional feature that contrary to gravity, permeability gradients can be implemented and vary in space. For example, consider drainage in a horizontal medium with a permeability gradient. Because the non-wetting fluid preferably penetrates the high permeability (pore size) regions, we expect that a form of gradient percolation would also describe the process, but with a different Bond number, B_k, now based on the permeability gradient[11].

Finally, when the fluids are flowing at a sufficiently large flow rate v, characterized by the capillary number $Ca=\mu v/\gamma$, viscous (μ) effects would act in a similar fashion reducing the front extent, but with a Bond number, B_μ, which is now related to the capillary number.

In case such that both heterogeneity and viscous effects act simultaneously fluid saturation profiles reflect the underlying heterogeneity permeability field[10,11]. This signature of heterogeneity leads to a possible suitable tool to determine the permeability field.

We present three-dimensional experiments (using an acoustic technique[12]) in bead packs and pore network simulations to test our conjecture on IPG for the two cases of buoyancy and permeability heterogeneity. The corresponding scaling for the case of viscous effects is also given. Then we analyse the noise correlations occuring on IPG fluid saturation profiles. We determine steady-state saturation profiles in heterogeneous media, where the permeability stratification is perpendicular to the flow direction. These steady-state experiments are then simulated with a continuum model and an extension to a correlated permeability field is also shown.

2 Invasion Percolation in a Hydrostatic or Permeability Gradient

Consider a porous medium (Fig1.) fully saturated with a wetting fluid (water). For a non-wetting fluid (oil) to displace this wetting fluid, requires that the interfacial capillary pressure at the pore throat, $\Pi=p_{nw}-p_w=2\gamma/a$, is overcome. At sufficiently low flow rates ($Ca<<10^{-6}$) the displacement is controlled by capillarity. As Π increases, pores of the lowest resistance, i.e. of the largest throat, are invaded, provided that they are connected to the injection inlet. This simple invasion rule is analogous to IP on a lattice leading to a fractal cluster of the non-wetting phase for an uncorrelated regular distribution of pore sizes. This pattern is isotopic over the 3D space. In reality, however, boundaries, body forces or heterogeneity can quench the development of this fractal cluster and lead to finite size effects.

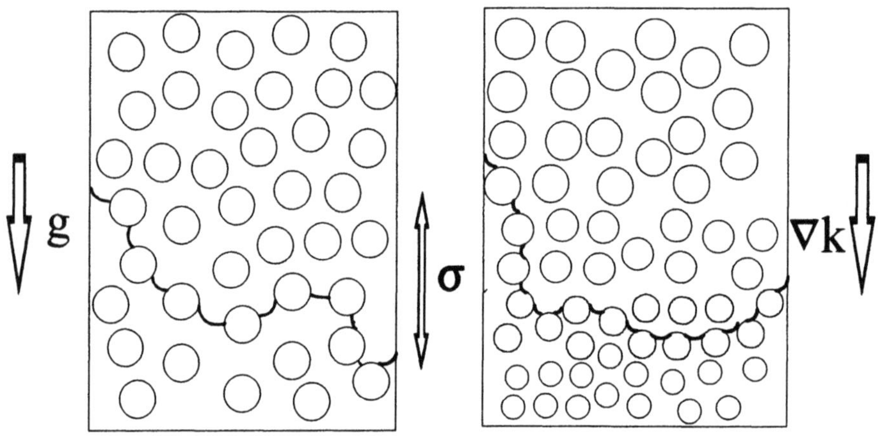

Fig. 1a: Displacement in a homogeneous medium in the presence of gravity. Top: Lower density non-wetting Fluid Bottom: Higher density wetting Fluid

Fig. 1b: Displacement in a heterogeneous medium with equal density fluids. Top: Non-wetting Fluid Bottom: Wetting Fluid

Let us first address this question for a homogenous porous medium in the presence of gravity (Fig. 1a). Here, displacement of a heavier wetting fluid by a lighter non-wetting fluid proceeds from top to bottom. Gravity has a stabilizing effect: due to the hydrostatic pressure gradient it is more favorable to invade smaller pores at the top than larger ones at the bottom; the overall effect is a limitation of the front extent in the vertical direction, so that the larger the density difference, the smaller the front width σ. Equivalently, consider two fluids of the same density but in a heterogenous porous medium of a constant permeability gradient (Fig. 1b). The invading non-wetting fluid fills the larger pores at the top, leading to the same kind of front extent limitation in the vertical direction as previously. It follows that hydrostatic pressure gradient and permeability gradient are expected to give the same type of front width.

These effects can be understood in the framework of IPG. We recall that GP is a lattice percolation process, where the occupation probability p is not constant, but decreases linearly with one of the lattice coordinates, thus giving rise to a finite length scale[5,7,8]. In IPG[7,8] the occupation probability p(x) is spatially varying; p is related to the capillary pressure Π via the pore size distribution[11]. In the absence of gradients, the fluid displacement is described by IP (where p is uniform in space), the front has the fractal structure of the percolation cluster and its extent is given by the correlation length ξ, which satisfies the percolation scaling[4] $\xi \sim |p-p_c|^{-\nu}$ (ν=0.88, in 3D, and p_c is the bond percolation threshold, which is a lattice property. The displacement is driven by an increase in Π and reaches breakthrough when p reaches p_c, i.e. when Π attains the capillary threshold Π_c. In the presence of gradients along the direction x, the displacement process is described by IPG, where p is continuously decreasing, and a front location x_c, where $p(x_c)=p_c$, can be identified. Clearly, the front location x_c varies with time, following the imposed pressure difference, P(t), which must increase for the displacement to proceed (e.g. $P(t)=\Pi_c+ \Delta\rho g_x x_c(t)$ for the case of buoyancy). The vicinity of x_c is a cross-over region, described by the theory of finite size effects in percolation[4]. In both GP and IPG, there is a region around x_c of finite extent σ, where percolation theory applies. Its scaling is given by[7,8]

$$\sigma \sim |\nabla p|^{-\nu/(1+\nu)} \tag{1}$$

In the stable case (negative values of dp/dx), IPG allows for the invading fluid concentration $S_o(x)$ to be determined[8] through

$$S_o(x) \sim (\delta p)^{\beta}\, \Pi_f[(x_c-x)/\sigma] \tag{2}$$

where the relation $(p-p_c)/\delta p=(x_c-x)/\sigma$ was implied, $\delta p \sim |\nabla p|^{-1/(1+\nu)}$, β=0.45 in 3D and the scaling function Π_f was computed in Ref. 8,11 The function Π_f is unique to the GP process. It accounts for the fractal nature of the front in the vicinity of p_c as well as for the approach to a compact pattern away from the front (near the inlet), where the invading saturation is described by the expression $S_o \sim (x_c-x)^{\beta}$.

Consider now the application of the above to gravity or permeability gradients. In either case, the link of the probability gradient to the Bond number is through the previous capillary pressure relationship shown above which leads to $p-p_c \sim (a/\gamma)(\Pi-\Pi_c)$, with typically $\Pi_c \sim \gamma/a$. In the case of gravity , hence $dp/dx \sim -B_g$. In the case of a permeability gradient, variations in capillary pressure result from variations in the pore size in the direction of the gradient (recall the permeability scaling $k \sim r^2$). Thus, $\Pi-\Pi_c \sim \gamma[1/r(x)-1/a] \sim B_k \Pi_c (x_c-x)/a$, where $B_k=-dr/dx \sim -d\sqrt{k}/dx$, hence the probability gradient scales as $dp/dx \sim B_k$. If the permeability is decreasing in the direction of displacement ($B_k > 0$)), invasion by the non-wetting fluid requires higher pressures, the permeability gradient acting to stabilize the front. In the opposite case ($B_k < 0$) invasion is facilitated, leading to capillary fingering. In the stable case, therefore, we shall have

$$\sigma \sim B^{-\nu/(1+\nu)} \tag{3}$$

The experiments were performed with porous media consisting of packed glass beads, the diameter of which ranged between 80μm and 1mm. For the gravity experiments, we used oil and various water mixtures of variable density difference ρ varying from 20 to 100 Kgm^{-3}, γ=25mJm^{-2}) and variable bead diameters (but of constant size in each experiment), resulting into a range of values of B_g from 5×10^{-4} to 10^{-1}. For the permeability gradient experiments, buoyancy was inhibited by using equal density fluids. The permeability gradient was generated by filling of the region with layers of filtered beads of monotonically changing size varying from 800 to 80μ m. The thickness of each layer varied such that the gradient of the bead diameter is constant and equal to B_k. Because of practical limitations, only one decade in B_k (2.4 10^{-3} - 1.44 10^{-2}) was covered in the heterogeneity experiments. The non-wetting fluid (lighter or equally dense fluid) was injected from top to bottom (or from high to low permeability) in a sample saturated with the wetting fluid (water solution) at a low constant flow rate (corresponding to a low capillary number, Ca=10^{-7}). The concentration profile $S_0(x)$ was recorded with the use of an acoustic technique described elsewhere[12]. Saturation measurements are indirectly obtained from the velocity variations of a sound wave along the sample. From the calibration curve and the accuracy of the relative velocity measurements (10^{-4}) we estimate the overall accuracy in saturation as better than 1%. A spatial resolution of 2 mm was implemented in order to keep track of precise variations along the length of the sample. This high accuracy technique allows for a precise determination of the average front position and of the cross-over regime. In the permeability gradient case all data were collected near the mid-point of the heterogeneity. Fig.2 shows the experimental saturation profile plotted versus the reduced position $(x_c-x)/\sigma$ for the case of strong buoyancy effects ($B_g=10^{-1}$) (over a thousand experimental points are plotted).

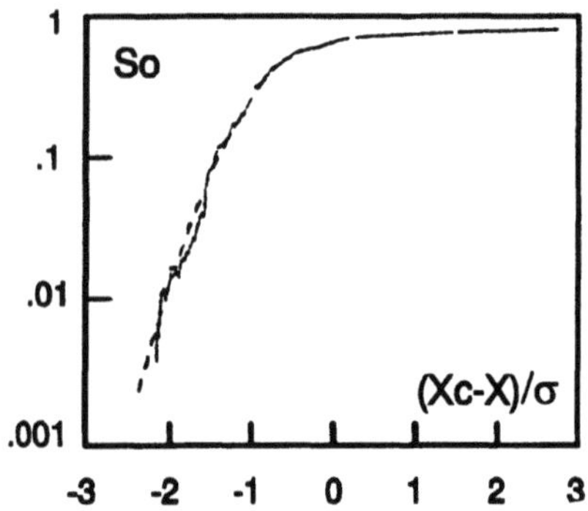

Fig.2: Non-wetting fluid saturation $\underline{S_0}$ vs the reduced space variable. The dashed line is the theoretical prediction.

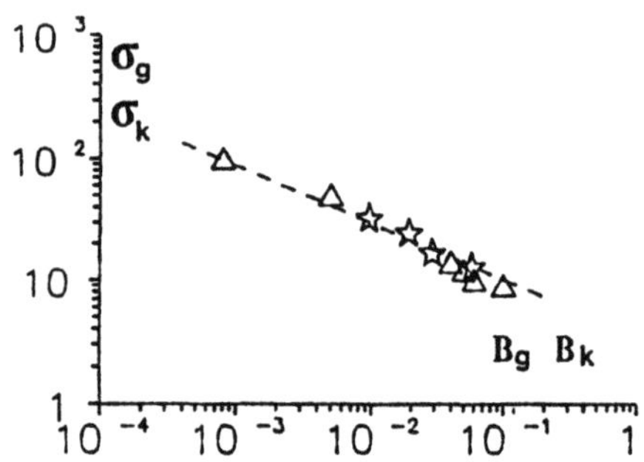

Fig.3: Tail width σ vs B_g (Δ) or B_k (⋆). The dashed line of slope -0.47 is the theoretical IPG prediction.

The small fluctuations of the concentration are reminiscent of the invasion percolation process. The dashed line through the data is the best fit of the cross-over function (2). The fit is excellent and demonstrates the validity of the theory in 3-D and over a much wider range than in earlier experiments[5,9] (note the use of a semi-logarithmic plot in Fig2). Corresponding profiles for the case of permeability gradient lead to good fits to the data[11]. From the experiments, the width can be determined as a function of either gravity or permeability Bond numbers, by matching with the theoretical function (2). A plot of the normalized front width vs the Bond number is shown in Fig.3. The dashed line of slope -0.47 is predicted from (1); both data sets (B_g and B_k) are consistent with this slope. We must point out that the B_k data required a shift (prefactor adjustment) to coincide with the B_g data, thus, the tail for a hydrostatic pressure gradient with B_g=0.1 is the same as that for a permeability gradient with B_k=0.025. The excellent agreement between theory and experiment confirms the validity of the IPG theory for either gravity or permeability gradients in 3-D. We note that this is the first time that such experimental confirmation has been obtained. However, while the case of gravity is very satisfactory, the agreement of the permeability gradient results with (2) and (3) is somewhat unexpected. GP and IPG apply to media of uniform lattice spacing, as is indeed the case with the gravity experiments. In the permeability heterogeneity experiments the spacing varied with the pore size, which is certain to change the scaling. To determine this scaling, a modification is necessary. To take into account the spacing variations, a mapping of the real variables (x,σ) to a uniform lattice of variables (X,Σ) is needed. Such a mapping is obtained from a stretching of the real lattice. The overall effect is the modifed scaling[11]

$$\sigma^{1+\nu} \sim -B_k^{-1-\nu}\,(1-x_c B_k)\,\ln(1-B_k) \qquad (4)$$

The theoretical prediction was tested in lattice models, using square and cubic lattices. The agreement obtained is quite satisfactory[11] for both 2D and 3D networks. We note that scaling (4) reduces to $\sigma^{1+\nu} \sim B_k^{-\nu}$, for $B_k \ll 1$.

Let us address now the question of viscous effects on the front extent, i.e. determine the appropriate Bond number B_μ. We recall that the characteristic number which compares vicous to capillary pressure drops is Ca. Depending on the viscosity ratio of the two fluids, two cases must be analysed: If the injected non-wetting (nw)fluid is much more viscous than the displaced, the pressure drop occurs mainly in the nw fluid. The fractal nature of the nw fluid pattern reduces the (relative) permeability of the nw phase in a maner characterised by the transport exponent t. A straightforward calculation then leads to $B_\mu \sim Ca^{(1+\nu)/(1+\nu+t)}$. In the opposite case, where the nw injected fluid is much less viscous than the displaced, the viscous drop occurs mainly in the wetting phase which has a Euclidian geometry; hence $B_\mu \sim Ca$. In both cases the front width σ should be is given by (3). Experiments and numerical simulations to test this theory are still in progress.

3 Noise of IPG Fronts

As noticed in Fig.2, the saturation profiles are very noisy. The average saturation led to the determination of the percolation features of the invading pattern. It is interesting to inquire what information is to be obtained from an analysis of the fluctuations. Specifically, we are concerned with the spatial correlation of the fluctuations of the saturation S_o as shown in Fig2. Noise analysis deals with the correlation function (covariance)[13] , $C(r)=<S_0(x)S_0(x+r)>$, or the semi-variogramm, $\Delta(r)=<[S_0(x)-S_0(x+r)]^2>=2[C(0)-C(r)]$. If the distribution of the noise is Gaussian, these correlation functions are sufficient to characterise the noise[13]. Let the (Hurst) exponent H be the characteristic exponent of the fluctuations. For $-1<H<0$, $C(r)\sim r^{2H}$ and the correlation vanishes at large r; for $0<H<1$, $\Delta(r)\sim r^{2H}$and the noise is correlated at all scales. In 1-D profiles, the latter case corresponds to the class of fractional Brownian motion (fBm) of exponent H, the former to fractional Gaussian noise (fGn) of exponent H'=H+1; the classical white noise is fGn with H'=.5 (H=-.5). The corresponding power spectrum of the noise (spectral density) scales as $l^{-(2H+1)}$, where l is the spatial frequency. Using simultaneously the power spectrum, C(r) and $\Delta(r)$ of the noise and of its graded function[13], we identified two regimes in the noise of Fig2: in the compact part of the curve ($S_0\sim1$) the trace is a white noise, whereas for small S_0 values, the trace is fBm of $H\approx.5$, i.e.the noise in the cross-over gradient percolation region is a regular Brownian motion, statistically self-affine. To account for this result, we recall that in the cross-over region, the IPG cluster is the percolation cluster near the percolation threshold (ξ is "infinite" for all sizes smaller than σ). Therefore, this noise must be related to the noise of a classical percolation cluster at p_c. In a separate study we have presented theoretical and numerical simulations to show that on a 3D lattice[14] the noise is fBm with $H\approx.5$. From this analysis, we obtain information on the correlations and nature of the percolation cluster itself, but not necessarily of the underlying porous medium, as long as the permeability of the porous medium satisfies[13] $H_p<-1/\upsilon$. In tracking the heterogeneity of the porous medium, we need to consider more correlated porous media. This is the purpose of the following.

4 Mapping Heterogeneous Permeability Fields

Capillary effects during immiscible flow, although significant at the microscale, they are typically neglected at larger scales in favor of viscous (or gravity) forces. The estimate of the length scale where they are appreciable, is given by $l_c\sim\sqrt{k}Ca^{-m}$, where the viscous correlation length exponent m is 1 or $\upsilon/(1+t+\upsilon)$, as already mentionned above (B_μ), although other exponents have also been proposed[15]. While this length scale is typically negligible compared to the geometric dimensions of a reservoir, capillarity can be nonetheless important in processes in heterogeneous reservoirs provided that the length scale of permeability variation, $l_k=\sqrt{k}/B_k$, be smaller than or equal to l_c. Such condition is likely to be satisfied near permeability discontinuities (end-effects) and in antipersistent or poorly correlated fields[13], where relatively sharp changes in permeability take place.

The importance of permeability heterogeneity on saturation profiles under static conditions has been known for a long time. Capillary continuity then dictates that the non-wetting saturation is larger in larger permeability regions. However, this simple condition fails when flow conditions (wether steady or unsteady) apply. Based on a continuum approach[10], rules were derived for the saturation profile, where a permeability heterogeneity in the direction of flow is encountered. The approach was also extended to countercurrent flows. The purpose of this part is to provide an experimental test of this approach, in determining the steady-state saturation profiles in heterogeneous media, where the permeability stratification is perpendicular to the flow direction. Immiscible displacement experiments were conducted in a long sample, consisting of a vertically placed rectangular glass bead pack of dimensions 30-4-2 cm, filled in different regions with different size glass beads according to a specified permeability heterogeneity. In one experiment, the bottom 21 cm were filled with large size beads (d_1=280μm) and the top 9 cm with small size beads (d_2=90 μm close packing was ensured by shaking). As the permeability is proportional to the square of the bead diameter the porous medium has an almost ten-fold permeability change somewhere in the middle. The fluids used were water as the wetting fluid $\mu_w=10^{-3}$ Pas, ρ_w=10 kgm^{-3}) and nonane as the non-wetting fluid (μ_o = 0.7 10^{-3} Pas, ρ_o=780 kgm^{-3}) with an interfacial tension between the fluids equal to γ=25 mJm.$^{-2}$

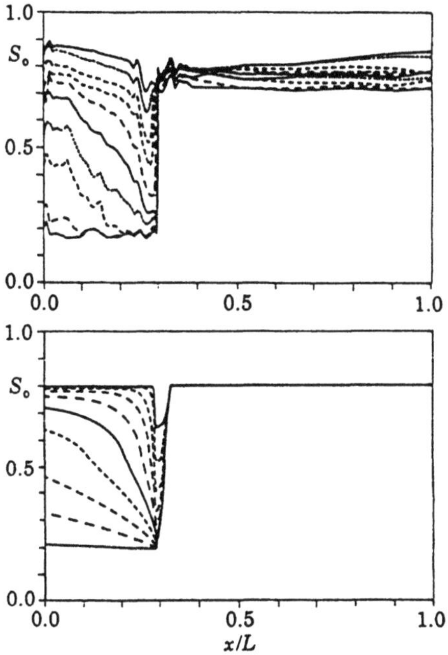

Fig.4: Top, experimental oil saturation profiles at 10 flow rates. Bottom, corresponding numerical simulations.

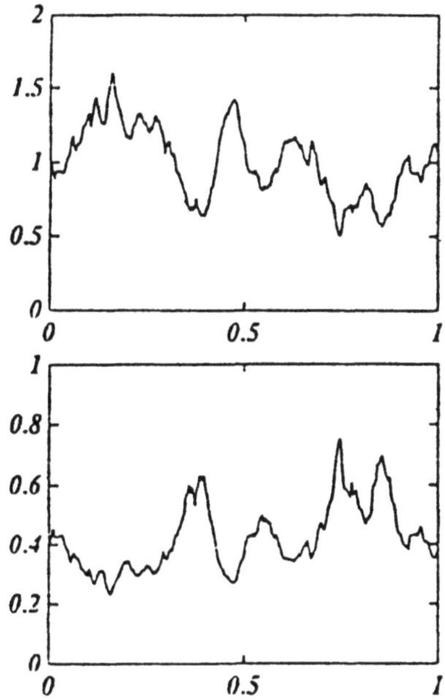

Fig.5: Top, H=.5 fBm permeability field. Bottom, corresponding saturation profiles for drainage at low Ca.

Experiments corresponded to primary drainage, thus we displaced water from an initially water-saturated sample by the lighter oil in the direction from top-to-bottom (in order to avoid gravitational instabilities). The flow rate q varied between 1.25 and 2700 cm/h, corresponding to a capillary number variation ($Ca=\mu_o q/\gamma$) between 9.7 10^{-8} and 2.0 10^{-4}. The same acoustic technique was also used to determine the saturation profile along the sample with an improved spatial resolution of 1 mm within the permeability variation zone (for a total of 150 sampling points).

Steady-state saturation profiles for the case of a permeability increase are shown in Fig4 (top), where the oil saturation S_0 is plotted vs the reduced sample length x/L for 12 different flow rates corresponding to more than three orders of magnitude variation in the capillary number. Typically, the lowest-rate displacement is first conducted, until a steady-state is established. Subsequent profiles were obtained by increasing the injection rate to the desired level and by continuing until another steady-state was reached. These profiles were very similar to those obtained by primary displacement at the same capillary number. A clear characteristic of the saturation response is its rapid decline before the permeability jump is encountered. The existence of this segment is more evident in the high Ca runs (although its magnitude is actually smaller). Upstream and downstream saturation values approach each other, particularly as the Ca increases. At low Ca, however, outlet boundary effects are felt in the profile, which, now resembles a saturation discontinuity in the region near the permeability jump. Numerical results are shown in Fig.4 (bottom) for the same conditions as in the experiment. A small noise was added to the permeability profile before integration to capture parts of small scale heterogeneity. The agreement between experiment and numerical results appears quite good, the numerical curves reproducing nicely all the salient features of the saturation profiles. For a more quantitative check, the depth of the saturation decline was plotted[10] as a function of the inverse capillary number Ca^{-1}. According to the analysis, this saturation decline should be a function of Ca^{-1} only. Numerical and experimental data agreed quite well for a variation of Ca spanning four orders of magnitude. Since the numerical analysis involves the (unknown) width of the region of heterogeneity, this represents a key adjustable parameter. However, it can be readily estimated by a shift along the horizontal axis. Best fit between theory and experiment was obtained for a width of the permeability jump of 2 mm, representing a few bead diameters, which is quite reasonable considering the uncertainty involved in the packing. The close agreement between experiments and the macroscopic model points out to the relevance of the latter. Indeed, this should be expected, as long as fractal regimes, described above are not involved.

The sensitivity of this saturation response to the permeability heterogeneity can in principle, be used to map permeability profiles. For example, the small, flow rate-dependent fluctuations of the profile in Fig.4 (top) is the signature of fluctuation of the permeability most likely due to packing defects. Hence, this can be used as an experimental tool to determine permeability variations in a direction perpendicular to the flow direction. It has been reported from porosity log[13] that many natural media have permeabilities, which can be correlated at all scales with an fBm trace of exponent H~.8 (close to the fracture exponent described in J.P.Vilotte and S.Roux lectures). Fig.5 (top) shows a permeability (actually a $\sqrt{k}$) field with H=0.8. Fig.5 (bottom) shows the saturation S_0 response obtained from the numerical simulation at steady-state at low values of Ca; the power spectrum of S_0 decreases as $l^{-2.51}$ close to the expect -(2H+1)=-2.6 value. It is, therefore, possible to identify the characteristics of the permeability profile from those of the measured saturation profile. However, we must note that for increasing Ca values the agreement deteriorates[10].

5 Conclusions

We have demontrated, both theoretically and experimentally, that a hydrostatic pressure gradient or a permeability gradient quench the front pattern of Invasion Percolation to a finite width σ, which is related to the respective Bond numbers via the percolation critical exponents. The analysis of the front in the fractal cross-over region reveals that the correlation of the noise of the 2-D saturation profile is of the fractional Brownian motion, as in ordinary percolation at p_c. Permeability heterogeneities perpendicular to the flow direction can also be tracked in non-wetting flow displacement leading to a promising tool for mapping of the permeability field.

This work was partly supported by NATO travel grant CRG 90 L053. MC, NR and DS would like to acknowledge support by GDR CNRS Physique des Milieux Heterogenes Complexes. BX and YCY were also supported by DOE Contract DE-FG22-90BC14600. All these sources of support are gratefully acknowledged.

References

1: R. Lenormand, J. Phys.Condens. Matter 2,SA79(1990); Lenormand, R., Proc. Rev. Soc. London. A423, 159 (1989).
2: L. Patterson, Phys. Rev. Lett. 52, 1621 (1984); T.A.Witten and L.M. Sander, Phys. Rev. B 27, 5686 (1983).
3: D. Wilkinson and J.F. Willemsen, J. Phys. A: Math.Gen.16, 3365 (1983).
4: J. Feder, Fractals Plenum, New York (1988).
5: J.P. Hulin, E. Clement, C. Baudet, J.F. Gouyet and M. Rosso,Phys. Rev. Lett. 61, 333 (1988); J.F. Gouyet, M.Rosso, E. Clement, C. Baudet and J.P. Hulin, in Hydrodynamics of Dispersed Media, J.P. Hulin, A.M. Cazabat, E. Guyon and F. Carmona (eds.) Elsevier, 179 (1990).
6: V. Frette, J. Feder, T. Jossang and P. Meakin, Phys. Rev.Lett.68, 3164 (1992).
7: D. Wilkinson, Phys. Rev. A30, 520 (1984).
8: B. Sapoval, M. Rosso and J.F. Gouyet, J. Phys. Lett 46, L149 (1985); J.F. Gouyet, M. Rosso and B. Sapoval, Phys. Rev. B 37, 1832 (1988).
9: A. Birovljev, L. Furuberg, J. Feder, T. Jossang, K.J. Maloy and A. Aharony, Phys. Rev. Lett.67, 584 (1991).
10: J. Chang and Y.C. Yortsos, Transport in Porous Media 5, 399 (1990); J. Chang and Y.C. Yortsos, SPERE, 285 (1992); M. Chaouche, N. Rakotomalala, D. Salin and Y.C. Yortsos, Europhys.Lett.21, 79 (1993); A.K. Stubos, C. Satik and Y.C. Yortsos, Int. J.Heat MassTransf. 36, 967 (1993).
11: M. Chaouche, N. Rakotomalala, D. Salin, B.Xu and Y.C. Yortsos (Phys.Rev.E)
12: J.-C. Bacri, M. Hoyos, R. Lenormand, N. Rakotomalala, A.Soucemariadin and D. Salin, J. Physique (France) III1, 1455 (1991); J.-C. Bacri and D. Salin, Geophys. Res. Lett.13, 326(1986).
13: Hewett, T.A., paper SPE 15386 presented at the 59 th Annual Technical Conference and Exhibition of the Society of Petroleum Engineers, New Orleans, LA (1986); M.B.Isichenko, Rev.Modern.Pys.64, 961 (1992) and references therein.
14: M. Chaouche, N. Rakotomalala, D. Salin, B.Xu and Y.C. Yortsos (in preparation).
15: Blunt, M., King, M.J., and Sher, H., preprint (1991).

Pattern Formation in Particulate Complex Fluids: A Guided Tour

Henri Van Damme, Emmanuel Lemaire, Y. Ould Mohamed Abdelhaye
A. Mourchid and P. Levitz

CRMD, CNRS and Université d'Orléans, F-45071 Orléans cedex 2, France

Abstract. Dispersion of colloidal particles in a liquid leads to the formation of complex fluids which are most often non-Newtonian and viscoelastic. When larger particles are used, plastic pastes may be obtained. Without dispersion medium, the particles just form a powder. Each of these systems exhibit peculiar flow or fracture behaviors which allows for a rich variety of fractal patterns, including viscous fingers, viscoelastic fracturing patterns, rough fracture surfaces, and granular displacement patterns.

Keywords. Non-Newtonian fluids, viscoelasticity, pastes, granular media, viscous fingering, viscoelastic fracturing, fracture

1 Particulate Fluids: Colloidal Suspensions, Gels, Powders and Pastes

The variety of pattern formation processes involving fluids is virtually unlimited. Mixing, evaporation and/or drying, chemical reactions, surface tension and temperature gradients are all ingredients which, alone or combined, may generate static or dynamic interfaces or heterogeneities. The fluid rheology interferes with the patterning process in almost every case. The purpose of this contribution is to show how the presence of colloidal or larger particles in the fluid may lead to specific patterns, by invasion or fracture, in which non linear aspects play an essential role.

The rheology of a particulate fluid is controled several factors: the size and shape of the particles, the range of the interaction, and the solid volume fraction. It is their combination which gives each particulate fluid its own behavior.

1.1 Dispersion of colloidal particles in a "simple" (in rheological respects) liquid such as water or a light hydrocarbon may lead to dramatic rheological changes. With independent, spherical and noninteracting particles (which is seldom the case), the process starts by a slow increase of the fluid viscosity with solid volume fraction Φ, according Einstein's law or one of its higher order version [1]. With spherical colloids interacting by long range repulsive forces of electrostatic origin, like with silica in water, the fluid very rapidly becomes non-Newtonian. Two cases may be observed: shear-thinning or shear-thickening. In the first case, the viscosity decreases as the shear rate is increased. In the second case the opposite happens. The first case is by far the most frequent in colloidal fluids. Concomitantly, the fluid often displays a yield stress, i.e. a minimum stress to be overcome before the fluid starts flowing. Both factors are sources of non-linearity. In simple shear flow, shear-thinning behaviour with a yield stress is described by the Herschel-Bulkley equation [2]:

$$\tau = \tau_y + \beta(d\gamma/dt)^n \tag{1}$$

$\tau = |\tau_{yx}|$, $\gamma = |\gamma_{yx}|$, τ_y is the yield stress and $n<1$ is the shear-thinning exponent. β has dimensions Pa s^n. Thus, the non-newtonian viscosity, $\eta = \tau/(d\gamma/dt)$, decreases as $d\gamma/dt$ increases. For $n=1$, eq.(1) describes a Bingham fluid, whereas for $\tau_y=0$ and $n<1$, it describes a simple shear-thinning fluid.

As the volume fraction is further increased, the non-Newtonian character is often coupled with the appearance of frequency-dependent viscoelastic properties and, more precisely, the onset and the increase of a storage modulus G' (real part of complex modulus). At still higher solid concentration, the medium may undergo a sol-gel transition and G' may overtake the loss modulus, G''. The solid volume fraction at which the fluid becomes non-Newtonian, viscoelastic and eventually undergoes a sol-gel transition is highly dependent on the interparticle interactions. Everything happens as if the volume fraction to consider was the effective volume fraction of the particles surrounded by their Debye screening length, ϕ_{eff} (Debye spheres). Thus, when the Debye length is very large, dramatic rheological changes may occur at extremely low real solid volume fractions ϕ (a few %), which in fact correspond to almost random close packing conditions for the virtual objects ($\phi_{eff} \cong 0.65$). This non-trivial dependence of rheological properties on real solid volume fraction is further reinforced when the particles are anisotropic (fibers, rigid platelets or large flexible flakes).

Another frequent feature of colloidal fluids is the time dependence of their properties, like in thixotropic or rheopectic behavior: at constant shear rate, the viscosity decreases or increases with time until a stationary state is reached. Both processes are associated with non zero structural relaxation times. These lead to an elastic response at high frequency, when the characteristic time of the flow event, τ_{fl}, approaches (from above) the time taken by the microscopic reorganization processes in the fluid, τ_{str}. This can be rationalized using the concept of Deborah number [2]:

$$De = \tau_{str} / \tau_{fl} \tag{2}$$

For a characteristic experimental length l and a velocity U, $\tau_{fl}=l/U$. The structural relaxation time is less obvious and is related to the reciprocal frequencies of the colloidal particles motion in the fluid. De may be interpreted as the ratio of elastic to viscous forces. For $De<<1$, viscous effects dominate, whereas for $De>>1$, the system will essentially behave as an elastic solid.

1.2 Fluids containing non-colloidal particles may have quite different properties. In the "dry" state (without suspension medium), the particles form non cohesive powder beds, with dilatant properties and friction-dominated flow behavior (see contributions by H. Herrmann et al., D. Bideau et al. and J. Duran et al. in this volume). Immersing the grains in a non-interacting liquid medium does not basically modify the behavior. Wet sand is of course more cohesive than dry sand (due to capillary forces), but it is still dilatant and contact forces still control its deformation. However, plasticity may be induced if the particles are wrapped in a layer of a soft and lubricating medium such as a polymer mix. A dramatic example of this is fresh cement or concrete which can withstand large forces without flowing but which, as soon as a little polymer is added, deforms under its own weight, even at a solid volume fraction of the order of 0.6 (i.e. not far from random close packing conditions) [3].

In the following, we will consider examples of pattern formation in each one of the media which were just briefly described: non-Newtonian colloidal fluids, viscoelastic gels, dry powders and plastic pastes. We will deliberately favor morphological and descriptive aspects.

2 Non-Newtonian Viscous Fingering

Viscous fingering in a Hele-Shaw cell (two transparent plates separated by a narrow gap) is one of the simplest pattern formation phenomenon involving liquids. The experiment consists in injecting a low viscosity fluid (fluid 1) into a high viscosity liquid (fluid 2) in a narrow gap, i.e. in conditions where flow is dominated by friction with the walls. Two square or rectangular glass plates separated by spacers a fraction of millimeter thick is suitable. In such conditions, and in the absence of net gravitational forces (horizontal cell), motion and deformation of the interface is controlled by the balance between viscous and capillary forces [4]. In each fluid, the flow velocity is governed by Darcy's equation and, if the fluids are incompressible, Laplace's equation also applies:

$$U_i = -(b^2/12\mu_i)\ \nabla P \ \text{ and } \ U_i = \nabla^2 P = 0 \qquad (3)$$

U_i is the velocity in fluid i averaged over the thickness of the cell, b; μ_i is the fluid viscosity. Equation 3 implies that any bump generated by fluctuations of the interface increases the pressure gradient in front of it, thereby accelerating further the more viscous fluid and increasing the size of the bump. The interface is basically unstable. Capillary forces enter in the game as stabilizing forces. The pressure jump across the interface between fluid 1 pushing fluid 2 is given by:

$$P_1 - P_2 = \Gamma_{12} K \qquad (4)$$

where Γ_{12} is the interfacial tension and K the local curvature of the interface. As a result, each bump (positive K) produces a pressure drop in front of it which decreases the gradient induced by the viscous forces. In terms of dimensionless group, the stability of the interface is controlled by two dimensionless numbers, the capillary number, $Ca=U\mu/\Gamma_{12}$, and the viscosity contrast, $A=(\mu_2-\mu_1)/(\mu_2+\mu_1)$. For an infinite viscosity contrast (A=1), the interface may be considered as an equipotential surface (P constant) and Ca is the only relevant parameter. In these conditions, the linear stability analysis performed by Chuoke et al. [5] for the boundary conditions of rectilinear flow shows that there is a critical wavelength, $\lambda_c=(\pi/\sqrt{3})bCa^{-1/2}$, below which all fluctuations are damped out by the interface tension, and a wavelength of fastest growing rate, λ_m, related to λ_c by $\lambda_m=3\lambda_c$.

Less theoretical work has been done on viscous fingering with non-Newtonian fluids [6,7]. What is clear is that for simple shear-thinning fluids (eq. 1, with $\tau_y = 0$), Darcy's law and Laplace's equation have to be replaced by:

$$U \sim \nabla P^{1/n} \text{ and } \mathrm{div}(|\nabla P|^{(1/n)-1}\nabla P) = 0 \qquad (5)$$

In these conditions, the recent linear stability analysis of Sader et al. [7] shows that shear-thinning enables perturbations to the interface with higher wave numbers to commence growing much earlier than with Newtonian fluids.

Fig. 1. Viscous fingering patterns obtained in a rectilinear Hele-Shaw cell with (a) a Newtonian oil, at small capilary number (Saffman-Taylor finger, adapted from ref. 8); (b) a Newtonian oil at large capillary number (adapted from ref. 10); (c) with a non-Newtonian clay mud.

Experimentally, the initial phase of viscous fingering with immiscible Newtonian fluids is the formation of ripples on the interface. The foremost ripple of size λ_c

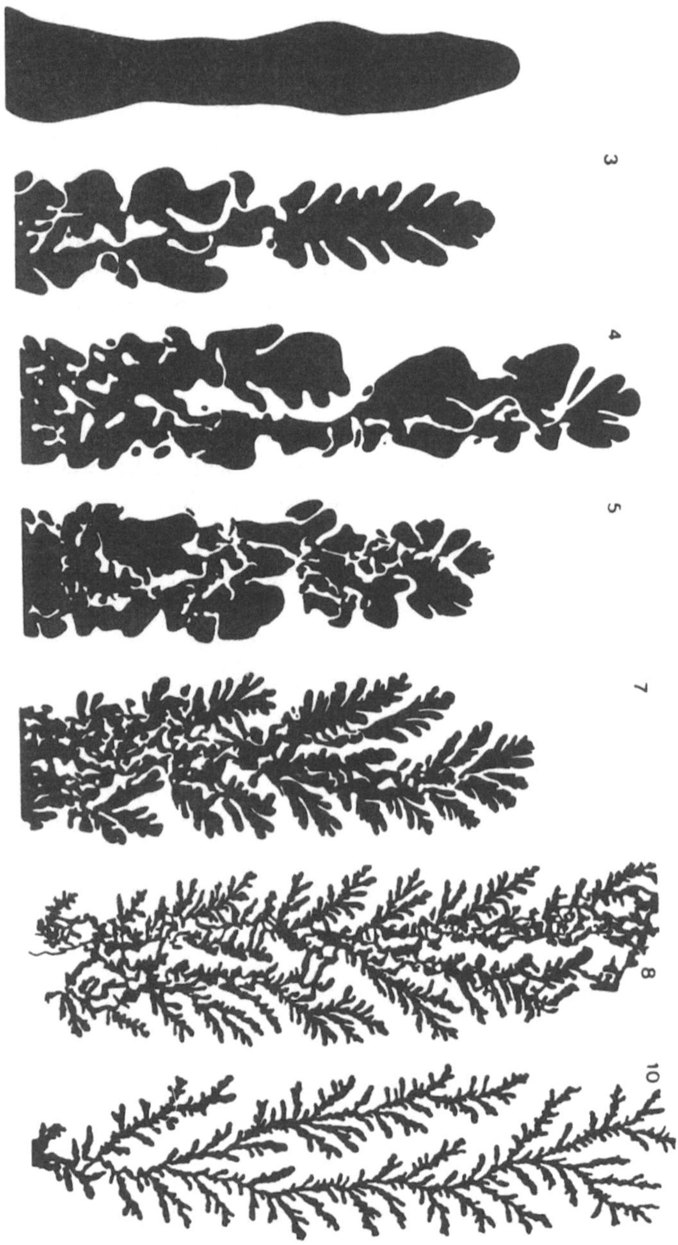

Fig. 2. A morphogenetic sequence leading to developed fractal viscous fingering patterns in non-Newtonian muds at increasing effective capillary number. Seven patterns, out of a sequence of ten, are shown (adapted from ref. 13).

soon overcomes the other and turns into a finger with smooth walls. This celebrated Saffman-Taylor finger [8] is an extremely robust pattern (fig.1a). In a broad range of viscosity and velocity conditions it retains its shape and the only thing which changes is its width relative to the cell width, $\lambda = l/w$ (finger width/cell width), which is a function of a modified capillary number, $1/B$ or $Ca' = 12(w/b)^2Ca$, which takes into account the aspect ratio (w/b=width/thicknes) of the cell. However, this "stable" regime of the interface instability becomes itself unstable when Ca' exceeds values of the order of 10^3 [9]. The finger exhibits a few simple instability modes, including tip-splitting and, if the control parameter is pushed far enough, it turns into a branched pattern which has a mass fractal character (fig.1b, [10]). However, this mass fractal character is not fully developed. The pattern boundary exhibits long straight parts which prevent it from having a surface fractal exponent covering the same range as the mass fractal exponent (larger low cut-off).

The rather robust behaviour of viscous fingers in newtonian fluids contrasts with their extremely unstable behaviour in shear-thinning fluids [11]. As soon as a non-Newtonian shear-thinning fluid is displaced, finger stability becomes extremely difficult to control, even at very low velocity, and one is rapidly led to a very tight branching cascade (Fig1c). An "effective" form of the control parameter, Ca'_{eff}, has to be defined, using the non-newtonian viscosity at a shear rate, averaged over the gap of the cell, of the order of 2U/b [12]. By doing so, one notices that shear-thinning properties allow for the exploration of extremely high values of Ca'_{eff}, larger than 106 [12]. The patterns obtained in this regime are fractals without loops, with a DLA-type structure. The absence of loops makes their "mass" fractal dimension (which describes the scaling of the pattern area) equal to their "surface" fractal dimension (which describes the scaling of the perimeter), in the same length range. That a mass fractal structure can be obtained is not obvious a priori since viscous fingering is primarily an interfacial instability. Mass fractality stems from the fact that a statistically uniform thickness is associated to the interface.

The morphogenesis of fractal non-Newtonian fingering pattern may be investigated in detail, starting from weakly shear-thinning fluids and low front velocities, and increasing progressively either the velocity or the shear-thinning properties, or both [13]. Figure 2 is a sequence of patterns obtained in that way, by injecting air at increasing pressure in a dilute colloidal clay suspension [12]. When doing so, one notices that a number of elements indicate that a single mechanism is operating all along the morphogenetic process of Fig.2. The first point is that the window of fractal behavior, with a low cutoff which is the branch width and a high cutoff which is the pattern width (both being identical in the starting Saffman-Taylor finger), opens continuously as Ca'_{eff} increases. In this window, the interface fractal dimension of the interface remains essentially constant. A second point is that simple Fourier filtering the more evolved patterns with increasingly strong low pass filters generate patterns which reproduce qualitatively in reverse way, the direct experimental sequence (Fig.3). This is clear suggestion that the morphological changes merely stem from an increasing number of spatial frequencies along the interface.

Fig.3. Low pass Fourier filtering of pattern number 7 of Fig.2 with increasingly strong filtering from right to left.

A third diagnosis element is the distribution of local curvatures, *K*, along the interface, which may be treated as a spatial series and analyzed by conventional signal analysis techniques [14]. In a Saffman-Taylor finger, only one spatial frequency (the positive curvature of the tip) is present. As the fractal regime develops, a quasi-continuous distribution of frequencies establishes itself, broadening towards high frequencies as the effective capillary number increases. This suggests that the interfacial instability process which leads to fractal viscous fingering patterns could be better described in terms of spatial turbulence than in terms of deterministic chaos [14,15]. Remarkably, the spectrum of the fully developed fractal tree is not a white distribution. The power spectrum exhibits a short plateau at low frequencies and a long high frequency side which obeys an inverse power law behavior over almost a decade. The exponent of this $1/f^{\beta}$ behavior of a self-affine set is related to the fractal exponent by $D_f=2-[(\beta-1)/2]$, which yields $D_f=1.20$ [14].

The analogy between fractal viscous fingering patterns and spatial turbulence may be pushed one step further by performing an analysis of the same type as that performed in the onset of thermally driven turbulence, in a layer of fluid heated from below. In the later case, one analyzes the ratio of the actual heat flow, mainly via turbulent convection, to the heat flow if conduction was the only heat transport mode (Nuselt number). The control parameter is the Rayleigh number. Over more than 7 decades, the relationship between Nu and Ra is of the power law type. The exponent is 1/3 in the weak turbulence regime, and 2/7 over about 5 decades of Ra in the hard turbulence regime [16] . The value of the exponents may be derived from scaling arguments. In the same spirit, one can look, in viscous fingering experiments, to the ratio of actual invading fluid flow through all the branches of the pattern to the fluid flow which would be observed if regular Saffman-Taylor through a single branch or finger was the only invading mode. This is a new dimensionless number, analogous to Nu. Our control parameter is still Ca'. The measure of the "flow" through the whole pattern is defined as the fraction of the cell width which is occupied by the invading fluid, δ, whereas the measure of the "flow" through a branch is just the relative branch width, λ. As shown in Fig.4, a good power law with exponent 1/3 is observed over more than 4 decades. The average relative branch width also follows a power law (Fig.4, [17]), with an exponent (-1/2.6) which is just a little different from the value (-1/2) predicted by Chuoke's analysis for immiscible Newtonian fluids. Thus, $\delta/\lambda \sim (Ca')^{1/3}$ and $\lambda \sim (Ca')^{1/2.6}$, which means that δ is essentially constant through the whole morphogenetic process which develops beyond Saffman-Taylor fingering. This is reminiscent of the observation of Hakim et al.[9], who showed that the occupancy probability of the cell in Newtonian viscous fingering experiments is concentrated over a constant fraction of the cell width (0.5). Both observations are clear indication that the branching behavior of complex fingering patterns, beyond the Saffman-Taylor regime, stems from a (yet hidden) scaling law.

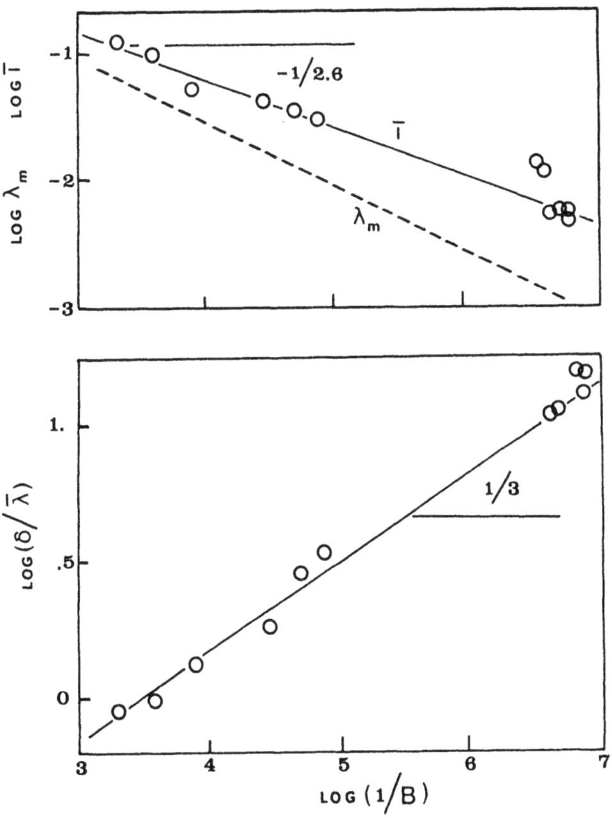

Fig. 4. Evolution of the relative "pattern flow" (a; see text) and relative finger width (b) as finction of the effective control parameter in the route to developed fractal viscous fingering

3 Viscoelastic Fracturing in 2D

As we mentionned in section 1.1, colloidal fluids also display viscoelastic properties, which show increasingly up as $De = \tau_{str}/\tau_{fl}$ increases. Thus, the use of viscoelastic fluids permits to think of a possible crossover from fractal growth by flow to fractal growth by fracture by increasing *De*. This can be achieved either by slowing down the rate of structural rearrangements, by concentrating the suspension, or by using faster invasion rates [18]. An example of transition from viscous fingering (VF) to viscoelastic fracturing (VEF) regime is shown in Fig.5. The transition is characterized by sharp morphological changes. The "fingers" now look clearly like cracks. They have a sharp tip. Branching occurs at right angle and, most important, *behind* the tips. In addition, the cracks are generally curved and branching occurs systematically on the convex side of the main crack. These elements show clearly that branching stems from the rise of a tensile stress parallel to the growth direction of the mother-cracks. As expected, the crossover from VF to VEF occurs for a critical value of *De*.

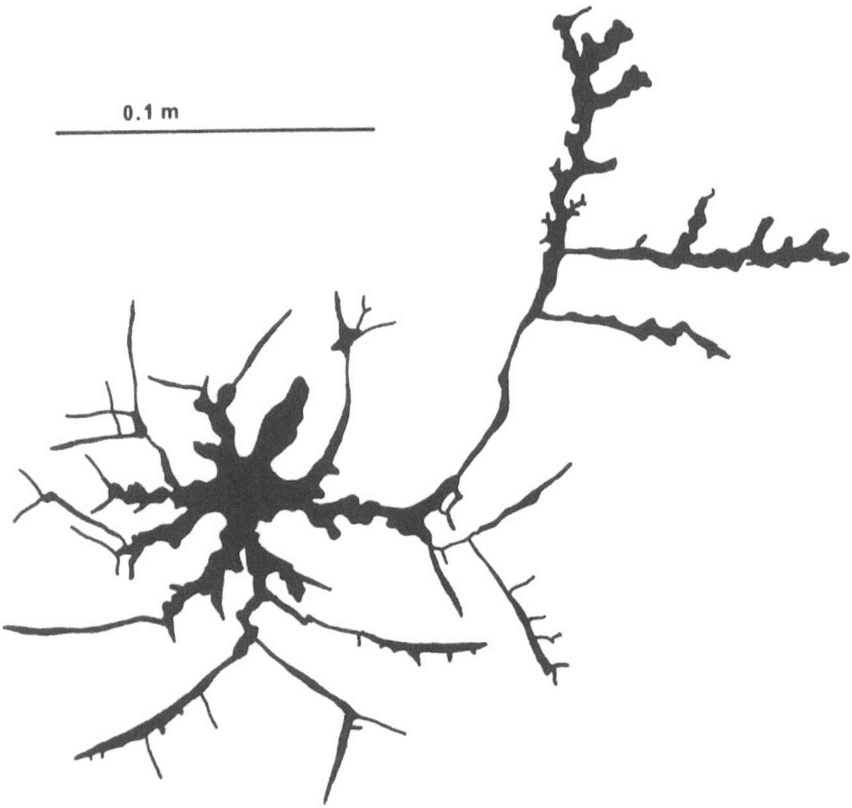

Fig. 5. Transition from viscous fingering (VF) to viscoelastic fracturing (VEF) in a single pattern. The process starts by VF, switches to VEF and then back to VF (top right)

In spite of these deep modifications, VEF patterns are still mass fractals, like VF patterns, but their mass fractal dimension, D_m, is systematically lower (1.4 to 1.5) than what is expected for DLA-type patterns ($D_m \cong 1.68$) as found for fractal viscous fingering in radial Hele-Shaw cells [19]. This lower value of D_m is in qualitative agreement with values obtained from simulation studies of crack propagation on bond networks using either deterministic [20] or random [21,22] bond-breaking models. Interfacial tension does not seem to influence D_m. It merely increases the average crack width.

An interesting variant of these experiments, which reinforces the difference between the VF and the VEF regimes, is a multiple injection experiment [23]. In VF, the driving force is the pressure gradient. When two fingers approach each other, the pressure gradient in the gap between the two tips vanishes as they approach each other. Thus, in the absence of inertial effects, in a multiple injection experiment within a single cell, VF patterns "repell" each other (Fig.6). On the contrary, VEF pattern "attract" each other because the stress produced by two approaching crack tips add to each other (Fig.7).

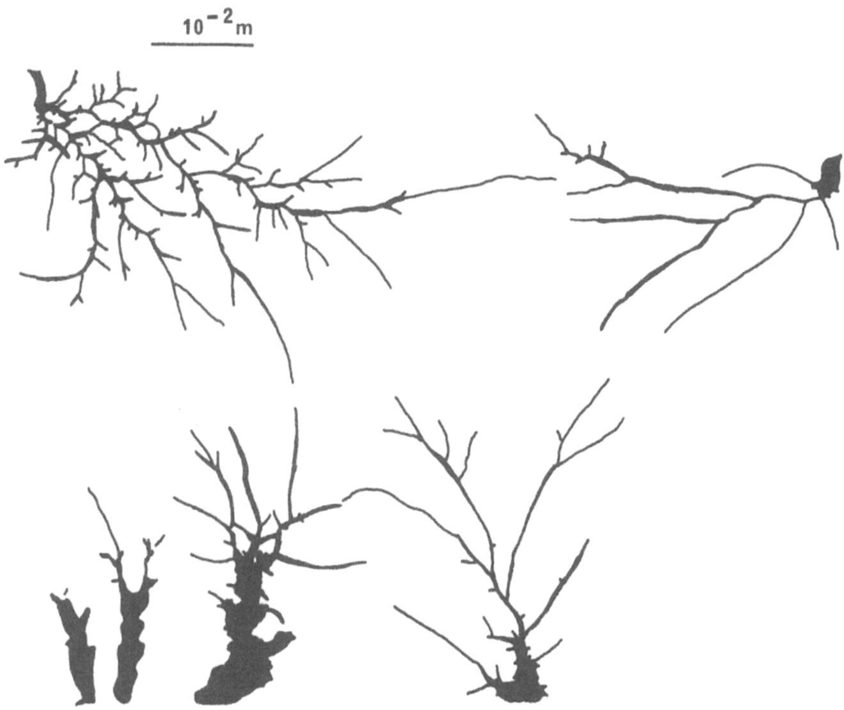

Fig. 6. Multiple injection experiment in the viscoelastic fracturing fingering regime.

4 Viscoelastic Fracturing in 3D

The experiments of section 3 are just a particular type of a general fracturation mode called hydraulic fracturing (HF), in which fracture is obtained by pressurizing a fluid into the medium to be fractured, generally a rock (see contribution by L. Petitjean in this volume). Besides these particular loading conditions, HF follows the same laws as ordinary fracture. In the classical analysis, it is assumed that the crack initiates when the tensile stress at the injection hole or cavity exceeds the tensile strength of the medium. At this point, the crack propagates as usually along the direction perpendicular to the smallest principal stress.

Fig. 7. Moulding of a multiple and branched crack obtained by hydraulic fracturing in a clay gel. The arrow indicates the injection direction of the syringe needle. Pattern width: ≅ 8 cm.

We performed our experiments with colloidal suspensions of a synthetic clay, Laponite, in 10^{-3} N NaCl aqueous solutions at pH 9. Laponite is a negatively charged magnesium silicate colloid with a surface area of the order of 300 m^2/g. It is made of 1 x 25 nm rigid platelets, with a surface charge of the order of 1 $e^-/2.5$

nm^2. It has a very strong gelling capacity. The concentration range investigated was between 1 and 5% (w/w). This corresponds to shear storage modulii, G', from 10^1 to 10^4 Pa, at a frequency around 1 rad s^{-1}, as measured with a controlled shear rheometer [24].

The pattern obtained by injecting a fluid epoxy resin into a 2% laponite suspension in a cylindrical glass beaker, without any stress besides gravitation, is shown in Fig.7. The pattern was obtained by letting the resin harden and washing the gel off. One obtains patterns which are clearly *fracture* patterns. The two important features are: (i) *multiple fracturing*, i.e. several cracks start from the injection hole; this is expected from the absence of well defined external stress. (ii) *branching*: each main crack splits into a large number of sub-cracks, slightly tilted with respect to the mother crack. This is more unusual. However, comparison of the estimated crack tip velocity (0.01 m/s) with the estimated velocity of sound (shear waves) in the fluid (v=0.1 m/s) reveals that the former is never far from the later. In other words, the fracture conditions are close to the expected instability conditions for crack propagation (ref). The conditions in which branching occur have still to be investigated more deeply.

5 The Fracture of Pastes

The fracture of fragile or ductile solids has been shown by several authors to produce self-affine rough surfaces [25-29] (see contribution by S. Roux in this volume). In ductile fracture with aluminium alloys [26], the fractal dimension of the fracture surface determined by the slit-island method was found to be in all cases close to 2.2, corresponding to a roughness exponent H=3-D of 0.8. On the other hand, for brittle fracture on a variety of materials including steel, graphite, porcelain, bakelite and plaster, K.J. Maloy et al. [29] found somewhat higher roughness exponents (lower local fractal dimension), with an average value of 0.87.

Viscoplastic fracture of a paste may be considered as the extreme case of ductile fracture (anyone who has ever prepared bread knows what it means). Experiments [30] were performed by separating along a vertical axis two horizontal 30x30 cm glass plates separated by a 1 cm thick layer of fresh ready-to-use mortar obtained by mixing lime particles with a polymer mix. This is tensile mode I fracture. The separation velocity was controlled in a range going from 1 to 10 mm s^{-1}. An example of fracture surface, showing their surprisingly realistic aspect with respect to natural profiles, is shown in Fig.8.

After hardening of the paste, the roughness of the fracture surfaces was analyzed by determining the roughness exponents or fractal dimensions of the one-dimensional profiles obtained by sawing the surface into vertical slices. No significant difference was observed between slow and fast fracture. The average value for the roughness exponents at different rupture velocities was found to be 0.88 ± 0.05. This value is remarkably close to the value obtained for brittle fracture by Maloy et al. [29].

The mechanism by which the layer of paste breaks is not yet clear. When this experiment is done with a colloidal fluid such as a clay suspension of much lower viscosity and modulus than the granular paste, it leads to a pattern which is, in fact, the result of viscous fingering due to air penetration from the boundary of the cell into the broadening gap. This leads to a pattern of smooth and rounded crests of uniform height where the valleys are all accessible from the boundary of the cell. This is not the case in the paste fracture surfaces, suggesting that fracture is not the result of quasi-2D fracturation by air entering from the boundary of the cell, but that it is initiated by cavitation or bubble growth at many places within the paste. The main question that these experiments raise is of course the origin of the universality of height correlations in brittle as well as viscoplastic fracture in disordered materials. The unifying concept of fracture by *damage* might be a clue.

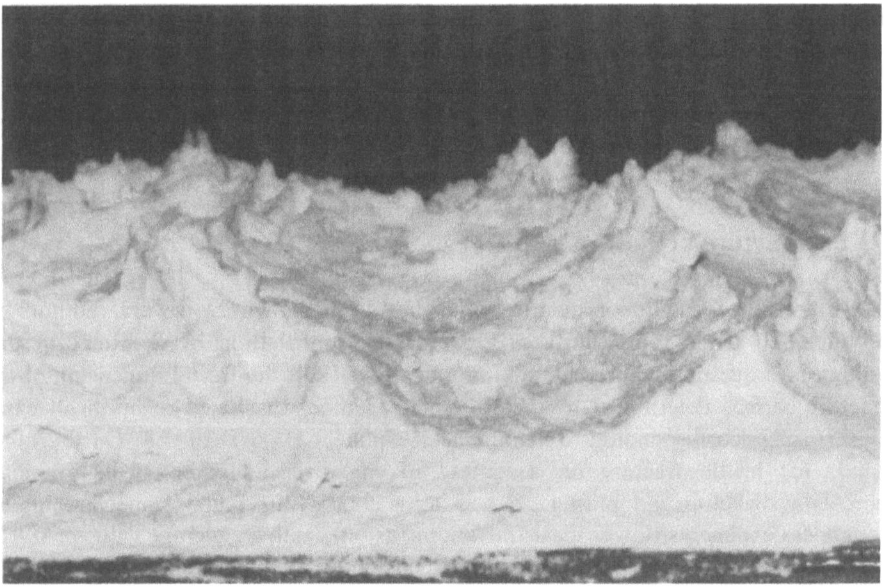

Fig. 8. Grazing view of the rough surface obtained by tearing away two plates separated by a layer of paste.

6 Granular Fingering in Powders

To what extent dry powders may be described as a class of non-Newtonian fluids with a yield stress is still under debate. Powders exhibit very peculiar phenomena such as arching but on the other hand, they also exhibit flow behaviors which may be interpreted in terms of fluid with a yield stress. As far as instabilities and patterns are concerned, a puzzling result is the occurence of what may be considered as a kind "viscous fingering" patterns in granular layers [32].

Fig. 9. Growth sequence of a fingering pattern obtained by injecting air in a powder.

The experiment is almost the exact replica of what has been described in section 2, with a radial Hele-Shaw cell. Air is injected into a layer of zirconia powder. The cell is filled by pouring the powder on the bottom plate, spreading it roughly and finally levelling it by gently sliding and rotating the upper plate until the desired spacing is reached. After filling, the cell is clamped on its four edges.

Two regimes are observed. At low inlet valve opening, the powder layer behaves as a rigid porous medium allowing only for gas permeation in its pores, all granular movements being blocked. On the other hand, at full valve opening, the particles are repelled by the gas flux, leading to a branched pattern of powder-free chanels remarkably similar to the fractal viscous fingering patterns obtained with non-Newtonian fluids in radial cells (Fig.9). Their mass fractal dimension is close to 1.6. The interface between the gas channels and the powder is in general very sharp (Fig.10), but its local morphology may be either straight or very wavy.

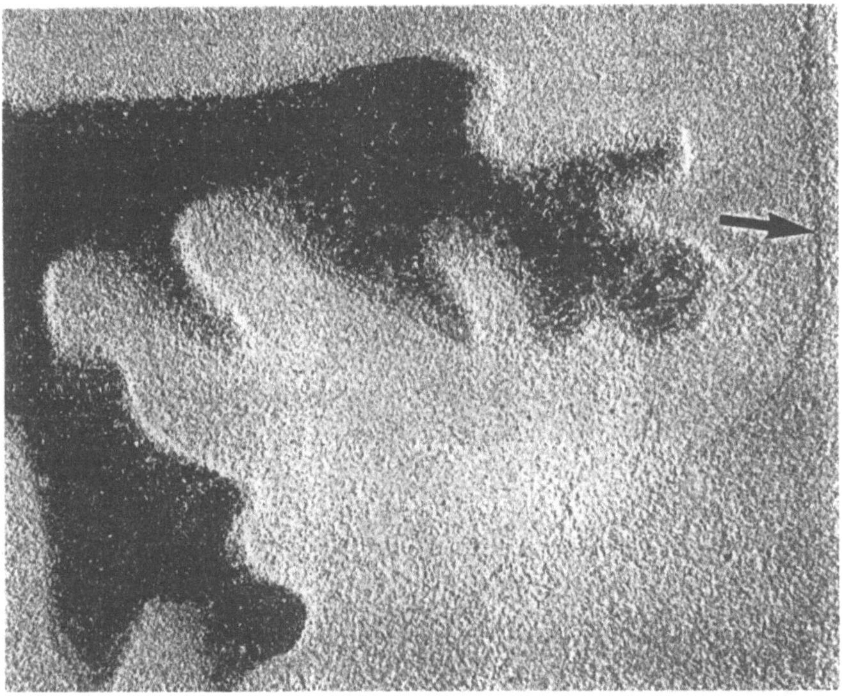

Fig. 10. Enlarged view of a "viscous" finger in a powder showing the interface between the compacted and the uncompacted zone (arrow).

The first question raised by these experiments is: how can a dilatant medium, confined in a narrow gap, flow? Two factors may contribute to make it feasible. The first is a deformation ("swelling") of the cell under pressure. The second, identified, stems from the fact that the initial state of the powder layer is not compact, due to the cell preparation method. Close examination of the patterns shows that, in fact, two regions may be detected in the invaded granular layer. The fingering pattern is surrounded by a compacted zone which is itself surrounded by the unperturbed, less compact, region (Fig.10). The interface between the compacted and the unperturbed regions is as sharp as that between the channels and the compacted region. The shape of the compacted zone is roughly circular. The distance between the branch tip and the boundary of the dense region is of the same order as the finger width, which is itself of the order of 20b, b being the cell spacing.

A second question is: what is the mechanism of wavelength (finger width) selection? In viscous fingering with miscible fluids, the mechanism of wavelength selection is related to energy dissipation or, more generally, entropy production [4,17,32,33]. It is likely that dissipation is also involved in the present case. More experiments on the influence of cell gap, particule roughness and flow rate should shed some light on this.

References

1: G.K. Bachelor, An Introduction to Fluid Dynamics, Cambridge Univ. Press (1988)
2: R.B. Bird, R.C. Armstrong and O. Hassager, Dynamics of Polymeric Liquids, Vol.1, Wiley New York (1987)
3: K. Kendall, in Tribology in Particulate Technology, eds. B.J. Briscoe and M.J. Adams (Adam Hilger, Bristol, UK, 1987), pp 91-102
4: G.M. Homsy, Ann. Rev. Fluid Mech., 19, 271 (1987)
5: R.L. Chuoke, P. Van Meurs and C.J. Van der Poel, J. Pet. Technol., 11, 64 (1959)
6: S.D.R. Wilson, J. Fluid Mech., 220, 413 (1990)
7: J.E. Sader, D.Y.C. Chan and B.D. Hughes, Phys Rev. E, 49, 420 (1994)
8: P.G. Saffman and G.I. Taylor, Proc. Roy. Soc. London, Ser. A, 245, 312 (1958)
9: P. Tabeling, G. Zocchi and A. Libchaber, J.Fluid Mech., 177, 67 (1987)
10: A. Arnéodo, Y. Couder, G. Grasseau, V. Hakim and M. Rabaud, Phys. Rev. Lett., 63, 984 (1989)
11: J. Nittmann, G. Daccord and H.E. Stanley, Nature, 314, 141 (1985)
12: H. Van Damme, E. Alsac, C. Laroche and L. Gatineau, Europhys. Lett., 5, 25 (1988)
13: H. Van Damme, E. Alsac et C. Laroche, C.R. Acad. Sc., serie II, 309, 11 (1989)
14: E. Lemaire, P. Levitz and H. Van Damme, C.R. Acad. Sc., serie II, 311, 1467 (1990)
15: V.K. Horvath et T. Vicsek, J. Phys. A: Math. Gen., 23, 1990, p L259-L264
16: B. Castaing, L. Kadanoff, A. Libchaber, E. Moses and G. Zocchi, J. Fluid Mech., 204, 1 (1989)

17: H. Van Damme and E. Lemaire, in Disorder and Fracture, J.C. Charmet, S. Roux and E. Guyon eds., NATO ASI Series, B235, Plenum Press, New York, pp 83-104 (1990)
18: E. Lemaire et H. Van Damme, C.R. Acad. Sci. Paris, série II, 3O9, 859 (1989)
19: E. Lemaire, P. Levitz, G. Daccord and H. Van Damme, Phys. Rev. Lett., 67, 2009 (1991)
20: H.J. Herrmann, J. Kertesz and L. De Arcangelis, Europhys. Lett., 10, 147 (1989)
21: E.L. Hinrichsen, A. Hansen and S. Roux, Europhys. Lett., 8, 1 (1989)
22: P. Meakin, G. Li, L.M. Sander, E. Louis and F. Guinea, J. Phys. A: Math. Gen., 22, 1393 (1989)
23: E. Lemaire and H. Van Damme, in preparation
24: A. Mourchid, P. Levitz, A. Delville and H. Van Damme, Les Cahiers de Rhéologie (Paris), X, 89, (1992)
25: B.B. Mandelbrot, D.E. Passoja and A.J. Paullay, Nature, 308, 721 (1984)
26: C.S. Pande, L.R. Richards and S. Smith, J. Mater. Sc. Lett., 6, 295 (1987)
27: D.L. Davidson, J. Mater. Sc., 24, 681 (1989)
28: E. Bouchaud, G. Laplasset and J. Planes, Europhys. Lett., 13, 73 (1990)
29: K.J. Maloy, A. Hansen, E.L. Hinrichssen and S. Roux, Phys. Rev. Lett., 68, 213 (1992)
30: E. Lemaire, Y. Ould Mohamed Abdelhaye, J. Larue, R. Benoit, P. Levitz and H. Van Damme, Fractals, in press
31: H. Van Damme, C. Laroche and L. Gatineau, Rev. Phys. Appl., 22, 241 (1987)
32: D.D. Joseph, Eur. J. Mech. B/Fluids, 9, 565 (1990)
33: L. Paterson, Phys. Fluids, 28, 26 (1985)

Solvable Models of Material Breakdown

P.L. Leath[1] and P.M. Duxbury[2]

[1] Department of Physics and Astronomy, Rutgers University, Piscataway, NJ 08855-0849.
[2] Physics/Astronomy Department and Center for Fundamental Materials Research, Michigan State University, East Lansing, MI 48824-1116.

Abstract. The history of the study of fracture of materials is briefly reviewed. Then the importance of analytically solvable models in understanding material breakdown is illustrated by a review of the work of Duxbury, Leath and Beale on simple analytically solvable models of fuse network breakdown in brittle systems. We then review recent work extending this analytically to include close pairs of clusters of defects or double clusters, which also exhibit the double-exponential failure distribution. Finally, a new analytic recursion method is presented for breakdown of systems with linear cracks, but a continuous distribution of breaking strengths. Remarkably, these systems exhibit an optimum sample size where the failure probability can, at low stress, be *reduced by many orders of magnitude* below that of a single bond.

Keywords. Breakdown, scaling theory, extreme statistics, solvable models

1. Introduction

The sample size dependence of the breakdown of materials is a very old subject dating back at least to the extensive experiments of Galileo [1], whose experimental set-up is illustrated in Fig. 1.

Figure 1. Galileo's sketch of an apparatus for measuring the breaking strength of a rectangular beam.

Galileo first discovered the proportionality of breaking strength to cross-section of a beam, and the inverse relation to the beam length for a given load E, as shown in Fig. 1. This relation, like that of force and displacement to stress and strain gives the primary relation between the macroscopic and microscopic breakdown properties of materials, but does not explain the variation of the breakdown or fracture stress with sample size. It was not until 1880, that the size effect in the breakdown stress was proposed by W.S. Chaplin [2] to be statistical in nature, varying from sample-to-sample depending upon the "weakest-link" in the system. But Chaplin's proposal was far ahead of its time and nearly forgotten until rediscovered and developed by A.A. Griffith [3] in 1920. Griffith's work had a profound effect on scientists, mathematicians and engineers. In mathematical statistics, there followed a classical period of the statistics of extremes, through the 1920's, 1930's and 1940's with major contributions from Dodd [4], Frechet [5], Fisher and Tippett [6], von Mises [7] and Gnedenko [8]. In 1939, the engineer W. Weibull [9] produced his marvelously successful ad hoc formula for the failure distribution. This period witnessed remarkably little communication between the mathematical statisticians, and scientists and engineers, whose progress evolved along separate tracks, and was marked by raging controversy because most scientists and engineers did not understand that it was only the extreme tail of the probability distribution (the weakest-link) that determined most of what was happening at the brittle fracture or breakdown point. Finally, in 1948, B. Epstein [10], made the connections between the statistics of extremes and real problems (such as brittle fracture) and finally the connections between the disciplines were begun to be made.

Today, the statistics of extremes finds applications in a great variety of fields including brittle fracture, dielectric and electrical breakdown, superconducting critical current, earthquake occurrence, probability of maximum floods, error rates in spacecraft receivers, air quality extremes, bacterial extinction time after innoculation, and the time to occurrence of cancer.

The fact that the sample size dependence of the average fracture strength usually scales to zero in the thermodynamic limit is a natural consequence of the dependence of fracture on extreme statistics. Nevertheless, it is only in the past decade that a more complete understanding of this scaling behavior has developed. After some early indications[11] in work by Smith, and Harlow and Phoenix, large scale numerical simulations on fuse networks of up to size 100x100 showed clear evidence for this effect. Using the key numerical method developed by de Arcangelis et al.[12], these numerical simulations, reported by Duxbury et al. [13,14] showed that, in the case of diluted two-dimensional fuse networks (see Fig. 2), the breakdown electric field decreased monotonically towards zero as the sample size increased. More importantly, they used a Lifshitz-type argument[13,14] to give a simple, analytic formula to explain the sample-size dependence. This "worst-defect-cluster" model is given in the next section.

2. Worst-Defect-Cluster Model

The worst-defect-cluster model is a weakest-link model. For the case of a randomly-diluted network, such as the one shown in Fig. 2, Duxbury et al.[13,14]

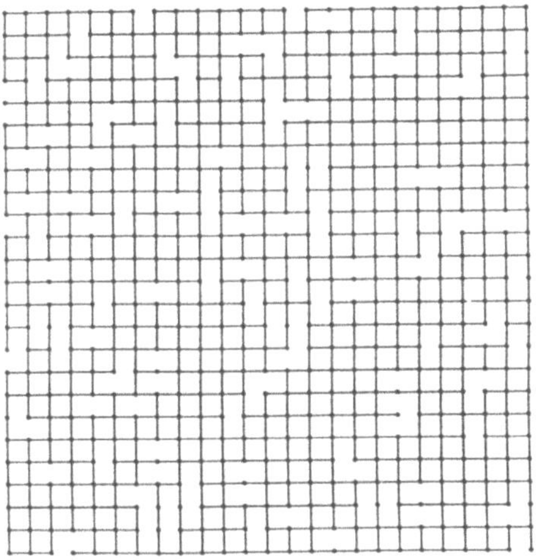

Figure 2. A diluted, square network of bonds with $p = 0.9$ occupancy.

guessed that an important "worst defect cluster" would be a linear cluster of adjacent missing fuses, as shown in Fig. 3, and calculated the size of the largest

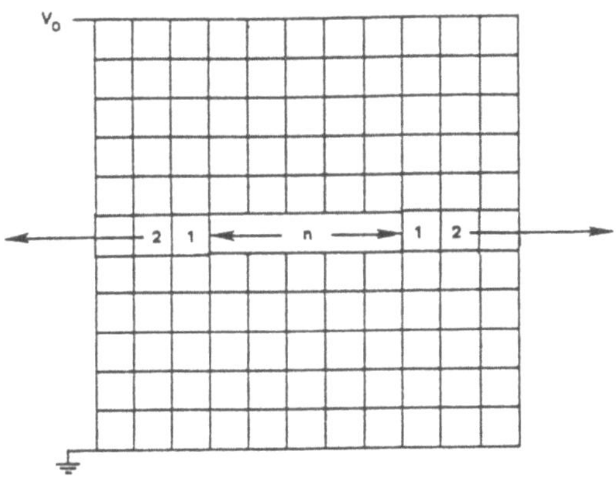

Figure 3. A linear cluster of $n = 4$ missing bonds. The arrows show the direction of crack propagation.

probable linear cluster in a sample of size LxL. In three dimensions, the equivalent model is to assume that the worst defect cluster is a disk or "penny-shaped" crack oriented perpendicular to the current flow. But the probability of n adjacent defects in an LxL sample is just,

$$P_n \sim L^2(1-p)^n, \tag{1}$$

since each bond is occupied independently, with probability p, and since there are L^2 possible placements of this cluster in the sample. The largest probable value n_{max} is then given approximately by setting Eq.(1) to one, so that

$$L^2(1-p)^{n_{max}} \sim 1, \tag{2}$$

or,

$$n_{max} \sim -2logL/log(1-p) \tag{3}.$$

Furthermore, if a large sample has a linear defect cluster of size n_{max}, the current per bond at the end of the defect is enhanced over that in the bulk (i) according to the formula [14,15],

$$i_{tip} \sim i(1 + Kn_{max}^{\alpha}), \tag{4}$$

provided n_{max} is large (Here $\alpha = 1/2$, and the result Eq.(4) is found by solving Laplace's equation in the presence of the defect). Therefore, the fuses at the defect tip break when $i_{tip} = 1$, corresponding to the critical current

$$i_c \sim \frac{1}{1 + Kn_{max}^{\alpha}} \sim \frac{1}{1 + K[\frac{-log(L^2)}{log(1-p)}]^{\alpha}}. \tag{5}$$

which means that i_c approaches zero logarithmically as L tends to infinity. Of course, this diluted perfect network is truly brittle in the sense that when the bonds at the defect tip (labeled 1 in Fig. 3) break, there is an even greater current enhancement on the bonds labeled 2 etc., so that the entire network is ripped apart (like a zipper) or cleaves. Duxbury et al.[14,15] showed that the result Eq.(4) also holds if the worst defect cluster is two adjacent cracks with a small surviving "neck" between them. In this case the exponent $\alpha = 1$. The numerical results [13,14] on the fully disordered lattices are shown in Fig. 4, and are fully consistent with the model result for i_c given by Eq.(5).

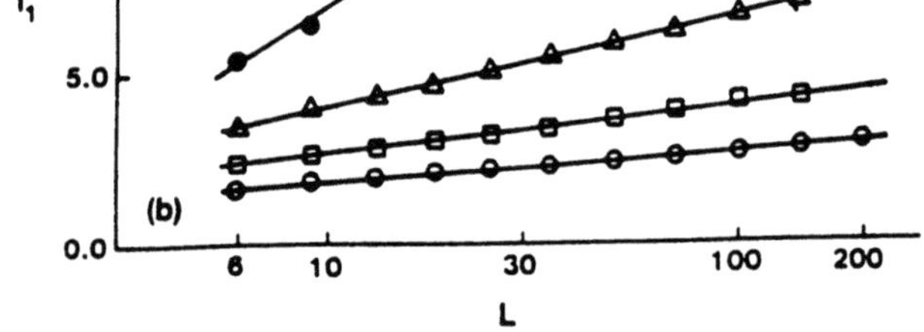

Figure 4. The inverse critical current density $(i_c/L)^{-1}$ vs. $log(L)$ for an ensemble of diluted LxL square lattices at $p = 0.90(\circ)$, $p = 0.80(\square)$, $p = 0.7(\triangle)$ and $p = 0.60(\bullet)$. The solid lines are straight lines, corresponding to Eq.(5) with $\alpha = 1$, from Ref. [14].

But the behavior of Eq.(5) depends upon the guess of the worst defect cluster in the sense that it was important that the probability of large defect clusters was exponentially small (decreases exponentially with increasing n as in Eq.(1)) and that the current enhancement at the defect tip grows algebraically with n (as in Eq.(4)). It may be that worse defect clusters other than those considered above exist, especially for large values of $(1-p)$, but the numerical results seem to show that whatever the worst defect clusters are, their probability of occurrence decays exponentially with n. If the probability of the worst defect cluster died out algebraically with n, for example, the sample-size dependence of i_c would be as a power law of sample size.

Several other "worst defect clusters" have been considered in the literature, the most prominant being the funnel defect [16,17] shown in Fig. 5a, which reduces to the double cluster [18,19] shown in Fig. 5b. These funnel type

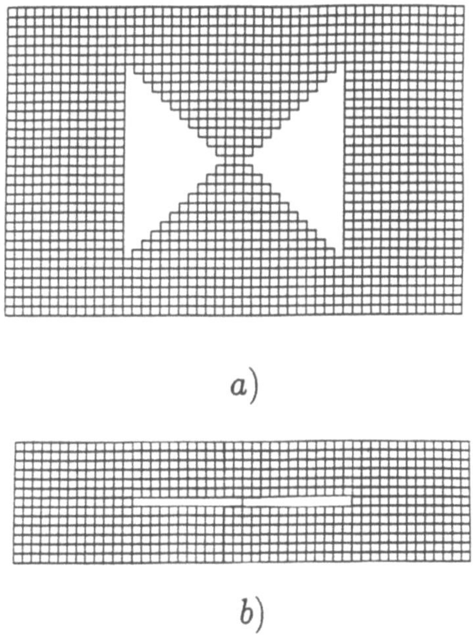

Figure 5. a) A funnel defect cluster; **b)** A double cluster.

defects also lead to the prediction of logarithmic size effects like that in Eq.(5), with the exponent α now dependent on the funnel angle.

The results of Duxbury et al. [13,14] for fuse networks were subsequently extended to the case of brittle fracture by Beale and Srolovitz[18], who found the essentially identical behavior in this case, with the current density enhancement replaced by the stress enhancement, the breakdown current density by the breakdown stress, the voltage by the strain and the electrical conductivity by the Young's modulus. However, the elastic breakdown problem is a tensor problem and allows for more freedom in the interplay between shear stresses and uniaxial stresses. Also others have successfully extended these calculations to include dielectric breakdown[20] and critical current in superconductors[21,22].

As to experimental verification of Eq.(5) on real materials, the concentration dependence (the $log(1-p)$ behavior) at the low concentration of defects has been verified semi-quantitatively many times on a variety of systems. Gilabert et al.[23] have observed the $log(1-p)$ behavior for the diluted fuse networks, several different groups [24] have seen it for dielectric breakdown and Hasselman and Fulrath[25] have seen a qualitatively similar effect in brittle fracture. But the decay of the breakdown stresses with increasing sample size is a true challenge for experimentalists because they must measure the average breakdown strength for ensembles of disordered samples prepared identically but differing in size by orders of magnitude. A far easier test for experimentalists to perform and one that is commonly done is to measure the failure probability distribution for an ensemble of samples of the same size. Since this test potentially contains the same size-effect information, as well as more information about fracture statistics, Duxbury et al. [14,26] next turned their attention to this quantity.

3. Failure Probability Distributions

The failure probability distribution $F(i, N)$ for the fuse network (or $F(\sigma, N)$ for the case of brittle fracture) is the probability that a sample of size N bonds breaks at applied current density i (or stress σ). Of course, this curve has a characteristic S-shape, as shown in Fig. 6, from the work of Duxbury et al. [14].

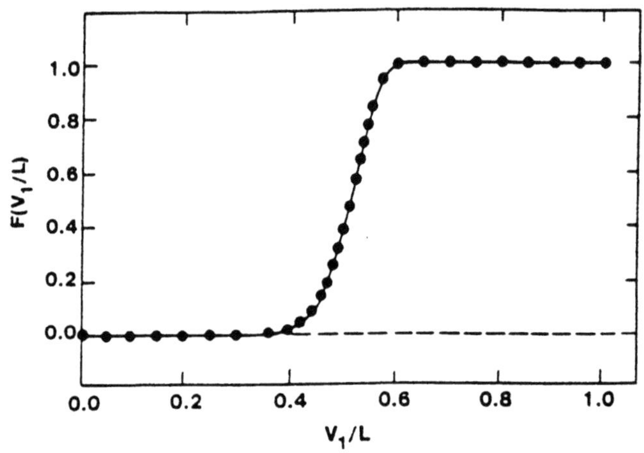

Figure 6. The failure distribution $F(e,n)$ vs. e, where the breakdown electric field $e = V_1/L$ is simply related to the applied critical current density i by the conductivity, for an ensemble of $1,500$, 50x50 square lattices with $p = 0.90$ bond occupancy, from Ref. [14].

A double-exponential analytic form for $F(i, N)$, namely,

$$F_{DE}(i, N) \sim 1 - exp[-cNexp(-ki^{-1/\alpha})] \tag{6}$$

was derived exactly by Duxbury et al. [14,26] for the case when the worst defect clusters are clusters of adjacent vacant bonds in a line or linear single clusters. In Eq.(6), c, k and α are parameters to fit to data or to be otherwise determined. But more generally a scaling argument was given as to the possible analytic form of $F(i, N)$ for large N. The argument goes as follows. Let $C(n, N)$ be the probability that there are no "worst defect clusters" of size greater than n [e.g. no linear cluster of adjacent vacant bonds greater than n] Then, the failure probability distribution is related to $C(n, N)$ by,

$$F(i, N) = 1 - C(n(i), N) \tag{7}$$

where $n(i)$ is the critical defect size necessary to cause breakdown for an applied current density i. Now consider a sample of size $N = L^2$ to be made up of M smaller samples of sizes $N' = L'^2$, so that $N = MN'$ as illustrated in Fig. 7.

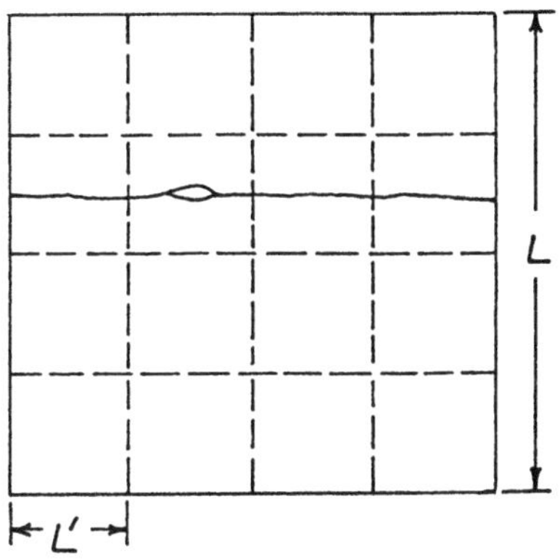

Figure 7. An LxL sample can be considered to be made up of M sub-samples of size L'xL'. If a crack is nucleated in any one L'xL' subsample, it cleaves the entire LxL sample.

Then the probability that there is no worst defect cluster in the entire sample of size N is the probability that none of the sub-units of size N' has a defect cluster, so that it is reasonable to assume the scaling form [27],

$$[C(n, N')]^M = C(a_m n + b_m, N), \tag{8}$$

which defines a renormalisation procedure for this problem. The b_m give the shift of the most probable breaking strength with increasing sample size, and the a_m coefficients give the narrowing or sharpening of the failure probability distribution with increasing sample size. As is discussed in Gumbel or Castillo [27] Eq. (8) has only two basic kinds of fixed points, namely, the Gumbel (a double-exponential) distribution

$$C_G(n, N) \sim exp(-cN exp(-k'n)), \tag{9}$$

and the Weibull distribution (for $\gamma > 0$)

$$C_W(n, N) \sim exp(-c'Nn^{-\gamma}), \tag{10}$$

which is called the Frechet distribution when $\gamma < 0$. Specifically, if the probability of the worst defect clusters dies out exponentially with n (as do linear clusters in a random medium see Eq. (2)) then the result is a Gumbel distribution, which for $n(i)$ given by Eq.(4) gives the double-exponential distribution (6). On the other hand, for worst defect clusters whose probability goes to zero as an inverse power of n, as may happen for example at the percolation threshold $p = p_c$, then the Weibull distribution (10) results, which with Eq.(4) gives the distribution,

$$F_W(i, N) = 1 - exp(-c'Ni^{\gamma}), \tag{11}$$

which is just the original form proposed by Weibull[9].

Duxbury et al. [14] showed that in their numerical simulations for an ensemble of randomly diluted LxL fuse networks, the failure distribution versus applied current density (or electric field) satisfied the double-exponential form better than the Weibull form, especially in the high reliability tail. This form can be checked in an unbiased manner by plotting the quantity,

$$A = log[-log(1 - F(i, N))/N], \tag{12}$$

either versus i^{-1} to test for the double-exponential form (6), or versus $log(i)$ to test for the Weibull form (11). Their plot [14] of A versus i^{-1} is shown in Fig. 8, which demonstrated the validity of the double-exponential form in this brittle case.

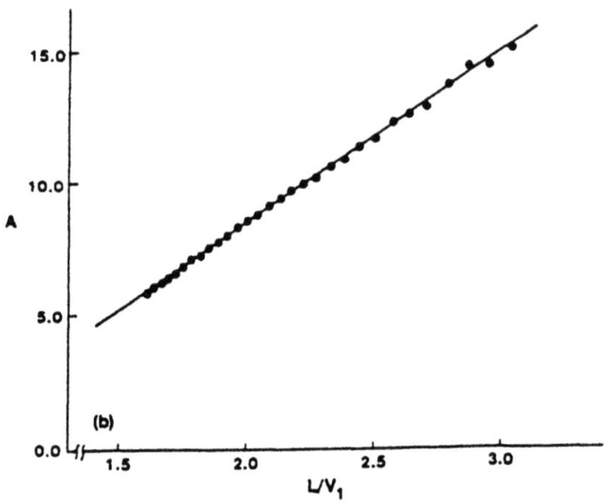

Figure 8. A plot [from Ref. 14] of A [as defined in Eq.(12)] vs. e^{-1} to test for double-exponential size scaling of the failure distribution, according to Eq.(6).

Also, from Eq.(6) and (11) one can easily deduce the size scaling for the double-exponential and Weibull forms. First, for the double-exponential form Eq.(6), the form of the i vs. N curve which holds $F(i, N)$ constant is,

$$i \sim (lnN)^{-\alpha} \tag{13}$$

And for the Weibull form (11) we similarly obtain,

$$i \sim N^{-1/\gamma}. \tag{14}$$

Thus, the Weibull form implies power-law size scaling and the double-exponential form implies logarithmic size scaling.

Finally, brittle fracture measurements on real materials have been done, for example by van den Born et al. [28], and have confirmed the usefulness of the double-exponential form, as is shown in Fig. 9.

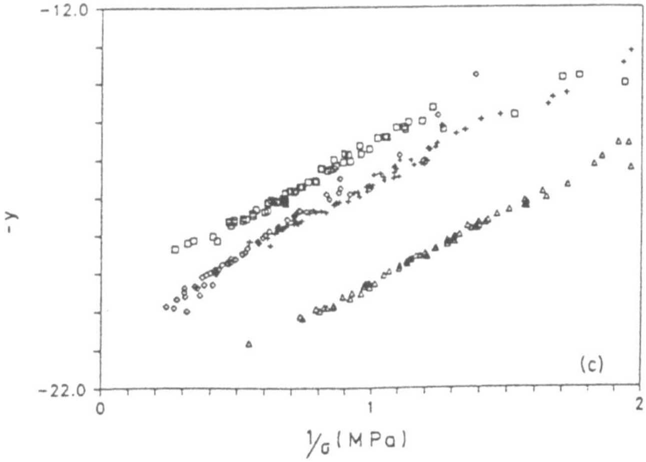

Figure 9. A plot [from Ref. 28] of $-y$ [the equivalent of A in Eq.(12)], for the case of brittle fracture, vs. σ^{-1} as a test of the double-exponential form (6), for the different series of experimental samples (with different porosities, and pore dimensions) of highly porous ceramics (silica extrudates).

4. Stochastic Transition-Matrix Method

Recently, we have extended the exact calculation of the failure probability distribution for the case of linear single cracks, Eq.(6), to the case of double cracks [29] or pairs of cracks separated by a single intact bond. The result is again the double-exponential form for brittle fracture with different coefficients corresponding to the different probability of double cracks.

The stochastic transition-matrix method (actually the matrix is really substochastic according to mathematical convention) is described briefly as follows. Consider the single cluster of n vacancies shown in Fig. 10a. Such a single cluster followed by an occupied bond, we shall denote by (100...00), where the 1 denotes the occupied bond, and the n zeros denote the cluster; and we shall denote the probability of occurrence of such a cluster by $\phi_{(100\ldots0)}$.

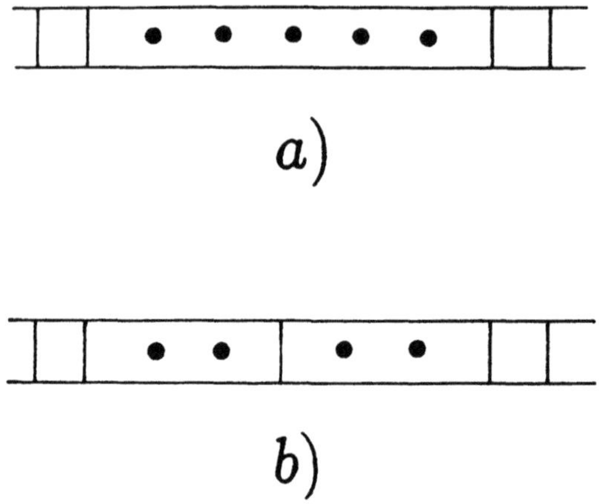

Figure 10. a) A single, linear cluster of vacant or failed bonds; **b)** a double, linear cluster of vacant or failed bonds, with one lone surviving bond in the center.

Then we define the $(n+1)$ component basis vector on site j to be.

$$\phi_j = \begin{pmatrix} \phi_{(1)} \\ \phi_{(10)} \\ \phi_{(100)} \\ \vdots \\ \phi_{(100..0)} \end{pmatrix}_j . \tag{15}$$

Thus we can relate ϕ_j to ϕ_{j+1} via the stochastic transition matrix according to the following formula.

$$\phi_{j+1} = M\phi_j = \begin{pmatrix} p & p & p & \dots & p \\ f & 0 & 0 & \dots & 0 \\ 0 & f & 0 & \dots & 0 \\ \vdots & \vdots & \ddots & \ddots & \vdots \\ 0 & 0 & \dots & f & 0 \end{pmatrix} \begin{pmatrix} \phi_{(1)} \\ \phi_{(10)} \\ \phi_{(100)} \\ \vdots \\ \phi_{(100..0)} \end{pmatrix}_j , \tag{16}$$

where $f = (1-p)$ is the probability of an unoccupied bond. The matrix M gives the probability for each of the $\phi_{(10..0)}$ on the next site $(j+1)$, given the $\phi's$ on the site j. Then if one goes l sites down the row from site i, one finds $\phi_l = M^l\phi_1$. Thus, if we impose periodic boundary conditions, such that $\phi_{L+1} = \phi_1$, then, since no $\phi_{(10..0)}$ are included corresponding to clusters of more than n vacancies, anywhere along the row (or in the entire sample if there are spiral boundary conditions at the end of each row) the probability $C_L(n)$ that in a sample of size L, there are only vacant clusters of size less than or equal to n is given by the trace of M^L,

$$C_L(n) = Tr(M^L) = \sum_k \lambda_k^L, \tag{17}$$

where λ_k are the eigenvalues of M. To determine the λ_k, consider the characteristic equation,

$$D_n = det|M/\lambda - I|, \tag{18}$$

where I is the identity matrix. We then write,

$$\frac{M}{\lambda} - I = \begin{pmatrix} a-1 & a & a & \dots & a \\ b & -1 & 0 & \dots & 0 \\ 0 & b & -1 & \dots & 0 \\ \vdots & \vdots & \ddots & \ddots & \vdots \\ 0 & 0 & \dots & b & -1 \end{pmatrix}, \tag{19}$$

where $a = p/\lambda$ and $b = f/\lambda = (1-p)/\lambda$. A co-factor expansion of the determinant D_n gives immediately the recursion relation,

$$D_n = -D_{n-1} + (-1)^n ab^n, \tag{20}$$

which gives,

$$(-1)^n D_n = (a-1) + ab + ab^2 + ab^3 + ... + ab^n = 0 \tag{21}$$

and finally the characteristic equation,

$$(\lambda)^{n+2} - (\lambda)^{n+1} + p(1-p)^{n+1} = 0. \tag{22}$$

Now since M is a primitive and non-negative matrix, the largest eigenvalue λ_{max} is real and distinct. But also it is easy to see from (22), that all the $\lambda's$ are less than one and that λ_{max} approaches 1 as $n \to \infty$. Therefore the trace of M^L in Eq.(17) can be approximated by λ_{max}^L. So, we set $\lambda = 1 - \epsilon$ in Eq.(22) and expand to lowest order in ϵ with the result,

$$\lambda_{max} = 1 - p(1-p)^{n+1} + O((1-p)^{2n}), \tag{23}$$

or from Eq.(17),

$$C_L(n) \sim (1 - pf^{n+1})^L \tag{24}$$

which is precisely the exact analytic form derived previously[14], and which gives the double-exponential failure distribution [14,29] of Eq.(6).

Then we used this stochastic transition-matrix approach to extend the calculation to the case of double clusters. That is in $C_L(n)$, we include the probability not only of single clusters as in Fig. 10a, but also that of all adjacent pairs of clusters (separated by a single intact bond) as shown in Fig. 10b. Thus we have calculated $C_L^{(2)}(n)$, the probability that there were no single or double cracks with greater than n total vacant bonds on both sides of any surviving bonds.

In this case the clusters to be considered are denoted by (11), (110), (1100), ...; (101), (1010), (10100)...; (1001),(10010), ...; ...(1000001), where no cluster or double cluster contains more then n zeros. Then the stochastic transition matrix is given, for example for $n = 4$, by

$$M_4 = \begin{pmatrix} p & 0 & 0 & 0 & 0 & p & 0 & 0 & 0 & p & 0 & 0 & p & 0 & p \\ f & 0 & 0 & 0 & 0 & 0 & 0 & 0 & 0 & 0 & 0 & 0 & 0 & 0 & 0 \\ 0 & f & 0 & 0 & 0 & 0 & 0 & 0 & 0 & 0 & 0 & 0 & 0 & 0 & 0 \\ 0 & 0 & f & 0 & 0 & 0 & 0 & 0 & 0 & 0 & 0 & 0 & 0 & 0 & 0 \\ 0 & 0 & 0 & f & 0 & 0 & 0 & 0 & 0 & 0 & 0 & 0 & 0 & 0 & 0 \\ 0 & p & 0 & 0 & 0 & 0 & p & 0 & 0 & 0 & p & 0 & 0 & p & 0 \\ 0 & 0 & 0 & 0 & 0 & f & 0 & 0 & 0 & 0 & 0 & 0 & 0 & 0 & 0 \\ 0 & 0 & 0 & 0 & 0 & 0 & f & 0 & 0 & 0 & 0 & 0 & 0 & 0 & 0 \\ 0 & 0 & 0 & 0 & 0 & 0 & 0 & f & 0 & 0 & 0 & 0 & 0 & 0 & 0 \\ 0 & 0 & p & 0 & 0 & 0 & 0 & p & 0 & 0 & 0 & p & 0 & 0 & 0 \\ 0 & 0 & 0 & 0 & 0 & 0 & 0 & 0 & 0 & f & 0 & 0 & 0 & 0 & 0 \\ 0 & 0 & 0 & 0 & 0 & 0 & 0 & 0 & 0 & 0 & f & 0 & 0 & 0 & 0 \\ 0 & 0 & 0 & p & 0 & 0 & 0 & 0 & p & 0 & 0 & 0 & 0 & 0 & 0 \\ 0 & 0 & 0 & 0 & 0 & 0 & 0 & 0 & 0 & 0 & 0 & 0 & f & 0 & 0 \\ 0 & 0 & 0 & 0 & p & 0 & 0 & 0 & 0 & 0 & 0 & 0 & 0 & 0 & 0 \end{pmatrix}. \tag{25}$$

By the same general technique, namely: the development of a recursion relation for $D_n = det|M_n/\lambda - I_n|$; then solving this recursion relation to find the characteristic equation for λ and; an expansion of this characteristic equation in powers of ϵ, leads to the exact asymptotic form,

$$C_L^{(2)}(n) \sim [1 - ((n+2)p^2 - p)(1-p)^{n+1}]^L, \tag{26}$$

which again gives the double-exponential form with modified coefficients. This was no surprise, since the qualitative arguments of section II above say that whenever the large defect clusters become less probable exponentially with size, the sample-size dependence will be logarithmic and thus the failure probability will be double-exponential in form.

But the arguments given here are for the case when the system is brittle such as for the diluted perfect network. On the other hand, when there is a distribution of breaking strengths, so that most cracks once started are stopped by the randomly occuring strong bonds, the system is said to be ductile or tough, and several authors have reported power-law size scaling [30]. So, recently, we have developed a recursion relation approach for this case, which is presented below.

5. "Ductile" Fracture

It has been known for some time that in heterogeneous materials with a broad distribution of bond breaking strengths that, at sufficiently low stress, the material is tough, the system is characterised by multiple-cracking as most cracks started are stopped by strong bonds that hold, and before fracture occurs a significant amount of "damage" has occured throughout the system[31,32]. Furthermore, Kahng et al. [32] pointed out that for sufficiently large sample sizes there is a crossover to brittle behavior in virtually all such ductile systems. They gave a qualitative "phase diagram" which is shown in Fig. 11 and which illustrates this point for the case of a uniform distribution of bond breaking strengths for $(1 - W/2) \leq \sigma \leq (1 + W/2)$. In particular, they predicted that the system was always ductile for $W = 2$, where the uniform distribution extends down to zero breaking strength.

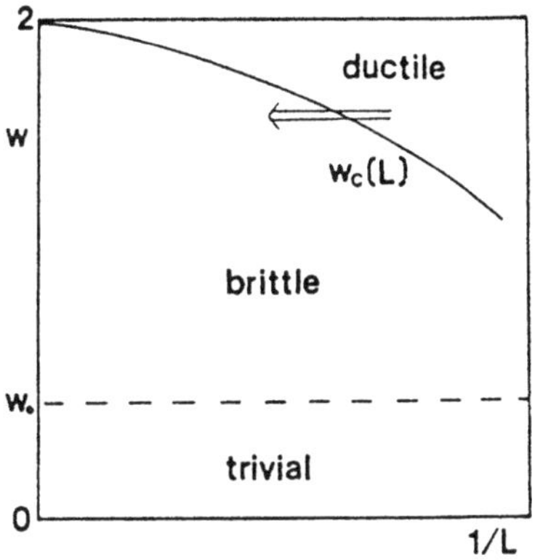

Figure 11. A sketch [Fig. 1 of Ref. 32] of their proposed phase diagram of ductile and brittle regions of a heterogeneous sample on a square lattice with a uniform distribution of bond breaking strengths of width $2W$ centered about $\sigma_0 = 1$.

Recently, we have been able to construct an exact recursion relation for the failure probability for linear cracking for the case of distributed breaking strengths [33,34] and find an amazingly deep minimum in the failure probability $F_L(\sigma)$ versus L for low values of applied stress σ, and for both the uniform distribution of bond breaking strengths,

$$f_u(\sigma) = \frac{1}{W} \quad \text{if} \quad 0 \leq \sigma \leq W \quad \text{and} \quad f_u = 0 \quad \text{otherwise,} \tag{27}$$

and for a local Weibull distribution,

$$c_w(\sigma) = \int_0^\sigma f_w(\sigma')d\sigma' = 1 - exp(-(\frac{\sigma}{\sigma_s})^m). \tag{28}$$

where σ_s, and m, are parameters and σ_s sets the stress scale.

We also assume a very local load sharing of the stress when a bond fails. In particular, the entire stress load that would have been on the entire length of a single crack is placed as a stress on the intact bonds on each end of the crack. That is, we use the model of Harlow and Phoenix [11,19], namely

$$\sigma_k = \sigma(1 + k/2) \tag{29}$$

where σ_k is the stress on an intact bond, σ is the applied stress, and k is the total number of failed bonds on both sides of the intact bond. This is illustrated in Fig. 12, where the value of k is shown for three particular bonds, and where the central lone bond has $k = 3$ because of the vacant bonds on both sides.

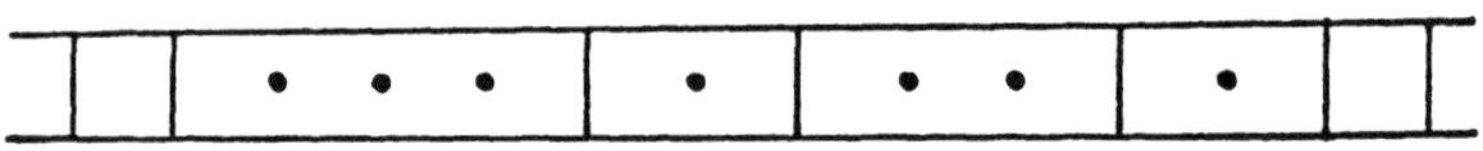

Figure 12. A typical sample of intact and vacant (or failed) bonds in a line. In the center of the line are three lone surviving bonds.

The goal of this calculation is to find $F_L(\sigma)$, the probability that a line of bonds of sample size L will fail when a stress σ is applied. But an easier problem to formulate is that for the survival probability ,

$$S_L(\sigma) = 1 - F_L(\sigma) \tag{30}$$

which is just the sum over the $2^L - 1$ surviving configurations, since $F_L(\sigma)$ is the failure probability of the one configuration corresponding to the failure of all the L bonds. In order to find $S_L(\sigma)$, we need first to note that the probability W_k that any particular bond surrounded by k failed bonds survives, or forms a "wall" to the further advance of the crack(s) on either side of it, is, with use of Eq.(29), given by,

$$W_k(\sigma) = 1 - \int_0^{(1+k/2)\sigma} f(\sigma_i)d\sigma_i \tag{31}$$

where $f(\sigma_i)$ is the probability that the bond has breaking strength σ_i.

To carry out this calculation, we separate the $2^L - 1$ survival configurations into four categories. First, we define a *lone bond* as an intact bond which is surrounded by failed bonds, as are the central ones in Fig. 12. Then, we let $\{A\}$ be the set of all survival configurations which contain only failed bonds and *lone bonds*, and which are bracketed on *each* end by *lone bonds.* The small n members of this set $\{A\}$ are $\{101, 1001, 10001, 10101, ..\}$. Second, the set $\{B\}$ is defined to be the same set as $\{A\}$ *except* that one (specified) end of the configuration *must* be failed. The small L members of this set (for the case where the left end is failed) are $\{01, 001, 0001, 0101, 00101...\}$. The complimentary set to this $\{B\}$ but with the failed ending on the right, has the same probability. Third, the set $\{C\}$ is again the same as $\{A\}$ *except* that *both* ends have failed, so the small L members of this set are $\{010; 0100, 0010; 01010...\}$. And, finally, we define, $\{p\}$ to be the set of all configurations containing no failed bonds $\{(.), 1, 11, 111, 1111...\}$, where (.) is the empty set (i.e. $L = 0$). Associated with these sets of survival configurations of size l, are the probabilities A_l, B_l, C_l and p_l respectively, and the corresponding generating functions are;

$$A(z) = \sum_{L=3}^{\infty} A_L z^L, \quad B(z) = \sum_{L=2}^{\infty} B_L z^L \quad \text{and} \quad C(z) = \sum_{L=3}^{\infty} C_L z^L \tag{32}$$

and,

$$p(z) = \sum_{L=0}^{\infty} (W_0)^L z^L = \frac{1}{1 - W_0 z}. \tag{33}$$

A careful study of the set of all survival configurations shows that its generating function $s(z) = \sum_L s_L z^L$ can be concisely written as

$$s(z) = C(z) + \frac{(1 + B(z))p(z)(1 + B(z))}{1 - p(z)A(z)}. \tag{34}$$

Furthermore, since $F_l = 1 - S_l$, we can write,

$$f(z) = \sum_{L=0}^{\infty} F_L z^L = \frac{1}{1 - z} - s(z), \tag{35}$$

where $F_L(\sigma) = F_L$ is the failure probability we seek. Equating the appropriate terms in Eq.(34), gives the exact recursion relation,

$$X_L = (1 + W_0)X_{L-1} - W_0 X_{L-2} - 2B_L + 2B_{L-1} - A_L + F_1 A_{L-1} - B_2 B_{L-2}$$

$$+ \sum_{m=1}^{L-4} B_{m+1}(B_{L-m-2} - B_{L-m-1}) + A_{m+2}(X_{L-m-2} - X_{L-m-3}). \tag{36}$$

Where $X_m = F_m + C_m$. Also, the equations for A_l, B_l and C_l can be expressed explicitly as

$$b_{L,l} = f_l W_l \delta_{L-l-1} + \sum_{r=1}^{L-l-2} b_{L-l-1,r} W_{l+r} f_l, \tag{37}$$

$$c_{L,l} = f_l W_{L-1} f_{L-l-1} + \sum_{r=1}^{L-l-3} c_{L-l-1,r} f_l W_{l+r}, \tag{38}$$

and

$$a_{L,l} = b_{L-1,l} W_l. \tag{39}$$

Where $A_L = \sum_l a_{L,l}$, $B_L = \sum_l b_{L,l}$ and $C_L = \sum_l c_{L,l}$. Here $a_{L,l}$ is the probability of a surviving state in the set $\{A\}$, *with* exactly l elements in the *last failed cluster* of the configuration (for example for $L = 8, l = 2$ one such configuration is 10101001). $b_{L,l}$ and $c_{L,l}$ are calculated in a similar way. Eqs. (36-39) give the exact recurrence relations for the failure probability ($F_L(\sigma)$) of a row or sample of size L at applied stress σ.

It is possible to directly generate and evaluate each of the $2^L - 1$ terms in $S_L(\sigma)$ on the computer, but that is a very slow process since the time required grows exponentially as 2^L. Nevertheless, with this brute-force enumeration, we have been able to evaluate $F_L(\sigma)$ for systems of up to size $L = 20$. However, the computer time required to evaluate the recursion relations grows at most as L^3, and in many cases of physical interest essentially exact results can be found in a computer time which scales linearly in L. In many cases, we are then able to carry out this analysis to $L = 5000$ on a SUN workstation. The numerical results are shown in Figs. 13-15.

First in Fig. 13, we show the numerical evaluation of $F_n(\sigma)$ vs. sample size L for the uniform distribution (Eq.(27)) of breaking strengths, and for three different values of the applied stress σ, namely, starting from the top curve of Fig. 13, $\sigma/W = 0.2, 0.1$ and 0.05, respectively. At higher stress the failure

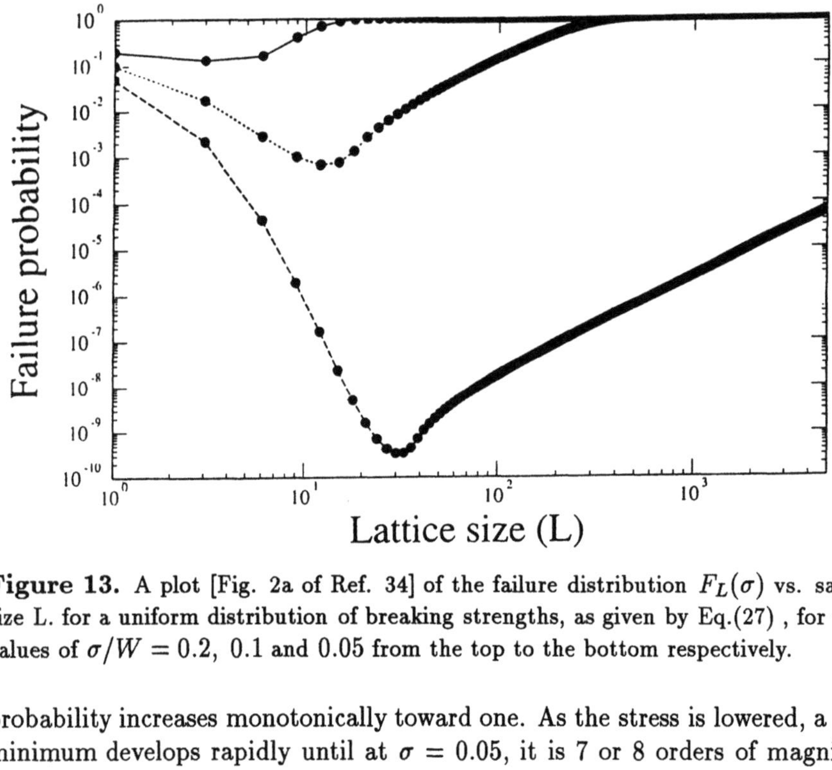

Figure 13. A plot [Fig. 2a of Ref. 34] of the failure distribution $F_L(\sigma)$ vs. sample size L. for a uniform distribution of breaking strengths, as given by Eq.(27) , for three values of $\sigma/W = 0.2,\ 0.1$ and 0.05 from the top to the bottom respectively.

probability increases monotonically toward one. As the stress is lowered, a deep minimum develops rapidly until at $\sigma = 0.05$, it is 7 or 8 orders of magnitude deep. This behavior indicates an optimum or strongest sample size which marks the crossover from a multi-crack region where many cracks are started but are almost always stopped by randomly occuring strong bonds, to the brittle region where *relatively* fewer cracks are stopped before a single crack begins to cleave the system apart. There are important applications of this result to real situations where there may be a strongly optimum sample size. Similar results are seen numerically for the local Weibull distribution [33,34]. We also checked for the stress dependence of $F_L(\sigma)$ for both the uniform and the local Weibull distributions [33,34], the results of which were similar for the two models, and

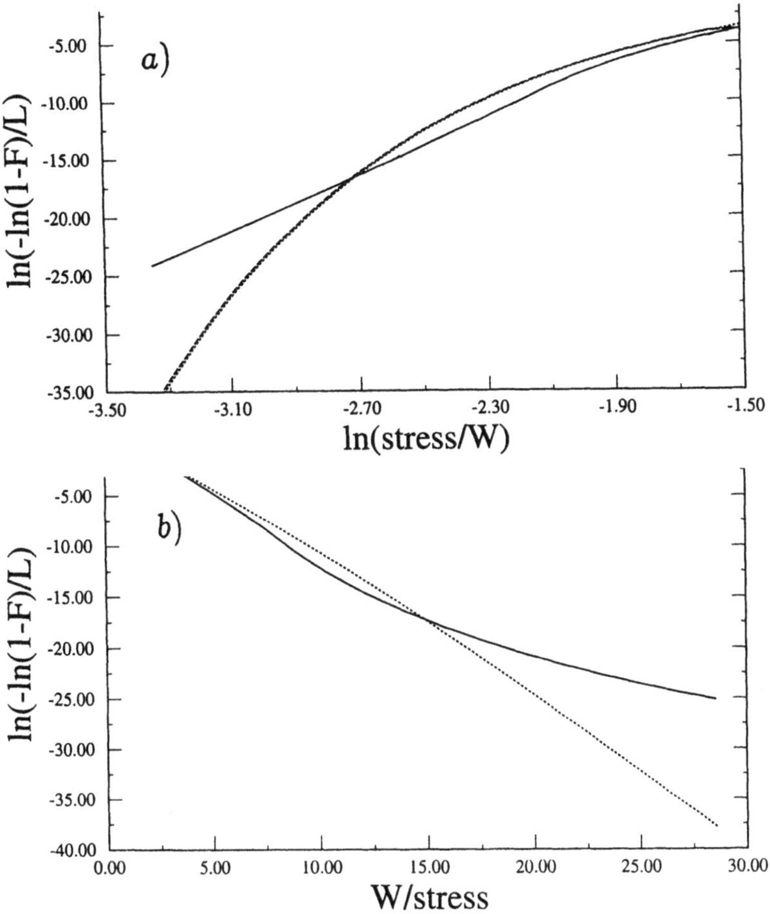

Figure 14. Plots of A [Eq.(12)] **a)** vs. $ln(\sigma/W)$ and; **b)** vs. W/σ, to test for Weibull and double-exponential distributions. In **a)**, data is for $L = 15$ (—); $L = 150$ (- -); and $L = 1500$ (- - -), and in **b)** for $L = 15$ (—) and $L = 150$(- - -).

produced results like that shown in Figs. 14a and 14b. Actually Fig. 14 shows the results with the approximation of no more than two adjacent lone bonds anywhere in the sample (i.e. - the terms in the iteration of Eqs.(37-39) with more then two $f's$ were ignored). Numerically, this was a good approximation to the exact result.

In Fig. 14a,b, we plot A of Eq.(12) vs. $ln(\sigma/W)$ to test for the Weibull distribution and vs. W/σ to test for the double-exponential distribution. The curves for $n = 150$ seem to show a reasonable fit to the double-exponential form, while for $n = 15$ there is a crossover from the Weibull-like behavior at high stress. Except for small samples at small stress, the data is fairly close to the double-exponential form. But for small samples at low stress (samples smaller than the minimum in $F_L(\sigma)$ vs. L - see Fig. 13), the behavior is more Weibull-like. This indicates that the multiple-cracking or crack-blocking is dominating the fracture process in this regime. It is clear that numerical simulations of more complex systems such as fuse and spring networks are not able to probe the behavior found here, and in particular are not able to find the deep minimum in Fig. 13, or often, to reliably distinguish between Weibull scaling and double-exponential scaling[34].

Finally, the sample size dependence of the average sample breaking strength $< \sigma_c >$ was measured. The results are shown for the uniform distribution case in Figs. 15a,b where $ln < \sigma_c >$ is plotted vs. $ln(n)$ in Fig. 15a to test for power-law scaling, and in Fig. 15b $< \sigma_c >^{-1}$ is plotted versus $ln(n)$ to test for

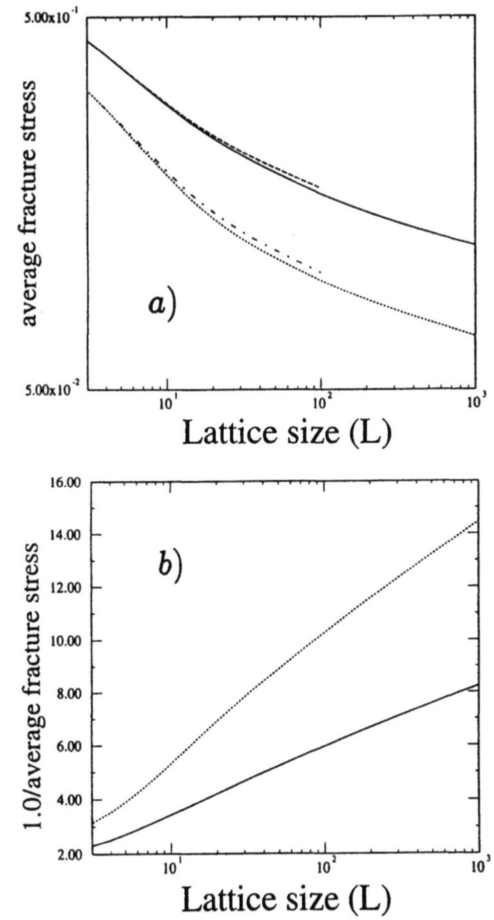

Figure 15 a) A log-log plot of the mean fracture stress vs. sample size L, and **b)** a semi-log. plot of the inverse mean fracture stress vs. L to test for power-law and logarithmic size scaling, respectively.

logarithmic size scaling. It is easy to see that the logarithmic size scaling works well for large samples, and the power law scaling works for a limited range of small samples. This is consistent with the regimes in which Weibull and double-exponential behavior were found above.

6. Conclusions

In conclusion, we have found that whenever there is brittle behavior (i.e. always for diluted homogeneous samples, and essentially always for sufficiently large heterogeneous systems), then the sample size dependence of the fracture stress is logarithmic and correspondingly, the failure distribution is double-exponential in form. But when multi-cracking dominates, as in the tough or ductile regime, there may be a region of power-law size-scaling and a Weibull-like failure distribution. In any event, there are very interesting physical phenomena in the fracture and breakdown behavior of disordered and heterogeneous materials which are of very great practical importance. Moreover, the interesting phenomena discussed here are very difficult to see in numerical simulations [34], so that the use of analytically solvable models are very important for their discovery and understanding.

Acknowledgements. PLL thanks the Department of Physics/Astronomy at Michigan State University and the Department of Theoretical Physics Oxford for their hospitality during his sabbatical year 1993, during which this work was done. PMD was supported by the DOE under grant DE-FG02-90ER45418, by the CMSC at MSU and by a Humboldt fellowship at the IFF-KFA Jülich. PMD also thanks the HLRZ at the KFA Jülich for its hospitality during his sabbatical year 1993/1994, during which this work was done.

References

[1] G. Galileo, *Discorsi*, Leiden (1638); for a review see *A history of the theory of elasticity and the strength of materials*, by I Todhunter, Cambridge University Press, 1886.
[2] W.S. Chaplin, Van Nostrands' Engineering Magazine **23**, 441 (1880); *and* Proc. Engineers Club of Philadelphia **3**, 15 (1882).
[3] A.A. Griffith, Phil. Trans. Roy. Soc. **221A**, 163 (1920)
[4] E.L. Dodd, Trans. Am. Math. Soc. **25**, 525 (1923).
[5] M. Frechet, Ann. de. la. Soc. de Math. (Cracow), **6**, 92 (1927).
[6] R.A. Fisher and L.H.C. Tippett, Proc. Camb. Phil. Soc. **24**, 180 (1928).
[7] R. von Mises, Rev. Math. Union Interbalkanique **1**, 141 (1936).
[8] B. Gnedenko, Ann. Math. **44**, 423 (1943).
[9] W. Weibull, Ing. Vetenskaps Akad. Handl. **Numbers 150 and 151** (1939).
[10] B. Epstein, J. Appl. Phys. **19**, 140 (1948); B. Epstein, J. Am. Stat. Assoc. **43**, 403 (1948); B. Epstein and H. Brooks, J. Appl. Phys. **19**, 544 (1948).
[11] D.G. Harlow and S.L. Phoenix, Int. J. Fracture **17**, 601 (1981); R.L. Smith, Proc. Roy. Soc. **A382**, 179 (1982).
[12] L. de Arcangelis, S. Redner and H.J. Herrmann, J. Phys. (Paris) Lett. **46**, L-585 (1985).
[13] P.M. Duxbury, P.D. Beale and P.L. Leath, Phys. Rev. Lett. **57**, 1052 (1986).

[14] P.M. Duxbury, P.L. Leath and P.D. Beale, Phys. Rev. **B36**, 367 (1987).
[15] Y.S. Li and P.M. Duxbury, Phys. Rev. **B36**, 5411 (1987).
[16] J. Machta and R.A. Guyer, Phys. Rev. **B36**, 2142 (1987).
[17] J. Helsing, J. Axell and G. Grimvall, Phys. Rev. **B39**, 9231 (1989).
[18] P.D. Beale and D. Srolovitz, Phys. Rev. **B37**, 5500 (1988).
[19] D.G. Harlow and S.L. Phoenix, J. Mech. Phys. Solids **39**, 173 (1991).
[20] P.D. Beale and P.M. Duxbury, Phys. Rev. **B37**, 2785 (1988).
[21] P.L. Leath and W. Tang, Phys. Rev. **B39**, 6485 (1989).
[22] W. Xia and P.L. Leath, Phys. Rev. Lett. **63**, 1428 (1989); P.L. Leath and W. Xia, Phys. Rev. **B44**, 9619 (1991).
[23] A. Gilabert, C. Vanneste, D. Sornette and E. Guyon, J. Phys. (Paris) **48**, 763 (1987).
[24] C. Laurent, E. Kay, N. Souag, J. Appl. Phys. **64**, 336 (1988); R.W. Coppard, L.A. Dissado, S.M. Rowland and R. Rakowski, J. Phys. Cond. Mat. **1**, 3041 (1989); P.M. Duxbury, P.D. Beale, H. Bak and P.A. Schroeder, J. Phys. **D23**, 1546 (1990).
[25] D.P.H. Hasselman and R.M. Fulrath, p367 in *Ceramic microstructures*, ed. by R.M. Fulrath and J.A. Pask, E. Krieger (Huntington,NY 1966).
[26] P.M. Duxbury and P.L. Leath, J. Phys. **A20**, L411 (1987).
[27] E.J. Gumbel, *Statistics of Extremes*, Columbia University Press (New York, 1958); E. Castillo, *Extreme Value Theory in Engineering*, Academic Press San Diego 1988.
[28] I.C. van den Born, A. Santen, H.D. Hoekstra and J. Th. M. de Hossen, Phys. Rev. **B43**, 3794 (1991).
[29] P.L. Leath and P.M. Duxbury, NIST Journal of Research, 1994 in press; P.M. Duxbury and P.L. Leath, Phys. Rev. **B** May 1994 issue.
[30] H.J. Herrmann, A. Hansen and S. Roux, Phys. Rev. **B39**, 637 (1989).
[31] M. Sahimi and J.D. Goddard, Phys. Rev. **B33**, 7848 (1986).
[32] B. Kahng, G.G. Batrouni, S. Redner, L. de Arcangelis and H.J. Herrmann, Phys. Rev. **B37**, 7625 (1988).
[33] P.L. Leath and P.M. Duxbury, Phys. Rev. **B** May 1994 issue.
[34] P.M. Duxbury and P.L. Leath, Phys. Rev. Lett. **72**, 2805 (1994).

Fracture and Other Breakdown Phenomena in Disordered Solids

Bikas K. Chakrabarti
Saha Institute of Nuclear Physics
1/AF Bidhan Nagar, Calcutta 700064, India.

We review briefly three kinds of breakdown phenomena occurring in disordered (in particular, percolating) solids; namely the Griffith fracture (mechanical breakdown) of percolating solids, Peierls' instability (thermal breakdown or 'melting') of percolating solids, and Zener breakdown in Anderson insulators (dielectric breakdown in quantum percolation). The theoretical results for the fracture exponents and of the fracture strength distribution, are compared with those available from experiments and from simulations. In absence of comparable simulations and experiments, the breakdown behaviour and instabilities in other cases remain speculative.

I. Introduction

With extreme perturbation, the mechanical or electrical properties of solids tend to get destabilised and failure or breakdown occurs. In fact, these instabilities in the solids often nucleate around disorder, which then plays a major role in the breakdown properties of the solids. The growth of these nucleating centres, in turn, depend on various statistical properties of the disorder [1,2], namely the scaling properties of percolating structures, its (fractal) dimensions, etc. [3]. These statistical properties of disorder induce some scaling behaviour for the breakdown of the disordered solids. Recently, there has been considerable developments in such studies [1,2].

We will discuss here the scaling properties of fracture of brittle percolating solids. For a brittle solid with a single linear micro-crack, Griffith [4] obtained the breakdown strength or the fracture stress by equating the elastic energy, released due to the extension of the micro-crack, with the surface energy of the newly grown surface of the crack. This procedure, therefore, maps the nonlinear (and irreversible) breakdown problem to the linear (and reversible) elastic problem for brittle systems. This idea has been extended [1] to percolating solids, where the pre-existing micro-cracks due to random vacancies, are of different sizes and shapes governed by the percolation statistics. The resulting scaling properties and the statistics of fracture strength are now quite well studied and established [1,2]. We give here a brief account of these developments.

The melting behaviour of a solid is another well-known thermal instability, which is again expected to be extremely sensitive to the presence of disorder in the solid. Since the instability, coming from thermal fluctuations, get more pronounced in lower dimensions (because of the failure to arrest such fluctuations), melting nucleates near the void surfaces of the disordered solids. The melting behaviour of a percolating solid (containing all kinds of voids with different sizes and shapes, determined by the percolation statistics) is thus expected to be quite interesting. However, in view of the difficulies with the melting theories of (perfect) solids, we disciss here the problem of Peierls' instability (due to thermal fluctuations of the lattice) [5] for the percolating solids.

Quantum particles, like electrons, often fail to diffuse through a percolating lattice (through which of course a classical particle diffuses away), because of coherent back scatterring (of the wave function) from the random geometry of the percolation cluster. Below three dimension, all the states on any percolating lattice gets (exponentially) localised and the electrons fail to diffuse through the disorderd (percolating) lattice. In three dimension, a mobility edge exists, below which all the states are localised, but extended above it. If the fermi level is below the mobility edge, it is in the (Anderson) insulating phase and if it is above the mobility edge, it is in the metallic phase. This Anderson transition (from insulator to metal) accross the mobility edge in a disodered solid (above two dimensions), is now well studied [6]. We discuss here the possibility of the (Zener-type) breakdown [7] of the insulating phase of the Anderson insulator (when the fermi level falls in the localised region, below the mobility edge), by applying strong electric field. We also compare such quantum breakdown problem with the classical dielectric breakdown problem of a metal loaded dielectric with metal concentration below the percolation threshold.

II. Griffith fracture of percolating solids

As is well known, sharp thin cracks in solids, present due to disorder, reduce the strength of the solids enormously as the local stress near a sharp notch may rise to a level several times that of the applied stress. Griffith, in 1920, equating the released elastic energy (assuming linear elastic behaviour upto the breakdown point;

brittle fracture), with the energy of the surface newly created (as the fracture occurs or the crack grows), arrived at a quantitative criterion for the equilibrium extension of the micro-crack already present within the stressed material [4]. If a crack of length l already exists in an infinite elastic continuum, subject to uniform tensile stress σ perpendicular to the length of the crack, then for the onset of the brittle fracture, Griffith equates (the differentials of) the elastic energy with the surface energy:

$$(\sigma^2/2Y)l^d \cong \gamma l^{d-1} , \tag{1}$$

where Y represents the elastic modulus, γ the surface energy density and d the dimension. Equality holds when no energy dissipation occurs (unlike the cases of plastic deformation or dynamics of cracks), and one gets

$$\sigma_B \cong k/\sqrt{l} \ ; \quad k \cong \sqrt{(Y\gamma)} , \tag{1a}$$

for the breakdown stress at (and above) which the existing crack of length l starts propagating and a macroscopic fracture occurs.

Fracture exponents

In a disordered solid, as modelled by the percolation model [3] with intact (non-broken) bond concentration p on a lattice, there occurs various vaccancy clusters (voids), or pre-existing cracks, of different sizes and shapes. The typical crack size in such a percolating solid being of the size of the correlation length [3] ξ_p, the application of the Griffith fracture criterion gives [1,8]

$$(\sigma^2/2Y)\xi_p^d \cong \gamma\xi_p^{d-1} \cong \xi_p^{d_B} \tag{2}$$

where $\xi_p \cong |\Delta p|^{-\nu_p}$; $Y \cong |\Delta p|^{T_e}$; $\Delta p \equiv (p - p_c)/p_c$, p_c being the percolation threshold and d_B the backbone dimension [3]. This assumes that the entire (ramified) backbone of the percolating cluster forms the surfaces and shares (equally) the surface energy. This gives [1]

$$\sigma_B \cong |\Delta p|^{T_B} \ ; \quad T_B = [T_e + (d - d_B)\nu_p]/2 , \tag{2a}$$

for the variation of the typical fracture strength σ_B of the percolating solid with the bonding probability p near the percolation threshold p_c, for a fixed size of the sample. The brittle fracture exponent

T_B is thus expressed in terms of (linear) elastic exponent T_e and the other exponents (and dimensions) representing the percolation cluster structures (statistics).

In fact, using the node-link-blob model [3] for the percolation cluster, one can easily derive a rigorous bound for the fracture exponent T_B [9, 1]. Near the percolation threshold p_c, the nearest-neighbour occupied bonds (or sites) form a statistically defined "super-lattice", made of tortuous "link-bonds" (of "chemical length" L_c) crossing at nodes seperated by an average distance ξ_p, the percolation correlation length (see Fig. 1).

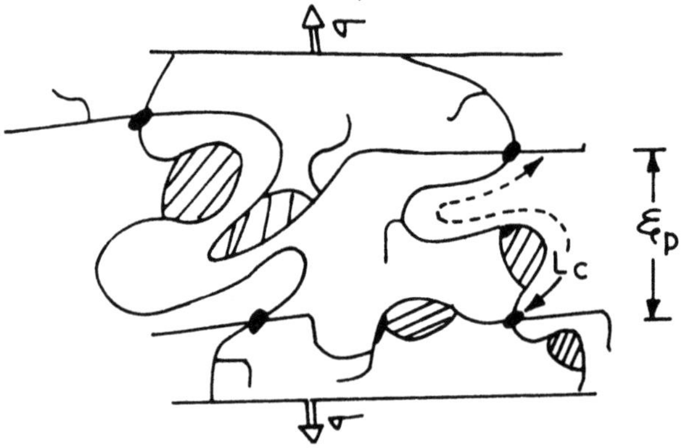

Fig. 1: A typical node-link-blob structure of the percolating backbone, which supports the external stress (σ). The dangling ends do not support any stress.

The external stress σ is shared by $\xi_p^{-(d-1)}$ number of parallel links, giving the stress per link $\sigma_L \cong \sigma\xi_p^{(d-1)}$. The total strain $\epsilon = \sigma/Y \cong \sigma\xi_p^{(T_e/\nu_p)}$ is shared by ξ_p^{-1} number of link-bonds, giving the strain per link-bond $\epsilon_L \cong \sigma\xi_p^{(1+T_e/\nu_p)}$. This gives the elastic energy per link-bond $\cong \sigma_L\epsilon_L \cong \sigma^2\xi_p^{(d+T_e/\nu_p)}$. If this entire elastic energy of the link is shared equally by all the $L_c \cong \xi_p^{d_B}$ (elemental) bonds of the link-bond, then the energy per bond is $\cong \sigma^2\xi_p^{[(d-d_B)+T_e/\nu_p]}$. Because of the assumption of equal sharing by all the bonds of the link, this gives an underestimate of the strain-energy per bond. If one now assumes a fixed energy threshold for each bond, the fracture will occur beyond $\sigma_B \cong |\Delta p|^{T_B}$; $T_B \geq [T_e + (d - d_B)\nu_p]/2$. If all the strain-energy is supported by the singly-connected bonds only, then $L_c \cong \xi_p^{1/\nu_p}$ [3], and one gets an over-estimate of the elastic energy per bond as $\cong \sigma^2\xi_p^{(d+T_e/\nu_p-1/\nu_p)}$. This gives the lower bound

: $T_B \leq [T_e + d\nu_p - 1]/2$. One thus gets the rigorous bounds for T_B [9, 1]

$$[T_e + d\nu_p - 1]/2 \geq T_B \geq [T_e + (d - d_B)\nu_p]/2 \ . \tag{3}$$

For $d \geq 6$, the links of the super-lattice network are indeed made up of singly-connected bonds [3], and therefore both the bounds become equal and equalities in (3) become exact. For systems with both bond-stretching and bond-bending forces, $\nu_p = 1/2$, $d_B = 2$ and $T_e = 4$ for $d \geq 6$, giving $T_B = 3$ for $d \geq 6$; and this is an exact result [1, 10]. Using the results for the elastic exponents d_B, ν_p and T_e in $d = 2$ and 3 [3, 12], one gets [1] $2.3 \leq T_B \leq 2.8$ and $2.5 \leq T_B \leq 2.9$ in $d = 2$ and 3 respectively. The experimental results for fracture stress of randomly cut lattices of perforated sheets of aluminium and copper (Benguigui et al [11]) give $T_B = 2.5 \pm 0.4$ in $d = 2$. The Monte Carlo simulation for percolating networks with both central and bond-bending forces (Sahimi and Arbabi [12]) give $T_B \cong 2.42$ and 2.64 in $d = 2$ and 3 respectively. All these results for the fracture exponent T_B are well within the above theoretical bounds. It may be mentioned that for continuum percolation, the above bounds for T_B become cosiderably different (larger; see [1]).

Fracture strength distribution

As is well known and established by now [4, 13, 14], the breakdown strength distribution of disordered solids is not quite self-averaging, as the statistics is governed by that of extreme events.

Let us assume that a solid of size L contains n number of (micro-) cracks, each with failure probability $f_i(\sigma)$, $i = 1, 2, ..., n$, under an applied stress σ. We further assume that the stress released region of each of the cracks are separate and do not overlap. If we denote the failure probability of the entire sample , under stress σ, by $F(\sigma)$, then [1]

$$1 - F(\sigma) = \prod_{i=1}^{n}(1 - f_i(\sigma)) \cong \exp[-\sum_i f_i(\sigma)] \cong \exp[-L^d g(\sigma)] \ , \tag{4}$$

where $g(\sigma)$ denotes the density of cracks weaker than stress σ (or the density of cracks which will start propagating at and above the stress σ). This is because, the sample survives, if each micro-crack within the sample survives.

a) **Weibull distribution:** For small bond dilution concentration, the probability $g(l)$ of a linear crack of length l is givewn by $g(l) \cong p^2(1-p)^l$. This gives [15]

$$g(\sigma) = \int_{l(\sigma)}^{\infty} g(l)dl \cong [p^2/\ln(1-p)](1-p)^{(k/\sigma)^2} , \qquad (5)$$

as the minimum length of the crack growing under stress σ is given by $l(\sigma) = (k/\sigma)^2$ from the Griffith law (1a). Note, as σ varies from 0 to ∞, $(1-p)^{(k/\sigma)^2}$ varies from 0 to 1 ($p < 1$); and this suggests a possible representation [15] $g(\sigma) \cong 1 - \exp(-c\sigma^m) \cong c\sigma^m$, with the (Weibull) modulus $m > 0$. This gives the Weibull failure strength distribution [4, 15, 1]

$$F(\sigma) \cong 1 - \exp(-cL^d\sigma^m) , \qquad (6)$$

giving

$$\sigma_B \cong L^{-d/m} , \qquad (6a)$$

for the variation of the typical strength of the sample (corresponding to a nonzero value of $F(\sigma)$), as a function of the sample size L.

b) **Large disorder: Scaling limit:** Very near the percolation threshold p_c , the probability of getting a vacancy cluster (crack) of linear size l will be given by [3]

$$g(l) \cong \exp(-l/\xi_p) \cong \exp(-\frac{k^2}{\sigma^2\xi_p}) , \qquad (7)$$

where the Griffith law (1a) has again been used to relate the minimum length l of the cracks which become unstable under stress σ. Using (1a) and (2), one gets $k^2/\xi_p = (Y\gamma)/\xi_p \cong |\Delta p|^{2T_B}$, where $T_B = [T_e + (d-d_B)\nu_p]/2$. This gives [cf. 1] the Gumbel distribution $F(\sigma)$ for the strength of the entire sample (Duxbury and Leath [16]):

$$F(\sigma) \cong 1 - \exp[-L^d \exp(-|\Delta p|^{2T_B}/\sigma^2)] . \qquad (8)$$

This gives

$$\sigma_B \cong |\Delta p|^{T_B}/(\ln L)^{1/2} \qquad (8a)$$

for the variation of the typical strength of the solid as a function of the sample size L.

Most of the computer simulation results for the fracture (see e.g. [12]) and for dielectric breakdown (see e.g. [13]) confirms the Gumbel (double exponential) distribution to be more appropriate than the Weibull distribution for the breakdown strength distribution of percolating solids. It should be noted, however, the Weibull form is actually derived also (see Section IIa) from a double exponential form after reducing one expontial function to a power law [15]. It may also be noted from the expression (8a), that the non-self-averaging distribution (8), coming from the statistics of extreme events (4), still allows for defining an average fracture strength σ_B of the disordered solid for any practical (finite) size L. Although σ_B vanishes in the thermodynamic limit ($L \rightarrow \infty$) due to the possibility of larger cracks present within the larger samples, the strength vanishes very slowly ($\sigma_B \cong (\ln L)^{-1/2}$) with the system size.

III. Peierls' instability ("melting") of percolating solids

In the last section, we have discussed the mechanical instability, under applied stress, of percolating or porous solids. Here, in this section, we intend to discuss another (mechanical) instability, induced by thermal fluctuations: namely the problem of Peierls' (melting-like) instability [5] of such percolating solids. In fact, extensive studies have been made for cooperative (continuous) transitions, like the magnetic transitions, on percolating lattices [17]. Thermally induced instability like the melting of a solid (or freezing of a liquid) is also a very common cooperative transition, dependent on the dimensionality (and is basically discontinuous in nature, in higher dimensions). The melting behaviour (or the question of thermodynamic stability at finite temperature) of a percolating solid or lattice is therefore quite intriguing, especially near the percolation point, where the dimensionality crossover takes place [18]. The melting properties of (low dimensional) fractals, like the colloidal aggregates, are quite interesting and have been studied recently [19]. Studies on discontinuous transitions of binary fluids in porous gels [20] might be a typical situation where such melting studies of percolating solids would be relevant. Molecular dynamic simulation of melting of an initially dilute (percolating) solid would be another example.

In the absence of a simple (and settled) theory of melting, let us consider the problem of Peierls' instability [5], induced by thermal fluctuations of the lattice, in a percolating solid. In particular, we intend to study the stability of the (correlations in) percolating lattice, against lattice fluctuations at any finite temperature [18]. For an effective harmonic system, applying the equipartition law to each independent lattice mode (phonon), with wave vector q, we have

$$w_q^2 < |u_q|^2 > \;\cong K_B T$$

or,

$$< |u_i|^2 > \;\cong\; K_B T \int_{q_l}^{q_u} (q^{d-1}/w_q^2) dq \;, \tag{9}$$

where u_q and w_q denote respectively the phonon amplitude and energy for the wave vector q (u_i being the lattice fluctuation at the site i), and q_u and q_l denote the upper and lower cut-off wave vectors respectively. As one approaches the percolation point, the modulus of elasticity decreases as $|\Delta p|^{T_e}$, giving $w_q \cong q|\Delta p|^{(T_e/2)}$ in the long wavelength limit. It may be mentioned that the elastic percolation point coincides with the (normal) percolation point for systems with (realistic) bond-bending forces. It is obvious from (9) that the integrand becomes vulnerable near the lower cut-off ($q_l \cong 1/L$, as the system size $L \to \infty$), indicating Peierls' instability of the solid with respect to long wavelength lattice fluctuations. The fluctuations become unbounded for $d \leq 2$ [5]. For higher dimensions, the lattice fluctuations of the infinite percolation cluster remain bounded below the 'melting' temperature T_m :

$$T_m(p \geq p_c) \cong |p - p_c|^{T_e} \;. \tag{10}$$

(The exponent T_e in (10) will be replaced by $(T_e - \beta/2)$ if one multiplies by the probability of getting the infinite percolation cluster or the percolation order parameter $\cong |\Delta p|^{\beta}$, for getting the lattice fluctuation, from (9), of the percolation cluster). The above expression indicates that the percolating solid will melt at a lower temperature determined by the reduced elastic modulus, and the melting temperature (in going from low temperature percolating solid to the high temperature melt) reduces monotonically (see Fig. 2) to zero ($T_m(p_c) = 0$ at the elastic percolation threshold) with the same exponent (T_e) as that of the elastic modulus. These results are for zero (or constant) external stress (σ).

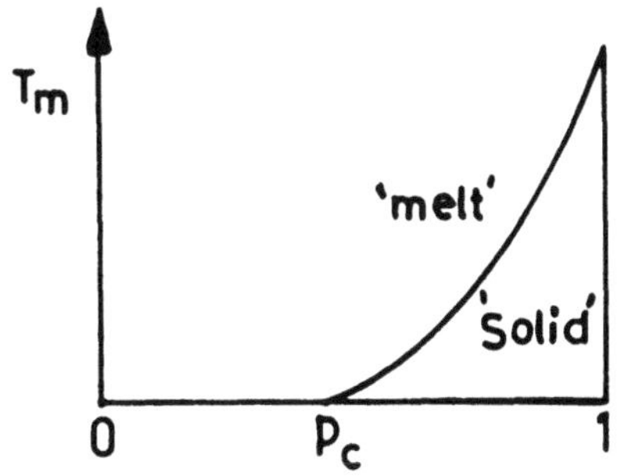

Fig. 2: Schematic phase diagram of melting of a percolating solid (at constant external stress or pressure).

In fact, at p_c the dimensionalities (as seen by the long wavelength fluctuations) crossover to the corresponding fractal values [3]. The dimensionality being reduced, the above mentioned Peierls' instability, induced by the long wavelength lattice fluctuations, becomes even more enhanced and the instability temperature drops to zero. For a percolating fractal (at $p = p_c$) of size L (for which the mass $M \cong L^{d_f}$), the average lattice fluctuation at a temperature T is given by eqn. (9) as

$$< |u_i|^2 > \cong K_B T q_l^{(d_f - z)} \cong K_B T M^{(2-d_s)/d_s} , \qquad (11)$$

where $q_l \cong L^{-1}$, d_f and d_s denote respectively the fractal and spectral dimensionality of the percolation cluster and $z(= 2d_f/d_s)$ denote the dynamical exponent for diffusive modes on the percolation cluster. This is because, d in (9) is replaced by d_f of the fractal, as the number of modes here are determined by d_f, and the dispersion for phonons (fractons) changes to $\omega_q^2 \cong q^z$. (If for anomalous diffusion [3] on the percolation fractal, random walk distance R_t scales with time t as $R_t \cong t^{\nu_w}$, $\nu_w < 1/2$, then $q^{-1} \cong \omega^{-\nu_w}$ or $\omega \cong q^z$ where $z \equiv 2d_f/d_s = 1/\nu_w (> 2)$; for phonons $\omega \rightarrow \omega^2 \cong q^z$). Unlike Euclidean objects, for which the lattice correlations are spontenously broken for dimensions $d \leq 2$ [5], fractals (like the percolating solid at the threshold) lose their correlations at any finite temperature if the spectral dimensionality d_s (and not the fractal dimensionality d_f) is less than two [19, 18]. Since, for percolating fractals d_s is much less than two ($d_s \cong 4/3$) [3], any infinite ($L \to \infty$) percolation cluster would have unbounded lattice fluctuation at any finite temperature. Thus, the vanishing melting temperature at the percolation threshold p_c ($T_m(p_c) = 0$) is doubly assured: It becomes zero as the elastic constant vanishes there (eqn. (10)), and also because of the Peierls' instability of the percolation fractal (at any finite temperature), as its spectral dimension is less than two.

It can also be argued [18] that on melting, the entropy (or latent heat) discontinuity would vanish at p_c as

$$\Delta S_m \cong \exp[-c/|\Delta p|^{(T_e-\beta)}] \ . \qquad (12)$$

This indicates, that the melting transition becomes continuous ($T_e >> \beta$) at the percolation threshold.

The above discussions on the thermal instability of percolating solids is for zero external stress. This instability or melting temperature would obviously be affected by the presence of external stress on the solid. In particular, the stress concentration at the sharp notches or at the sharp points of the void surfaces in the percolating solid, would help destabilisation at lower temperatures.

IV. Zener breakdown in Anderson insulators

Dielectric breakdown in metal loaded dielectrics, or in (classical) insulators containing a random fraction p of conductors near but below the percolation threshold p_c , has been widely studied [1]. For $p > p_c$, the conductivity of the composite grows as $|\Delta p|^{t_c}$, where t_c is the classical conductivity exponent [3]. For $p < p_c$, the breakdown voltage (for a fixed sample size) or the breakdown field E_B grows with the minimum gap and the minimum gap, in turn, decreases with the inverse percolation correlation length ξ_p , giving $E_B \cong |\Delta p|^{\nu_p}$ [1].

One can consider now the equivalent problem of breakdown in Anderson insulators [1]. In fact, such quantum percolation (or the Anderson transition) problem belongs to a different university class, compared to that of classical percolation [21] : As discussed in the introductory section (I), the lower critical dimensionality for quantum percolation seems to be two [6], compared to its unit value in classical percolation. For quantum percolation in three dimension, a mobility edge ϵ_c exists, and the system makes transition from insulator (for systems with fermi level $\epsilon_f < \epsilon_c$) to a conductor (for $\epsilon_f > \epsilon_c$), and in the metallic phase ($\epsilon_f > \epsilon_c$) the conductivity grows as $|\epsilon_f - \epsilon_c|^{t_q}$; $t_q \neq t_c$.

The above comparisons in conductivity properties of classical and quantum percolation are for cases with conductor concentration above the percolation threshold ($p > p_c$) and with the fermi energy above the mobility edge ($\epsilon_f > \epsilon_c$) respectively. Like the complementary problem of classical breakdown in insulators or dielectrics with nonpercolating conductors ($p < p_c$), as discussed earlier, we consider here the problem of quantum insulation breakdown in Anderson insulators for $\epsilon_f < \epsilon_c$, by applying strong electric field ([1, 22]; see however [23] for an alternative view).

In fact, this quantum breakdown in Anderson insulators (due to the tunnelling of the exponentially localised states, to the extended states above the mobility edge) is similar to the Zener breakdown in narrow gap semiconductors. In the case of Zerner breakdown [7], the energy bands get tilted along direction of the external field E and this reduces effectively the width of the gap or of the barrier between the bands. The breakdown occurs when the exponentially localised states within the band gap (localised within the lattice constant) can tunnel through the (tilted) barrier with reduced width. This happens when the energy gained by the electron (with charge e) in travelling the distance of a lattice constant a, is equal to the band gap ϵ_g : $eEa = \epsilon_g$; or in other words, when the effective width $w = \epsilon_g/(eE)$ of the gap between the (tilted) bands become of the order of the lattice constant a, within which the states in the band gap are (exponentially) localised. Interband tunnelling occurs for $w \leq a$. The Zener breakdown field E_B for a semiconductor with band gap ϵ_g , therefore, scales linearly with the gap : $E_B = \epsilon_g/(ea)$.

In the case of Anderson Insulators or quantum percolation for $d > 2$, the states are extended for energies above the mobility edge ϵ_c , below which the states are exponentially localised [6, 21] within a distance ξ_q , the localization length: $\xi_q \cong |\Delta\epsilon|^{-\nu_q}$; $\Delta\epsilon \equiv \epsilon_f - \epsilon_c$. If we assume that the nature of the states and the mobility edge remain unchanged in presence of the external electric field E (see Fig. 3), and that the localised electrons can gain energy from the field for at least a travel of distance ξ_q along the field, then the bands of all the (localised and extended) states get tilted in the direction of the field and the effective width w of the mobility gap $\Delta\epsilon$ decreases

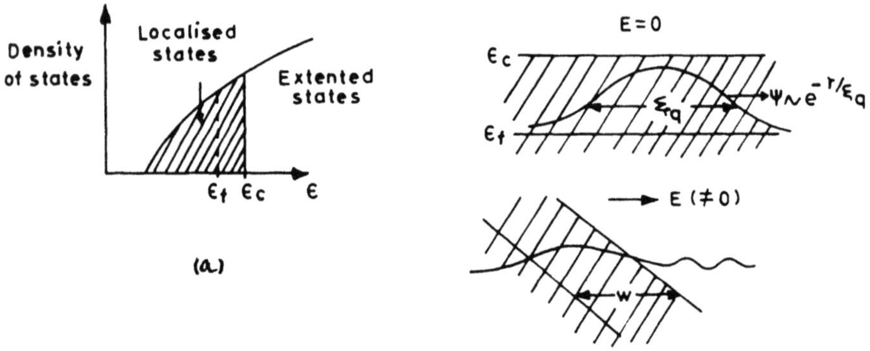

Fig. 3: Schematic density of states for Anderson insulators ($d > 2$) shown in (a), where the fermi level ϵ_f is below the mobility edge ϵ_c (metallic phase for $\epsilon_f > \epsilon_c$). For strong electric field E, the band of (localised) states get tilted and tunnelling occurs when the effective width $w(= |\Delta\epsilon|/eE)$ of the mobility gap $\Delta\epsilon$, is less than or equal to the localisation length ξ_q .

with the increase of the field : $eEw = \Delta\epsilon$, or $w = \Delta\epsilon/eE$. Breakdown due to tunnelling to extended states (above the mobility edge) then occurs for $w \leq \xi_q$, giving the breakdown field

$$E_B = \Delta\epsilon/e\xi_q \cong |\Delta\epsilon|^{T_q};\ T_q = 1 + \nu_q\ . \tag{13}$$

This breakdown field may also be viewed as the field at which the localised states make a cross-over from an exponential to a power law type behaviour. Unlike the Zener breakdown in semiconductors, where the breakdown field scales linearly with the band gap, the breakdown field in Anderson Insulators scales almost quadratically ($\nu_q \cong 0.9$ [6, 21], giving $T_q \cong 1.9$ in $d = 3$) with the mobility gap ($\epsilon_c - \epsilon_f$).

Since the tunnelling probability $g(E)$ decreases exponentially with the width w of the effective barrier, $g(E) \cong \exp(-w/\xi_q) \cong \exp(-\Delta\epsilon/eE\xi_q) \cong \exp\ (-|\Delta\epsilon|^{T_q}/E)$. The failure probability of the entire sample (of size L) under field E would then be given, from eqn. (4), by

$$F(E) \cong 1 - \exp[-L^d \exp(-|\Delta\epsilon|^{T_q}/E)]\ . \tag{14}$$

This Gumbel form for the strength distribution for Zener breakdown in Anderson insulators may be compared with that for fracture (8). It then gives

$$E_B \cong |\Delta\epsilon|^{T_q}/(\ln L) , \qquad (14a)$$

for the variation of the typical breakdown field (for and above which $F(E)$ is significant) with the sample size L.

V. Concluding remarks

As discussed in section II, the statistics (scaling properties etc.) of (static) fracture properties of disordered solids are quite well-studied and established. However, not much is known about the fracture dynamics. The precursor effects of macroscopic failure or fracture are not yet very well studied, although there has been some indication [24] that a ratio of the longitudinal and transverse elastic modulus approaches an universal value as the macroscopic fracture point is reached. Such studies of precursor effects associated with the dynamics of fracture are extremely important for predictive capablities (before the occurrence of the catastrophy). The melting properties of initially dilute percolating solid (section III) is quite interesting. However, no result of any molecular dynamic simulation of melting of such percolating solids, is available yet. The reverse problem of freezing of the melts to such disordered solids or fractals, is even more intriguing. The question of electrical breakdown in quantum percolation or Zener breakdown in Anderson Insulators (section IV) is obviously quite complicated and needs further study.

Acknowledgements: I am grateful to my collaborators Subhrangshu Sekhar Manna, Abhijit Mookerjee, Barnana Pal, Purusattam Ray and Asok K. Sen, with whom the studies mentioned above were developed. I am grateful to Paul Leath for several useful comments.

REFERENCES

[1] B. K. Chakrabarti, *Mechanical & Electrical Breakdown in Disordered Solids*, Rev. Solid State Sc. **2** (1988) pp. 559-579

[2] H. J. Herrmann and S. Roux (Eds.), *Statistical Methods for the Fracture of Disordered Media*, North Holland, Amsterdam, (1990);

R. Englman and Z. Jaeger (Eds.), *Fragmentation, Form and Flow in Fractured Media*, Ann. Israel Phys. Soc. (1986);
M. Sahimi,*Brittle Fracture in Disordered Solids*, Physica A **186** (1992) pp. 160-182;
D. J. Bergman and D. Stroud, *Physical Properties of Macroscopically inhomogeneous media*, in Solid State Physics, Vol. 46 (Eds. H. Ehrenreich & D. Turnbull; Academic Press, N. Y.; 1992) pp. 147-269

[3] D. Stauffer, *Introduction to Percolation Theory*, Taylor & Francis, London (1985)

[4] B. R. Lawn and T. Wilshaw, *Fracture of Brittle Solids*, Cambridge Univ. Press, Cambridge, (1975)

[5] R. Peierls, *Surprises in Theoretical Physics*, Princeton Univ. Press, Princeton, N. J. (1979) p. 85

[6] P. Lee and T. V. Ramakrishnan, *Disordered Electronic Systems*, Rev. Mod. Phys. **57** (1985) pp. 287-337

[7] J. M. Ziman, *Principles of the Theory of Solids*, Cambridge Univ. Press, Cambridge, (1965) pp. 163-168

[8] P. Ray and B. K. Chakrabarti, J. Phys. C **18** (1985) L185

[9] P. Ray and B. K. Chakrabarti, Phys. Rev. B **38** (1988) 715

[10] D. J. Bergman, in Engllman and Jaeger (Eds.) [2]

[11] L. Benguigui, P. Ron and D. J. Bergman, J. Physique **48** (1987) 1547

[12] M. Sahimi ans S. Arbabi, Phys. Rev. B **47** (1993) 713

[13] P. Duxbury, in Herrmann and Roux (Eds.) [2]

[14] P. L. Leath, in this Proceedings

[15] P. Ray and B. K. Chakrabarti, Solid State Commun **53** (1985) 477;
S. Z. Zhang, C. W. Lung and K. L. Wang, Phys. Rev. B **42** (1990) 6631

[16] P. M. Duxbury and P. L. Leath, J. Phys. A **20** (1987) L411; P. M. Duxbury, P. D. Beal and P. L. Leath, Phys. Rev. Lett. **57** (1986) 1052; Phys. Rev. B **36** (1987) 367

[17] see e.g., R. B. Stinchcombe, in *Phase Transition & Critical Phenomena*, Vol. 7, Eds. C. Domb & J. L. Lebowitz, Academic Press, N. Y. (1983) p 151

[18] B. K. Chakrabarti and A. K. Sen, J. Phys. Cond. Matt. **2** (1990) 9183

[19] B. Pal, S. S. Manna and B. K. Chakrabarti, Solid State Commun **64** (1987) 1309 ;
Z. Wang, C. D. Gong, A. Holz and C. S. Ting, J. Phys. Cond. Matt. **1** (1989) 2575

[20] S. Hirotsu, Y. Hirokawa and T. Tanaka, J. Chem. Phys. **87** (1987) 1392

[21] see e.g., B. Shapiro, in *Percolation structures and Processes* , Eds. G. Deutscher, R. Zallen and J. Adler, Hilger, Bristol, (1983) p. 367;
Y.Shapir, A. Aharony and A. B. Harris, Phys. Rev. Lett. **56** (1986) 976;
Y.Meir, A. Aharony and A. B. Harris, Europhys. Lett. **10** (1989) 275

[22] A. Mookerjee, B. K. Chakrabarti, I. Dasgupta and T. Saha, Physica A **186** (1992) 258

[23] see A. K. Sen, in this proceedings.

[24] M. Sahimi and S. Arbabi, Phys. Rev. Lett. **68** (1992) 608

Spring-Network and Finite-Element Models for Elasticity and Fracture

A. Jagota and S.J. Bennison

Central Research and Development,
E.I. DuPont de Nemours & Co.,
Experimental Station,
Wilmington, Delaware 19880-0356, USA.

Abstract This paper discusses computational models for elastic deformation and fracture of brittle, disordered materials. In particular, we compare models based on networks of springs and finite-element discretizations. When the length scale of elements in the numerical model corresponds to discrete material elements, spring-network approaches are valid. Their validity as models for a continuous medium is examined by comparing to a finite-volume discretization and by studying (a) the ability to capture uniform strain, and (b) the ability to model crack propagation in an isotropic, homogeneous, brittle material. As models of continuous media, spring-networks exhibit several numerical artifacts. For such problems, we conclude that finite-element methods can be used to the same degree of accuracy or better than spring-network models.

Keywords spring-network, finite-element, fracture, disorder, brittle, elasticity.

1 Introduction

Like many mechanical processes in materials, fracture is strongly influenced by defects, i.e., cracks. Much of the work on the study of fracture is based on the physics of single cracks [1]. For materials with few major flaws and relatively little disorder, this approach has been very successful in predicting failure and in improving materials using the theory of elasticity and defined properties such as the fracture toughness K_{Ic}. When materials are highly disordered and failure is preceded by the growth and propagation of defects distributed over the volume of the material, continuum damage mechanics approaches can be used to characterize damage evolution [2,3]. However, these can rarely be extended to final failure of the material, nor do they contain specific information about the relationship between material microstructure and fracture behavior.

In recent years, much attention has been focussed on the fracture of materials where either approach fails. This is of great practical importance; for example, constant efforts to improve mechanical properties of brittle materials invariably lead to failure modes that involve multiple fracture events, e.g., fibre-reinforced ceramic matrix composites [4] and residually-stressed ceramic composites [5]. Statistical issues assume greater importance: disorder in the material; nucleation, growth, and interaction of separate cracks; ultimate localization of distributed damage to failure, etc.

Two broad aims in the study of such materials are (a) to discover general scaling relationships between, for example, strength, sample size, and disorder, and (b) to understand the effect of material microstructure on fracture.

Although considerable progress has been made in the first of the goals using analysis [6]-[10], most models remain of necessity rather simple. Computational methods have been developed and used widely for such studies. Being far less restrictive than analytical models, they have played a crucial role in exploring various regimes of behaviour. Although several flavours of the models exist, which we shall discuss in the following section, a common thread is the use points in space in a regular or irregular configuration, connected by springs. The network putatively models an elastic material, and individual breakage of springs models initiation and propagation of flaws [7], [11]-[18].

Such spring models have also been used to study more explicitly the effects of material microstructure on fracture, for example, fracture of polymers [19], composites [20,21], concretes [22]-[24], transformation toughening [25,26], and micro-crack shielding [27]. Finite-element methods [28] have also been used extensively in the study of fracture, albeit usually in the context of single cracks [29,30]. However, this is not a restriction of the method, and recent studies have employed it to study the behaviour of disordered materials [31,32].

In this paper we discuss and compare spring-network and finite-element methods for elasticity and fracture[1]. A particular concern is the validity of the models as representations of brittle, elastic materials. We shall distinguish between two types of problems:

P1 where the discreteness of the model matches the discreteness of the material, for example, granular materials [33] and sintering [34],and

P2 where the length scale of disorder in the material does not match that of the model, which is then interpreted as serving as a discretization of a continuum.

In the first case, the validity of the model is established *a priori*. For example, discrete spring-network models for granular materials [33,34] have an obvious correspondence with the material being modelled. The problem of the existence and nature of continuum models for the material, a *homogenization* issue, is then separate from the question of the validity of the computational model. However, this question remains in the more general second case, **P2**. We shall discuss this issue by examining the nature of the discretization for spring-network and finite-element models, and by studying the performance of the two approaches

[1] see also [15] for a related discussion on spring-network models, and [14,35] for comparative discussions of spring-network and finite-element models.

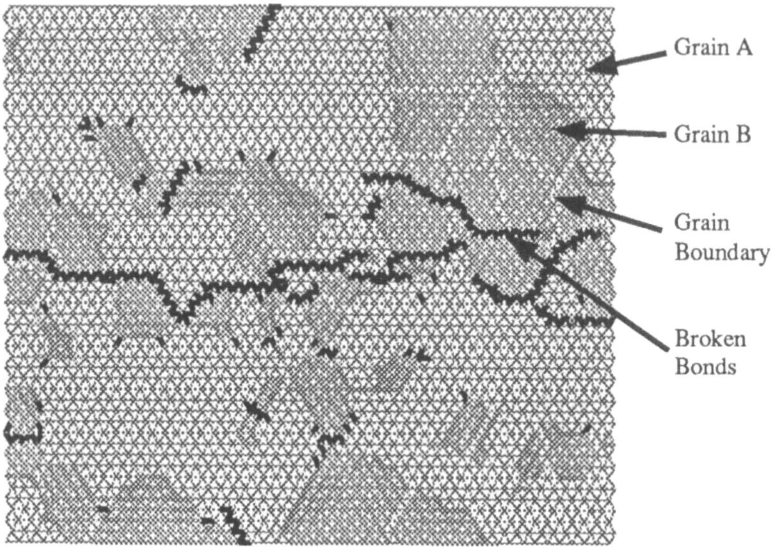

Figure 1: Beam based spring-network model of a two-phase, residually-stressed, brittle material strained vertically to failure. Residual stress disorder causes distributed damage which interacts and coalesces as the material is strained.

with respect to two problems: (a) the ability to model a state of uniform strain, and (b) the ability to model crack propagation in an isotropic, homogeneous, and brittle, solid.

The remainder of the paper is organized as follows: Section 2 reviews briefly the varieties of spring-network models, and their relationship to finite-element and other numerical techniques. In section 3 we examine the validity of the spring-network discretizations formally by attempting to derive discrete equations from the continuum governing equations, and by examining the behavior of the models with respect to the conditions stated in the previous paragraph. The overall conclusions are presented in section 4.

2 Computational methods

In this paper we will restrict our attention to the modelling of brittle linear-elastic solids in two dimensions (plane strain), although the conclusions extend to higher dimensions and to non-linear materials. This section reviews briefly the varieties of models that have been used.

2.1 Spring-network models

Figure 1 shows an example of a spring-network model of a two-phase (residually stressed) microstructure. It consists of a triangular grid of points connected

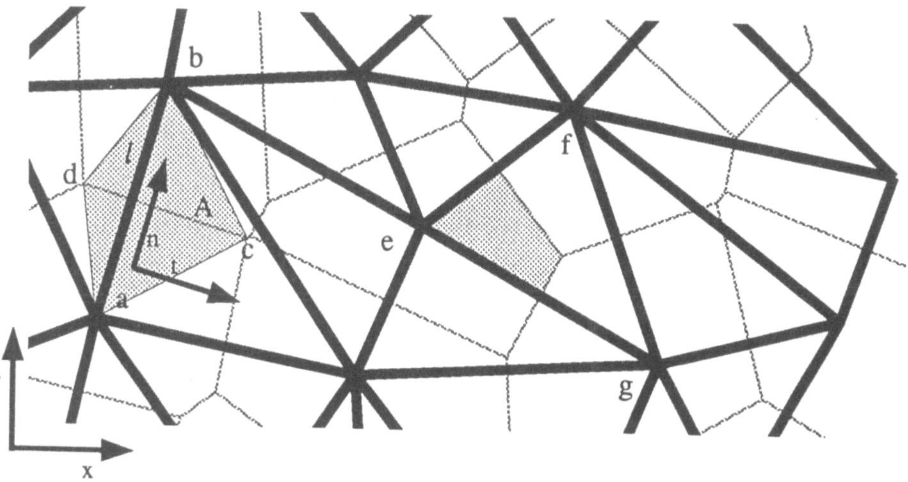

Figure 2: Part of a random spring-network in two-dimensions. Random points, e.g., a and b, are connected by springs (thick black lines) if they share a face of the Voronoi cell diagram of the collection of points (thin, light lines).

by one-dimensional spring elements. This network serves as a discretization for the materials being studied. Such a network is typically deformed by externally applied displacements or forces and all internal points deform in a manner that minimizes the total stored elastic energy, or equivalently, maintains force and moment equilibrium for every point. Different models vary in the specific expressions used to write the stored elastic energy, or, again equivalently, in the constitutive relations used for individual springs. Disorder is typically introduced by varying the properties of individual springs. Prior to their use for the study of fracture, such models were used extensively in the study of elastic percolating networks [36]-[41]. They have also been developed independently to study deformation and fracture of elastic continua [17,42]

Let E be the stored elastic energy, $\mathbf{u}$ be the displacement vector of each point, $\mathbf{F}$ be the force vector in each element, ω be the rotation at a point, and $\mathbf{M}$ the moment transmitted in an element. Also, let N^e be the number of springs and l the length of each spring, and let the superscripts a and b refer to the two points at the ends of an element[2] (figure 2). We will discuss below four types of models:

1. **Central force**[15,40] Each element between points transmits only a force $\mathbf{F} = F_n\mathbf{n}$ alligned along the element (unit vector $\mathbf{n}$), which is then characterized by a spring constant k^n: $F_n = k^n(\mathbf{u}^\mathbf{b} - \mathbf{u}^\mathbf{a}) \cdot \mathbf{n}$. The total elastic

[2] Subscripts denote cartesian vector or tensor components. Superscripts denote either indices identifying points in the lattice, or identify a spring constant. A comma represents a derivative with respect to a spatial coordinate: $u_{i,j} = \partial u_i/\partial x_j$.

energy is given by

$$E = \frac{1}{2}\sum_{i=1}^{N^e} k^n \left((\mathbf{u^b} - \mathbf{u^a})\cdot\mathbf{n}\right)^2 \tag{1}$$

2. **Born**[34,43] Elemental forces are no longer restricted to be along the element, and a second spring constant k^t is introduced as relating the tangential force to relative tangential displacement: $F_t = k^t(\mathbf{u^b} - \mathbf{u^a})\cdot\mathbf{t}$, where $\mathbf{t}$ is normal to $\mathbf{n}$. Then,

$$E = \frac{1}{2}\sum_{i=1}^{N^e}\left[k^n \left((\mathbf{u^b} - \mathbf{u^a})\cdot\mathbf{n}\right)^2 + k^t \left((\mathbf{u^b} - \mathbf{u^a})\cdot\mathbf{t}\right)^2\right] \tag{2}$$

   The Born model is *not* rotationally invariant; rigid-body rotation of a network stores energy [34,44].

3. **Granular or beam**[23,34] The kinematics of points in the network are now desribed by a rotation in addition to displacement, and elements transmit moments in addition to forces. This introduces a new relationship between moment and relative rotation for every element. Here, we shall restrict our attention to a special case of this model where the rotational stiffness is small. Then the total energy is

$$E = \frac{1}{2}\sum_{i=1}^{N^e}\left[k^n \left((\mathbf{u^b} - \mathbf{u^a})\cdot\mathbf{n}\right)^2 + k^t \left((\mathbf{u^b} - \mathbf{u^a})\cdot\mathbf{t} - l(\omega^b + \omega^a)\right)^2\right] \tag{3}$$

   Moments are negligible, and the extra kinematic variables for each point, the rotations, serve to render the model rotationally invariant.

4. **Bond-bending**[16,44] Here, springs resist motion along $\mathbf{n}$ as in the central force model. Additionally, changes in the angle between triads of points e, f, and g: θ^{feg} (figure 2), contribute to the stored elastic energy, and

$$E = \frac{1}{2}\left[\sum_{i=1}^{N^e} k^n \left((\mathbf{u^f} - \mathbf{u^e})\cdot\mathbf{n}\right)^2 + \sum_{i=1}^{N^\theta} c\left(\delta\theta^{feg}\right)^2\right] \tag{4}$$

   where c is a parameter that characterizes resistance to change in angle, and N^θ is the number of angles to be considered.

For a given network, subjected to applied displacements or loads at the boundary, the internal field of displacements and forces can be determined using several techniques. Enforcing equilibrium for every point leads to a set of simultaneous linear algebraic equations [34,45] which are solved for the unknown point displacements (and spins). Alternatively, the displacements can be determined iteratively by minimizing the energy [13], or dynamic relaxation [33].

Other techniques, such as transfer-matrix solution [46], or the use of lattice Green's functions to reduce three-dimensional problems to two dimensions [47] can also be used. The techniques have competing advantages with respect to speed, problem sizes handled, parallel implementations, etc. which we will not discuss here.

Fracture is typically modelled by specifying a limit to the linear constitutive equations for each element, for example, by simply removing an element if its stored energy exceeds a critical value[3], e.g., for the beam model we use:

$$\frac{F_n^2}{2k^n} + \frac{F_t^2}{2k^t} > E^c \qquad (F_n > 0) \tag{5}$$

A simulation proceeds by incrementally increasing the boundary conditions, and iteratively removing such bonds as achieve criticality, until an equilibrated state is achieved.

For problems of type **P1**, the various spring constants and criticality conditions are determined directly by the material being modelled. For problems of type **P2**, one must relate these parameters of the spring-network model to continuum properties, for example, for an isotropic brittle elastic material the Young's modulus Y, Poisson's ratio ν, and fracture toughness K_{Ic}. This we shall discuss in the following section.

In more general terms, we may think of spring-network models as a discretization of higher than 1-dimensional materials by a network of one-dimensional elements characterized by element constitutive equations, and a breaking condition for each element.

2.2 Finite-element and other methods

Finite-element methods [28] begin with a continuum boundary-value problem consisting of governing equations (usually expressed in a *weak* integral form) operating over a defined domain V with appropriate boundary conditions on the surface S. For example, for small-strain deformation one uses the principle of virtual work:

$$\int_V \sigma_{ij}\delta\epsilon_{ij}dV = \int_S T_i\delta u_i dS + \int_V b_i\delta u_i dV, \tag{6}$$

where σ_{ij} is the stress tensor, ϵ_{ij} is the strain tensor:

$$\epsilon_{ij} = (u_{i,j} + u_{j,i})/2, \tag{7}$$

T_i are boundary tractions, and b_i are body forces. The domain V is discretized into finite-elements and constitutive relations between σ_{ij} and ϵ_{ij} are introduced, for example for linear elasticity:

$$\sigma_{ij} = 2\mu\epsilon_{ij} + \lambda\epsilon_{kk}\delta_{ij}, \tag{8}$$

where μ and λ are the Lamé constants. In each element the displacement $\mathbf{u}$ is approximated by interpolating functions. The global integral equation (6) is thus

[3] Stochastic breaking conditions have also been used successfully [13,19]

reduced to a set of simultaneous algebraic equations with nodal displacements $\{U\}$ as the unknowns:

$$[K]\{U\} = \{F\}, \tag{9}$$

where $[K]$ is the stiffness matrix and $\{F\}$ is the vector of equivalent nodal forces. These equations can be solved directly or iteratively for $\{U\}$. As referred to in the Introduction, these methods have been used widely to study the behavior at the tip of a single isolated crack. Crack propagation has been simulated using process zone models [30], element removal [48], or separation of element boundaries [32].

In some respects, the spring-network model resembles finite-difference [49] or finite-volume [50] methods: the discretization consists of points connected by one-dimensional elements. However, whereas for finite-difference, finite-volume, and finite-element techniques the continuum equations are explicitly discretized, such a connection has not been provided for spring-network models.

3 Validity of the models

In this section we examine the validity of these models for the two types of problems **P1** and **P2** defined in the Introduction. For problems of type **P1** the spring-network model, by definition, has a direct correspondence with the discrete nature of the material. The issue then is of the appropriate effective continuum model for the material, one of ***homogenization***. For problems of type **P2**, finite-element, finite-difference, and finite-volume methods are explicitly derived as discretizations of the continuum. The question is how the spring-network model is related to these and whether it is valid as a discretization.

3.1 P1 problems; homogenization

Consider a network of springs of size L with springs of characteristic size l. When $L \gg l$ the network behaves as an effective continuum. For example, a large triangular lattice with idential springs modelled by the beam method has an effective Young's modulus and Poisson's ratio given by[34]:

$$\nu = \frac{k^n - k^t}{3k^n + k^t} \tag{10}$$

$$Y = \frac{\sqrt{3}}{4l}((3-\nu)k^n + (1+\nu)k^t). \tag{11}$$

That for large networks there corresponds an effective continuum has often been used as justification for their use. For regular grids using explicit relationships such as (10) and (11) spring properties can be related to continuum properties. However, even for regular and isotropic lattices, the entire range of Poisson's ratio is not accessible. When the network is irregular or random, the effective properties cannot be computed exactly but can be bounded [51]-[54]. Also, the appropriate effective continuum for the beam model is not that of standard linear elasticity because of the presence of spins, but some version of micro-polar elasticity [55]-[58]. In general therefore, the homogenization principle cannot be used to justify the validity of spring-network models for problems of type **P2**.

The situation is further complicated when studying fracture, for a moving crack senses the local structure of the network. The triangular grid, for example, is elastically isotropic in two dimensions but a crack moving through it senses three distinct preferred directions for propagation. This issue will be discussed further in the following subsection. The homogenization problem on its own is very important but is not the subject of this paper and we shall leave its discussion at this point.

3.2 P2 problems; discretization

The main issue to be resolved in this paper then remains as the following: for problems of type **P2**, are spring-network models valid discretizations of a continuum ? We shall address this issue firstly by attempting to derive the spring-network expressions for energy as discrete forms of continuum expressions, and secondly by examining the ability of the various methods to model (a) a uniform state of strain, and (b) crack propagation in an isotropic, homogeneous, brittle material.

3.2.1 Discretization

We seek, in two-dimensions[4], to derive the spring-network expressions for energy in the network (1-4) as a discretization of the continuum expression:

$$E = \frac{1}{2}\int_V \sigma_{ij}\epsilon_{ij}dV \tag{12}$$

to the discrete, spring-network expressions (1-4).

To this end, consider again figure 2 which consists of a discretization based on a set of points placed randomly in two-dimensions[5]. A random grid is chosen because (a) regular grids are not isotropic with respect to crack propagation, and (b) to examine the nature of the discretization using an arbitrary lattice. The Voronoi cell network [59] of the points is used to define the springs; any two points that share a face of a Voronoi cell are connected by a spring. This procedure provides a unique method for generating a spring-network for given points. It also provides a method to generate a corresponding finite-element mesh for direct comparisons between the two methods, e.g., points *a-c-d* form a triangular element (figure 2).

Let us first attempt discretization for the beam model. Consider points *a* and *b* in figure 2. We assign to spring *a-b* the region defined by the polygon *a-c-b-d*. Such a partition of space as an interpretation of the spring-network model has

[4] We note, without demonstration, that in one-dimension the spring-network discretization is *identical* to a finite-element discretization using linear elements, but is *different* from a finite-difference discretization of the governing differential equation.

[5] subject to being separated by a specified minimum distance.

been previously proposed [15,60] and is similar to methods use to derive finite-volume approximations for fluid flow [50]. If the length of the common Voronoi edge is A, the area of a-c-b-d is $Al/2$. Using (8), we write (12) as a sum over all such areas (assuming constant strain in each):

$$\frac{1}{2}\int_V (2\mu\epsilon_{ij}\epsilon_{ij} + \lambda\epsilon_{kk}\epsilon_{mm})dV = \frac{1}{2}\sum_{i=1}^{N^e}\frac{Al}{2}(2\mu\epsilon_{nt}\epsilon_{nt} + \lambda\epsilon_{kk}\epsilon_{mm}) \tag{13}$$

where, to evaluate the strain in each region, we have switched to a local coordinate system along (n) and tangential (t) to the direction of the spring. Part of the strain can be determined as:

$$u_{n,n} = \frac{(u_n^b - u_n^a)}{l}; u_{t,n} = \frac{(u_t^b - u_t^a)}{l} \tag{14}$$

However, the other two terms needed to define the strain (7), $u_{t,t}$ and $u_{n,t}$, cannot be determined using only two points. This points to the basic flaw in the attempted discretization. To complete the argument, if we disregard the two terms that have not been determined, and introduce particle rotations to make the model rotationally invariant, the energy can now be written as:

$$E = \frac{1}{2}\sum_{i=1}^{N^e}\frac{A}{2l}\left[(2\mu+\lambda)(\delta\mathbf{u}\cdot\mathbf{n})^2 + 2\mu\left(\delta\mathbf{u}\cdot\mathbf{t} - l(\omega^b+\omega^a)\right)^2\right] \tag{15}$$

where $\delta\mathbf{u} = \mathbf{u}^b - \mathbf{u}^a$. This is in the form of (3), and suggests that for irregular grids the spring constants should scale as (A/l) when being used to model a homogeneous material.

To relate the continuum (12) and bond-bending (4) expressions for energy, consider in figure 2 the points e,f, and g. Without loss of generality we may fix our coordinates on point e, set its displacement to zero, and choose the x axis to coincide with e-g. Defining the displacement gradient $F_{ij} = u_{i,j}$, we find

$$u_i^f = F_{ij}\alpha_j^f;\; u_i^g = F_{ij}\alpha_j^g \tag{16}$$

where α_j^f and α_j^g are the vectors that connect points f and g, respectively, to point e. In component form:

$$u_{x,x}x^g + u_{x,y}y^g = u_x^g \tag{17}$$
$$u_{y,x}x^g + u_{y,y}y^g = u_y^g \tag{18}$$
$$u_{x,x}x^f + u_{x,y}y^f = u_x^f \tag{19}$$
$$u_{y,x}x^f + u_{y,y}y^f = u_y^f \tag{20}$$

These are four simultaneous equations and can be solved for the unknown components of strain in terms of point displacements and coordinates. For the argument to be made in this section, it suffices to consider the special case of a square lattice, for which:

$$\epsilon_{xx} = \frac{u_x^g}{l}; \epsilon_{yy} = \frac{u_y^f}{l}; \epsilon_{xy} = \frac{1}{2}\left(\frac{u_y^g}{l} + \frac{u_x^f}{l}\right) \tag{21}$$

Then, the contribution to the strain energy made by the region under consideration is:

$$\frac{A}{4l}\left[2\mu\left((u_x^g)^2+(u_y^f)^2+\frac{1}{2}((u_y^g)^2+(u_x^f)^2+2u_y^g u_x^f\right)+\lambda\left((u_x^g)^2+(u_y^f)^2+2u_x^g u_y^f\right)\right] \tag{22}$$

Equation (4) can be expressed in a similar form, and for a square lattice the contribution to the energy from the same springs is

$$\frac{k^n}{2}\left((u_x^g)^2+(u_y^f)^2\right)+c\left((u_x^f)^2+(u_y^g)^2+2u_x^f u_y^g\right) \tag{23}$$

Comparing the discretization (22) with the bond-bending expression (23) we find a close but *not* exact similarity; the term containing $u_x^g u_y^f$ is missing in the bond-bending expression.

The point of these exercises is that discretizations based on geometrical constructions used in spring-networks are possible, but that they differ from the standard spring-network expressions.

3.2.2 Test problems

The previous section has raised some doubts about the validity of standard spring-network expressions as discretizations of continuum expressions for energy. However, alternative discretizations could be found. Moreover, we have not addressed the question of the validity of these models for fracture. To do this in a model-independent manner we propose to use two tests:

1. The ability of an *arbitrary* network to model uniform strain (the *Patch* test of finite-element methods [28]), and

2. The ability to model crack propagation in an isotropic, homogeneous, brittle material.

These we propose as minimum conditions to assure no lattice effects on the results of a computation where the network is disordered or has some structure.

Triangular and random spring-networks have been studied using the beam model. By dividing the Voronoi cells into triangles as shown in figure 2, e.g., *a-c-d* and *b-d-c*, we also obtain a corresponding finite-element mesh based on constant-strain triangles. This allows a direct comparison of the two methods. To conduct the first test, we solve for the internal displacements with uniform applied displacement in the y direction on the boundaries and uniform properties of the material. Plotted in figure 3 are the distributions of local strain. The spring-network model has been solved using the beam model and viscous relaxation; we have not tested the bond-bending case. We plot the distribution of $((u_y^b-u_y^a)/(y^b-y^a))/(\Delta U/L)$ for each spring, where $(\Delta U/L)$ is the macroscopic applied strain. For the finite-element results we plot the distribution of ϵ_{yy} in each element. Also shown are the results for a regular triangular spring-network. It passes the patch test; each spring stretches in accord with

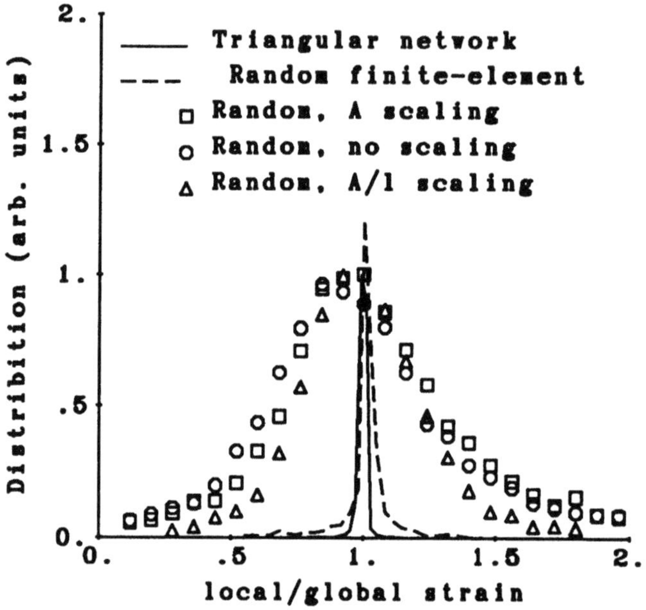

Figure 3: Distributions of local strains under uniform applied displacement.

the macroscopic strain. However, this is a consequence of the symmetry of the network and not a test of the discretization. The random finite-element mesh is able represent uniform strain; the small deviations are due to elements near the boundary. However, the spring-network model for the random grid is unable to model uniform strain whether or not the spring constants are scaled as A/l as suggested by (15). Clearly, the spring-network and its dual finite-element model correspond to different physical materials. We emphasize that this is of concern if the problem of interest is of type **P2**; for problems of type **P1** the distributions presented in figure 3 would represent some basic information about the statistical variation in local deformations. For **P2** problems, if disorder were now introduced onto the spring network, it would be difficult to separate the effect of the lattice from the effect of the disorder on the results obtained. Although we have not tested the bond-bending model, it is worth considering the case of a regular and square bond-bending network. This is incapable of modelling an isotropic, elastic medium, whereas the corresponding finite-element mesh can.

We turn now to the second test. A long crack in a triangular grid, alligned with one of the three preferred directions, models a crack in a continuum well [15]. In particular, the critical energy breakage criterion (5) can be related to the fracture toughness, e.g., for the beam model:

$$E^c \approx \frac{K_{Ic}^2 l^3}{k^n} f(k^n/k^t) \tag{24}$$

which we present without derivation, although it also follows from dimensional arguments. However, it is known that short cracks show anomalous behaviour;

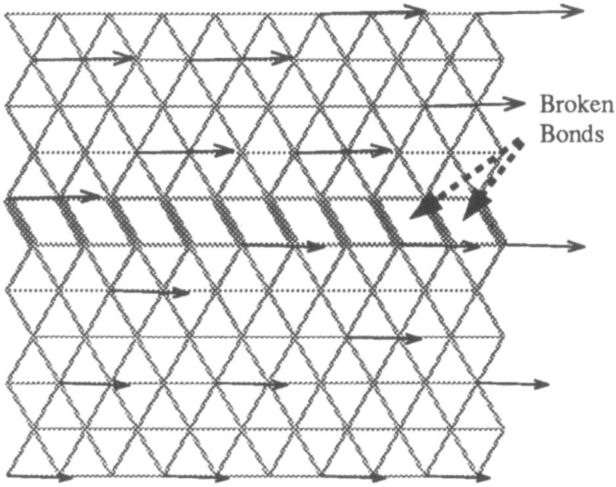

Figure 4: Alternate breakage of bonds in a triangular lattice under far-field shear.

cracks propagate in unexpected directions (see, for example, figure 4 in [15]). Moreover, when the principal strains in a triangular lattice do not lie along the three preferred directions, cracks propagate in an unexpected manner (figure 4, see also figure 2 in [11]). In the finite-element simulation we remove elements when the local energy density exceeds a critical value. Figure 6 shows a comparison of crack propagation in a random spring-network grid and a random finite-element grid. Note that in the random spring-network, the crack leaves behind several unbroken springs that artificially bridge the crack. The finite-element model does not suffer from this problem.

The issue of removal of springs or finite-elements as a model for fracture is clearly a difficult one to resolve, especially for nucleation and growth of short cracks. Our point here is that the finite-element method is capable of solving all the problems of type **P2** that the spring-network method can handle to comparable and usually better accuracy in terms of both the validity of the discretization and the propagation of cracks.

4 Discussion and conclusions

This paper has examined spring-network and finite-element methods for elasticity and fracture. For problems of type **P1**, where the discreteness of the model matches the discreteness of the material, the spring-network models are valid and appropriate. For problems of type **P2**, where the spring-network is being used as a discretization of a continuum, we show that there are deficiencies, both as models of elasticity and fracture. Finite-volume methods can be used to obtain a discretization based on a geometrical grid like that used for a spring-network, but the expressions do not match those of any of the standard spring-network

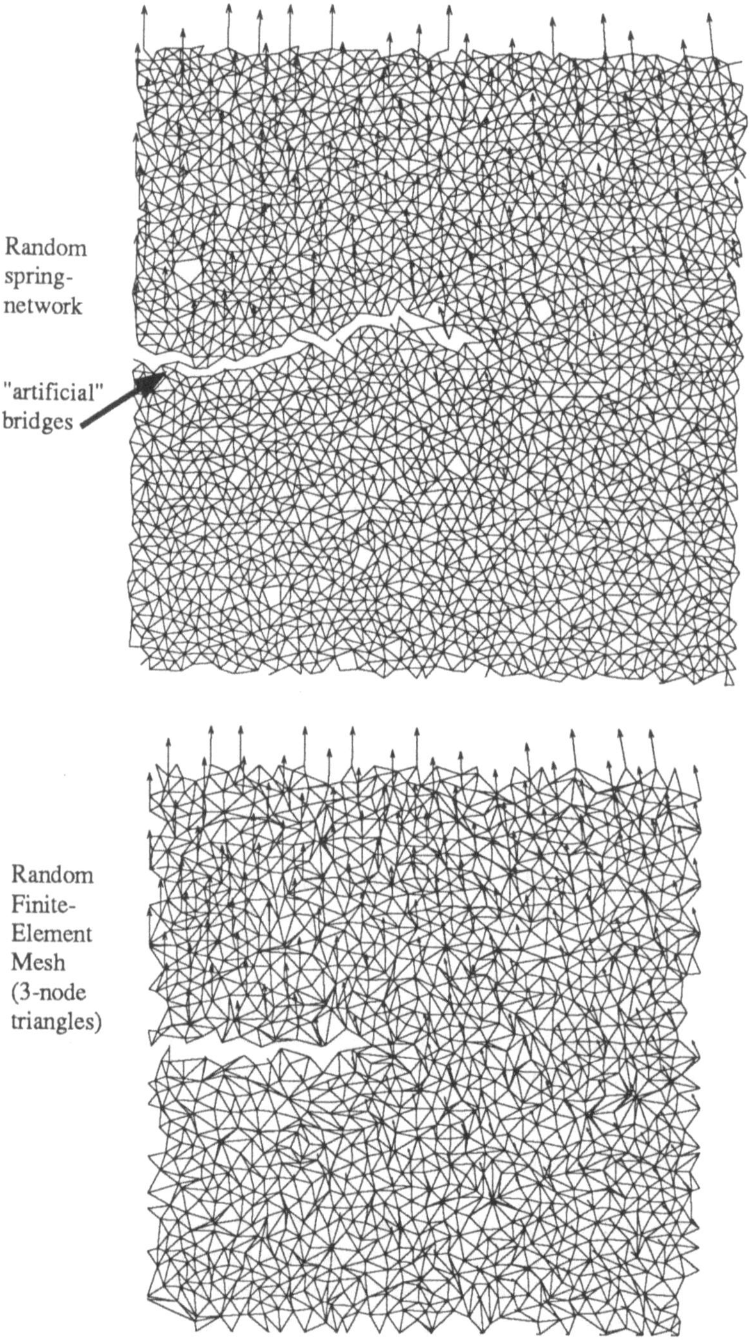

Figure 5: Comparison of crack propagation in a spring-network and a finite-element mesh.

models, nor is it clear that they can be interpreted as springs. Spring-network models show several anomalous fracture modes and must be used with care. This may not be an issue when dealing with large disorder and when the goal is to study universal scaling relationships, but may strongly influence results in other applications. Finite-element methods can be used to obtain discretizations corresponding to any spring-network model, and perform with the same accuracy or better.

We have studied the two models only in the context of isotropic linear elasticity. Finite-element methods are naturally extended to nonlinear material behavior, anisotropy, and different breaking rules. A complete reformulation would be necessary to solve, for instance, plasticity problems using spring-networks notwithstanding efforts to use spring-networks on an *ad-hoc* basis [61,62][6].

Acknowledgements

The authors gratefully acknowledge several useful discussions with Drs. E. Schlangen, W.A. Curtin, and E.R. Fuller Jr.

References

[1] B.R. Lawn, *Fracture of Brittle Solids*, Second edition, Cambridge University Press, New York (1993).

[2] M. Kachanov,*J. Eng. Mech. Div. Trans. ASCE, EM5*, 106, 1039 (1980).

[3] R. Talreja,*Mechanics of Materials*, **12** 165-180 (1991).

[4] A.G. Evans,*J. Am. Ceram. Soc.*, **73**[2]187-206(1990).

[5] J.L. Runyan, S.J. Bennison,*J. Eur. Ceram. Soc.* **7** 93-99 (1991).

[6] P.M. Duxbury, P.L. Leath,*J. Phys. A*, **20** L411-L415 (1987).

[7] P.M. Duxbury, Y. Li,in *Disorder and Fracture*, J.C. Charmet et al. (eds.), Chapter 8, p 141, Plenum Press, New York (1990).

[8] P.M. Duxbury, Y. Li,in *Random Media and Composites*, R.V. Kohn, G.W. Milton (eds.), SIAM, New York, chapter 8, p. 115, (1989).

[9] B.K. Chakrabarti,*Reviews of Solid State Science*, **2** [4] 559-579 (1988).

[10] W.A. Curtin,*J. Mech. Phys. Solids*, **41** [2] 217-245 (1993).

[11] D.J. Srolovitz, P.D. Beale,*J. Amer. Cer. Soc.*, **71** [5] 362-369 (1988).

[12] G.N. Hassold, D.J. Srolovitz,*Phys. Rev. B*, **39** [13] 9273-9281 (1989).

[13] P. Meakin,*Science*, **252**, 226-234 (1991).

[14] S. Roux,in *Statistical Models for the Fracture of Disordered Media*, H.J. Herrmann, S. Roux (eds), (87-114), Elsevier Science Publishers B.V. (North-Holland) (1990).

[15] W.A. Curtin, H. Scher,*J of Mat. Res.*, **5**, [3], 535-553, (1990).

[16] M. Sahimi, S. Arbabi,*Phys. Rev. B*, **47**, [2], 713-722 (1993).

[17] T. Kawai,*Seisan Kenkyu*, (Inst. Ind. Sci., University of Tokyo), **29**, 208-210, (1977).

[18] A. Kikuchi, T. Kawai, N. Suzuki,*Comp. Struct.*, **44** [1-2] 469-480 (1992).

[19] Y. Termonia, P. Smith,*Macromolecules*, **20** 835-838 (1987).

[6] If individual springs are allowed to "yield", the resulting network models anisotropic plasticity.

[20] Y. Termonia, *J. Mat. Sci.*, **24** (1989).
[21] L. Monette, M.P. Anderson, S. Ling, G.S. Grest, *J. Mat. Sci.*, **27** [16] 4393-4405 (1992).
[22] N.J. Burt, J.W. Dougill, *J. Engng. Mech. Div. ASCE*, 103, 365-376, (1977).
[23] E. Schlangen, J.G.M. Van Mier, *Int. J. Dam. Mech.*, **1**, 435-454 (1992).
[24] E. Schlangen, J.G.M. Van Mier, *Mat. Res. Soc. Symp. Proc.*, **278** 153-158 (1992).
[25] W.A. Curtin, H. Scher, *J. Mat. Res.*, **5** [3] 554-562 (1990).
[26] R. LeSar, A.D. Rollett, D.J. Srolovitz, *Mat. Sci. Engng.*, **A155**, 267-274, (1992).
[27] W.A. Curtin, K. Futumura, *Acta Met. Mat.*, **38** [11] 2051-2058 (1990).
[28] O.C. Zienkiewicz, R.L. Taylor, *The Finite Element Method*, Fourth Edition, McGraw-Hill (London) (1989).
[29] S.N. Atluri (editor), *Computational Methods in the Mechanics of Fracture*, North-Holland, New York (1986).
[30] A.R. Ingraffea, V. Saouma, *Fracture Mechanics of Concrete*, G.C Sih, A. DiTommaso (editors), EAFM series, Martinus Nijhoff publishers, Dordrecht, (171-222) (1985).
[31] S. Ghosh, S.N. Mukhopadhaya, *Comp. Meth. App. Mech. Engng.*, **104** 211-247 (1993).
[32] M. Ortiz, S. Suresh, *J. App. Mech.*, **60**, 77-84, (1993).
[33] P.A. Cundall, O.D.L. Strack, *Géotechnique* **29**, 47-65, (1979).
[34] A. Jagota, G.W. Scherer, *J. Amer. Cer. Soc.* (1993), **76** [12] 3123-3135 (1993).
[35] E. Schlangen, *Experimental and numerical analysis of fracture processes in concrete*, Ph.D. Thesis, Technical University of Delft, Delft, The Netherlands (1993).
[36] D. Stauffer, *Introduction to Percolation Theory*, Taylor and Francis, (London and Philadelphia), (1985).
[37] S. Feng, P.N. Sen, *Phys. Rev. Let.*, **52**, [3], 216-219, (1984).
[38] E. Duering, D.J. Bergman, *Phys. Rev. B*, **37** [16] 9460-9476 (1988).
[39] D.J. Bergman, E. Duering, *Phys. Rev. B*, **34**, [11], 8199-8201 (1986).
[40] M. Sahimi, S. Arbabi, *Mat. Res. Soc. Symp. Proc.*, **207** 201-228 (1991).
[41] A.R. Day, K.A. Snyder, E.J. Garboczi, M.F. Thorpe, *J. Mech. Phys. Sol.*, **40**, [5], 1031-1052 (1992).
[42] A. Hrennikoff, *J. App. Mech.*, A169-A175, (1941).
[43] M. Born, K. Huang, *Dynamical Theory of Crystal Lattices*, Oxford University Press, New York (1954).
[44] P.N. Keating, *Phys. Rev.* **145** [2] 637-645 (1966).
[45] W. McGuire, R.H. Gallagher, *Matrix Structural Analysis*, John Wiley and Sons, (New York), (1979).
[46] S. Roux, A. Hansen, *Europhys. Let.*, **6** [4] 301-306 (1988).
[47] R. Thomson, S.J. Zhou, A.E. Carlsson, V.K. Tewary, *Phys. Rev. B*, **46** [1] 10613-10622 (1992).

[48] R. Becker, A. Needleman, S. Suresh, V. Tvergaard, A.K. Vasudevan,*Acta Metall.*, **37** [1] 99-120 (1989).

[49] L. Lapidus, G.F. Pinder,*Numerical Solution of Partial Differential Equations in Science and Engineering*, John Wiley and Sons (New York) (1982).

[50] V. Selmin,*Comp. Meth. App. Mech. Eng.*, 102, 107-138 (1993).

[51] A. Jagota, K.R. Mikeska, R.K. Bordia, *J. Amer. Cer. Soc.*, **73**, [8], 2266-2273 (1990).

[52] M. Ostoja-Starzewski, C. Wang,*Acta Mech.*, **80** 61-80 (1989).

[53] M. Ostoja-Starzewski, C. Wang,*Acta Mech.*, **84** 47-61 (1990).

[54] M. Ostoja-Starzewski,*Mechanics of Materials*, **16** 55-64 (1993).

[55] J.T. Jenkins,in *Modern Theory of Elasticity and Applications*, J. Wu, T.C.T. Ting, D.M. Barnett (eds.), SIAM, Philadelphia, 368-377 (1991).

[56] C.S. Chang, C.L. Liao,*Int. J. Sol. Struct.*, **26**, [4], 437-453, (1990).

[57] C.T. Sun, T.Y. Yang,*J. Elast.*, **5** [1] 45-58 (1975).

[58] Z.P. Bažant, M. Christensen,*Int. J. Sol. Struct.*, **8**, 327-346, (1972).

[59] F.P. Preparate, M.I. Shamos, *Computational Geometry, An Introduction*, Springer-Verlag, Berlin (1985).

[60] M. Ostoja-Starzewski,*Mechanics Research Communications*, **14** [4] 255-262 (1987).

[61] R. Gibala, A.K. Ghosh, D.C. Van Aken, D.J. Srolovitz, A. Basu, H.Chang, D.P. Mason, W. Yang,*Mat. Sci. Eng.*, **A155** 147-158 (1992).

[62] S. Roux, A. Hansen,*J. Phys. II, France*, **2** 1007-1021 (1992).

Ginzburg-Landau Form Description for Steps on Creep Curves

Mulugeta Bekele[1,2] and G. Ananthakrishna[3]

[1] Department of Physics, [3] Materials Research Centre, Indian Institute of Science, Bangalore 560 012, INDIA
[2] On Leave From Department of Physics, Addis Ababa University, Addis Ababa, Ethiopia

Abstract: We consider the model proposed earlier by us for explaining the phenomenon of jumps on creep curves[1]. The model consists of three types of dislocations namely the mobile, the immobile and those with clouds of solute atoms and some transformations between them, leading to a coupled set of nonlinear differential equations for the densities of the dislocations. The model reproduces a large number of experimentally observed features [1,2]. The mathematical mechanism has been shown to be Hopf-bifurcation with respect to several physically relevant drive parameters. The earlier analysis had demonstrated the existence of a pair of complex conjugate slow modes and a fast mode [2]. Here, we present a mathematical analysis of adiabatic elimination of the fast mode and obtaining a Ginzburg-Landau form representation of the order parameter beyond the Hopf bifurcation point upto quintic terms [3]. The present analysis shows that over two ranges of one of the parameters we find a first order transition consistent with our earlier approximate solutions [2]. These ranges are on either side of the mid range of the parameter where we find a second order transition. This method also allows us to map the theoretical parameters to the experimentally observed macroscopic quantities.

1. Introduction

There are various forms of instabilities in plastic flow of which conceptually simplest form [4,5] arises when the material is subjected to a creep test wherein a force is applied and the response in the form of elongation of the material is measured. Under normal conditions, there is no instability and one finds a smooth strain-time curve. However, under certain metallurgical conditions, one sees steps on creep curve suggesting a form of instability [4-6]. A better known form of the instability, but conceptually more difficult, appears when the material is deformed under tensile deformation [4-6]. Here, the material is subjected to a predetermined response namely a constant rate of deformation, and the force or the stress developed in the sample is sought to be measured. Here again under normal conditions, one finds a smooth stress-strain curve. When the system is in the regime of instability, stress-strain curve exhibits several load drops. Each of the load drops is related to the formation and propagation of dislocation bands [7]. It is in the latter type of testing where the plastic instability manifests easily. The phenomenon is refered to as the Portevin-Le Chatelier effect (PLC) or the jerky flow. It is seen only in a window of strain rates and temperature. The phenomenon is seen in several metals such as commercial aluminium, brass, alloys of aluminium and magnesium [8]. The phenomenon of steps on creep curve, which is the subject of the present discussion is seen in much fewer instances and arises only in a window of temperature and solute atom concentration, although both these instabilities have a common physical origin [1,2,4-6,11]. The microscopic origin of the instability lies in the interaction of dislocations with mobile point defects and is refered to as dynamic strain ageing. This induces a negative strain rate charecteristic. The basic idea was formulated by Cottrell [9]. In most of the present theories, the negative strain rate sensitivity is used in one form or the other [10]. None of the models attempt to understand the emergence of the full temporal oscillatory behaviour of the instabilties. In this sense all the known phenomenological models are static. Until a decade ago, there were no models which looked at the problem from the point of view of far-from-equilibrium transition, wherein the new temporal order could be described as a cooperative phenomenon [1,2]. Further, no effort has been made to understand the connection between the interaction of the defects participating at the microscopic level and the macroscopically measured quantities such as the onset of jerky flow at a critical strain, the strain rate range over which it occurs etc. An attempt to understand the problem in the above perspective was made by our group several years ago [1,2,11]. We proposed a theory involving three types of dislocations and some transformations between them. Even though the spatial inhomogeneous structure was ignored and only the temporal oscillatory state was sought to be described, the model and its extensions to the case of constant strain rate test, proved to be very successful in that it could explain most of the experimentally observed features [1,2,11]. The basic idea could be summarized by stating that limit cycle solutions arise due to nonlinear interaction between three different types of dislocations, suggesting a new mathematical mechanism for the instabilty. Description of the phenomenon which includes the initiation and propagation of the bands during the PLC has also been recently attempted [12].

Since the introduction of bifurcation theory into this field several years ago by our group [1,2,11], several other groups have also undertaken similar lines of attack [7,12,13]. In the process, we feel that finer aspects of the dynamical systems have been glossed over.

For instance, one often finds that causual remarks are made about fast and slow modes without actually going through the proceedure of demonstrating the existence of such modes and their elimination. Under adiabatic elimination, the resulting modes which serve as order parameter variables are very complicated functions of the original modes. Yet, hand waiving arguments have been used in building models which we believe are technically suspect.

The purpose of this article is at least two fold. First, we wish to demostrate the technical aspects of adiabatic elimination and then derive the order parameter equation having the time dependent Ginzburg-Landau form. Second, to relate the theoretically introduced quatities to the macroscopically measured quantities. The latter will help us establish correspondence between the two. It should also help us to pick out the effect of various microscopic mechanisms on the macroscopic quantities. In addition, we can compare these results with the approximate solutions for the limit cycles obtained earlier [2].

2. A Model for Steps on Creep Curve

We start with a brief summary of the model. The details of the model can be found in the original references [1,2]. The model consists of the mobile dislocations and the immobile dislocations and another type which mimics the Cottrell's type, which are dislocations with clouds of solute atoms. Let the corresponding densities be N_m, N_{im} and N_i respectively. The rate equations for the densities of dislocations are:

$$\dot{N}_m = \theta V_m N_m - \beta N_m^2 - \beta N_m N_{im} + \gamma N_{im} - \alpha N_m \,, \tag{1}$$

$$\dot{N}_{im} = k\beta N_m^2 - \beta N_{im} N_m - \gamma N_{im} + \alpha' N_i \tag{2}$$

$$\dot{N}_i = \alpha N_m - \alpha' N_i . \tag{3}$$

The dot here refers to the time derivative. The first term in Eq.(1) is the rate of production of dislocations due to cross glide with a rate constant θ. V_m is the velocity of the mobile dislocations which in general depends on some power of the applied stress σ_a. The second term refers to two mobile dislocations either annihilating or immobilizing with rate constants $(1-k)\beta$ and $k\beta$ respectively (see $k\beta N_m^2$ in Eq. 2). The third term also represents the annihilation of a mobile dislocation with an immobile one. The fourth term represents the remobilization of the immobile dislocations due to stress or thermal activation (see γN_{im} in Eq. 2). The last term represents the immobilization of mobile dislocations either due to solute atoms or due to other pinning centers. α refers to the concentration of the solute atoms which participate in slowing down the dislocations. Once a mobile dislocation starts acquiring solute atoms we regard it as a new type of dislocation, namely the Cottrell's type N_i. This process is represented as an incoming term in Eq.(3). As they acquire more and more solute atoms they will slow down and eventually stop the dislocation entirely. At this point, they are considered to have transformed to N_{im}. This process is represented by the loss term in Eq.(3) and a gain term in Eq.(2). These equations can be cast into a dimensionless form by using scaled variables:

$$
\begin{aligned}
x &= N_m(\beta/\gamma), y = N_{im}(\beta/\theta V_m), z = N_i(\beta\alpha'/\gamma\alpha), \tau = \theta V_m t, \\
a &= (\alpha/\theta V_m), b_0 = (\gamma/\theta V_m), \text{ and } \mathrm{c} = (\alpha'/\theta \mathrm{V_m}),
\end{aligned} \tag{4}
$$

as

$$\dot{x} = (1-a)x - b_0x^2 - xy + y, \tag{5}$$

$$\dot{y} = b_0[kb_0x^2 - xy - y + az], \tag{6}$$

$$\dot{z} = c(x-z). \tag{7}$$

Eqs.(5- 7) are coupled set of nonlinear equations which support limit cycle solutions for a range of parameters a, b_0 and c, all of which are physically relevant [1,2]. a refers to the concentration of the solute atoms, b_0 refers to the reactivation of immobile dislocations and c to the time scales over which the slowing down occurs. The dependence on stress and temperature appears through V_m. The instability domain of the parameters can be calculated and approximate closed form solutions for the limit cycles can obtained [2]. Using the Orowan equation which relates the rate of change of strain($\dot{S}$) to dislocation density and the mean velocity: $\dot{S} = bN_mV_m$, with b as the Burger's vector, steps on the creep curve follow automatically since the densities of dislocations are oscillatory. Several experimental results are reproduced[2].

3. Reductive Perturbative Approach

For the sake of completeness, we outline the reductive perturbative approach to problems of formation of new states of order in far-from-equilibrium situations [16,17]. Transitions occuring in these systems are quite analogous to equilibrium phase transitions [18]. The general idea is to construct a 'potential like function' for the 'order parameter' like variable in the neighbourhood of the critical value of the drive parameter. This then would permit the use of the methods developed in equilibrium phase transitions for further analysis. For the sake of clarity, we will deal with the Hopf bifurcation. In this case, we know that near the critical value of the drive parameter, a pair of complex conjugate eigenvalues and another real negative eigenvalue exist for the linearized system of equations around the steady state. The instability sets in once we cross the critical value of the drive parameter. Consider the situation as we approach the critical value from below. In this case, the real part of the complex eigenvalues approach zero value and hence these modes get slower and slower. In contrast, the effect of the change in the drive parameter on the real negative eigenvalue is negligible. Thus, this eigenvalue corresponds to a fast mode. For this reason, the slow modes are the ones which determine the formation of new states of order. Equivalently, these modes enslave the fast modes. The reductive perturbative method is a method where the ensalving is carried out in a systematic way. The method involves in first finding the eigenvectors corresponding to the slow modes and expanding the nonlinear terms in power series of the deviation from the critical value of the drive parameter. The procedure is illustrated below. In this particular case, it just happens that there is a less need to go into the eigenframe of all the three modes since one of the original variables, namely x, itself turns out to be nearly the fast mode. However, we have carried out the full procedure also, which will be reported elsewhere.

Below we collect the pertinent details required for the reductive perturbative approach used for obtaining the Ginzburg-Landau form upto quintic term for the order-parameter variable. We emphasize that it is essential to go to the quintic order due to the fact that we have a first order transition for a range of parameter values. For the rest of the analysis we set $k = 1$. Consider the Eqs.(5-7). It can be shown that there is a domain of instability for the parameters a, b_0 and c. The range of values of b_0 is $\sim 10^{-4} - 10^{-2}$. Thus, we find that $b_0 \ll 1$ and $b_0 \ll a$. Under these conditions we can eliminate one full equation adiabatically. This aspect becomes transparent if the above equations are written in terms of variables which are deviations from the steady state. There is only one steady state defined by:

$$x_a = z_a = \frac{1 - 2a + [(1-2a)^2 + 8b_0]^{1/2}}{4b_0}, \text{and}, y_a = 1/2 \qquad (8)$$

Defining new variables which are deviations from the steady state

$$X = x - x_a, Y = y - y_a, \text{and}\, Z = z - z_a, \qquad (9)$$

Eqs.(5-7) become

$$\dot{X} = -(\alpha X + \chi Y + b_0 X^2 + XY), \qquad (10)$$

$$\dot{Y} = -b_0[\Gamma X + \delta Y - aZ - b_0 X^2 + XY], \qquad (11)$$

$$\dot{Z} = c(X - Z). \qquad (12)$$

Here the dot denotes differentiation with respect to τ and

$$\alpha \equiv a + 2b_0 x_a + y_a - 1, \chi \equiv x_a - 1,$$
$$(13)$$
$$\Gamma \equiv y_a - 2b_0 x_a, \delta \equiv x_a + 1$$

Now, we can rescale the time-like variable by $\tau' = b_0\tau$. Using the dot to denote differentiation with respect to τ' , we get

$$b_0\dot{X} = -(\alpha X + \chi Y + b_0 X^2 + XY), \qquad (14)$$

$$\dot{Y} = -[\Gamma X + \delta Y - aZ - b_0 X^2 + XY], \qquad (15)$$

$$\dot{Z} = \frac{c}{b_0}(X - Z). \qquad (16)$$

It is clear that in the limit of small b_0, $|X| \to \infty$, unless the left hand side vanishes identically. This is equivalent to saying that X has a very small time constant $\sim b_0$. Thus, we can eliminate X in favour of the other two variables. This leads to only two equations with X = X(Y) :

$$\frac{dY}{d\tau} = -b_0[aX + 2(x_a + X(Y))Y - aZ], \qquad (17)$$

and

$$\frac{dZ}{d\tau} = c[X(Y) - Z] \tag{18}$$

where

$$X(Y) \equiv (1/2b_0)\{-(\alpha + Y) + [Y^2 + 2(\alpha - 2b_0\chi)Y + \alpha^2]^{1/2}\}. \tag{19}$$

The two coupled equations found after the adiabatic elimination (Eqs.17-18) have two fixed points: $(0,0)$ and $-[x_a(\alpha - b_0 x_a, 1)]$. The $(0,0)$ is the fixed point of physical interest. Writing these equations as a matrix equation where the nonlinear part appears separately from the linear part, we get

$$\frac{d\vec{R}}{d\tau} = L\vec{R} + \vec{N} \tag{20}$$

where

$$\vec{R} \equiv \begin{pmatrix} Y \\ Z \end{pmatrix}, \tag{21}$$

$$L \equiv \begin{pmatrix} c_0 & ab_0 \\ -c\chi/\alpha & -c \end{pmatrix}, \tag{22}$$

with $c_0 \equiv b_0(a\chi/\alpha - 2x_a)$ and the nonlinear part, $\vec{N}$, given by

$$\vec{N} \equiv \begin{pmatrix} -ab(X + \chi Y/\alpha) - 2b_0 XY \\ c(X + \chi Y/\alpha) \end{pmatrix} = \begin{pmatrix} \sum_{n=2}^{\infty} \xi_n Y^n \\ \sum_{n=2}^{\infty} \zeta_n Y^n \end{pmatrix}. \tag{23}$$

Note that the last expression for $\vec{N}$ in Eq.(23) gives its two elements as as power series expansions in Y with coefficients ξ_n and ζ_n as functions of a, b_0 and c.

We now find the region where the Hopf bifurcation occurs leading to a limit cycle solution for Eq.(20). This is the case when the linear part of the matrix equation drives the system away from the steady state (fixed point) while the nonlinear part acts as a kind of 'stabilizing' agent (system being dissipative) such that the combined effect ultimately leads to a new steady state: a limit cycle. We analyse the stability of the fixed point as a function of the parameter c. The eigenvalues($\lambda_{\pm}$) of the matrix L are

$$\lambda_{\pm} = \frac{1}{2}[c_0 - c \pm \sqrt{((c_0 - c)^2 - 8b_0x_ac)}] \tag{24}$$

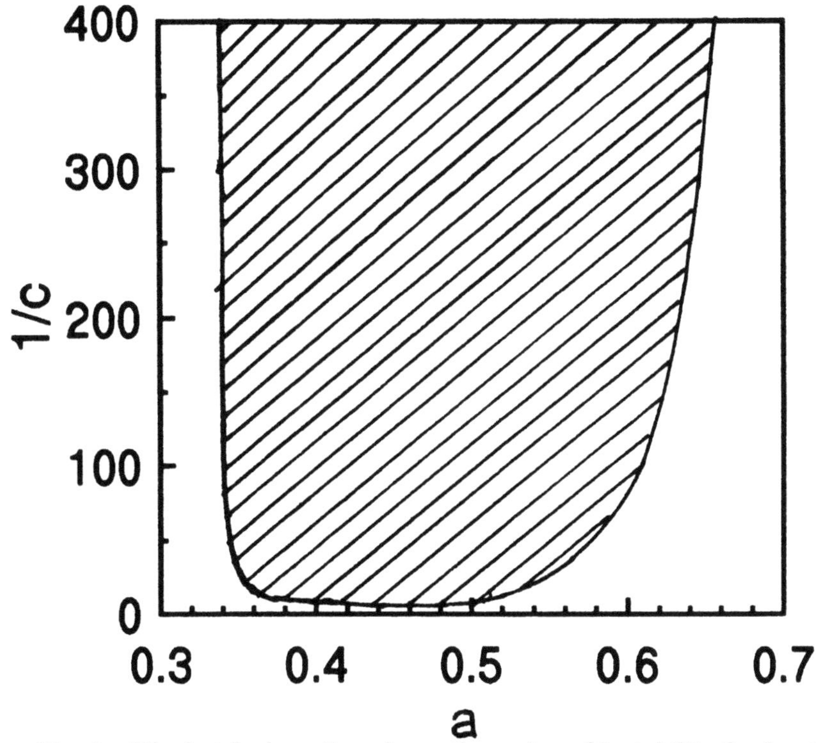

Fig. 1 : The hatched portion shows the region of instability for $b = 10^{-4}$

The fixed point becomes unstable when c is less than c_0 (c is non-negative) and the discriminant of Eq.(24) is negative giving a pair of complex conjugate eigenvalues. This holds when the inequality

$$(3a-1)(1-2a^2) \geq 2b_0(2+a)^2 \tag{25}$$

is satisfied. In particular, for values $b_0 \ll a, b_0 \ll 1$ which were considered in ref.[2], this inequality gives, approximately, the following bounds on a :

$$\frac{1}{3} \leq a \leq \frac{1}{\sqrt{2}}, \tag{26}$$

for fixed value of b_0. The instability region in the $\frac{1}{c}$ versus a plane for a fixed value of $b_0(= 10^{-4})$ is shown in Fig.1.

To get approximate analytical solution of Eq.(20), we follow the reductive perturbative approach [16-18] similar to that used by Mashiyama *et al.* This method is based on the Hopf bifurcation theory that the unstable modes grow continuously out of the critical eigenmodes giving rise to limit cycle solutions. The drive parameter (in our case c) should be taken near the critical value (here c_0). The procedure leads to a time-dependent Ginzburg-Landau(TDGL) equation. From this equation, the nature of the transition to the new steady state (limit cycle), its amplitude and frequency can be calulated.

We choose $c = c_0(1-\epsilon)$ with $0 < \epsilon \ll 1$ and write the matrix L as a sum of two matrices, $L = L_0 + \epsilon L_1$, where L_0 is the matrix L evaluated at $c = c_0$ and

$$L_1 \equiv \begin{pmatrix} 0 & 0 \\ c_0\chi/\alpha & c_0 \end{pmatrix}. \tag{27}$$

The eigenvalues of L_0 are

$$\lambda_{0\pm} = \pm i\omega, \tag{28}$$

where $\omega \equiv \sqrt{(ab_0c_0\chi/\alpha - c_0^2)}$.

We now express $\vec{R}$ as a sum of successive powers of $\epsilon^{\frac{1}{2}}$ and introduce a slowly varying time scale $\tilde{\tau}(\equiv \epsilon\tau)$ replacing $\frac{d}{d\tau}$ by $\frac{d}{d\tau} + \epsilon\frac{d}{d\tilde{\tau}}$. Then Eq.(20) becomes

$$\left(\frac{d}{d\tau} + \epsilon\frac{d}{d\tilde{\tau}}\right)\left(\epsilon^{\frac{1}{2}}\vec{R}_1 + \epsilon\vec{R}_2 + \epsilon^{\frac{3}{2}}\vec{R}_3 + ...\right) =$$
$$(L_0 + \epsilon L_1)\left(\epsilon^{\frac{1}{2}}\vec{R}_1 + \epsilon\vec{R}_2 + \epsilon^{\frac{3}{2}}\vec{R}_3 + ...\right) + \vec{N}. \tag{29}$$

This equation can be solved successively by equating terms of equal power in ϵ on both sides of the equation. Considering $\mathcal{O}(\epsilon^{\frac{1}{2}})$terms, we get the matrix equation

$$\frac{d\vec{R}_1}{d\tau} = L_0\vec{R}_1, \tag{30}$$

whose solution is given by

$$\vec{R}_1(\tau) = Ae^{i\omega\tau}\vec{r} + A^*e^{-i\omega\tau}\vec{r^*} \equiv \begin{pmatrix} Y_1(\tau) \\ Z_1(\tau) \end{pmatrix}. \tag{31}$$

where A and A^*(complex conjugate of A) are the coefficients while $\vec{r}$ and $\vec{r^*}$ are the right eigenvectors of the matrix L_0 , i.e.,$L_0\vec{r}_i = \lambda_i\vec{r}_i$. Similarly, we define the left eigenvectors $\overline{s}$ and $\overline{s}^*$ such that $\overline{s}_i\mathbf{L}_0 = \overline{s}_i\lambda_i$. (Note that $\overline{s}_i\vec{r}_j = 0$, for $i \neq j$). Next, $\mathcal{O}(\epsilon)$ terms of Eq.(29) give

$$\frac{d\vec{R}_2}{d\tau} = L_0\vec{R}_2 + \vec{N}_1, \tag{32}$$

where

$$\vec{N}_1 \equiv \begin{pmatrix} \xi_2 Y_1^2 \\ \zeta_2 Y_1^2 \end{pmatrix}. \tag{33}$$

Let the solution for $\vec{R}_2$ be

$$\vec{R}_2 = Be^{i\omega\tau}\vec{r} + B^*e^{-i\omega\tau}\vec{r^*} \equiv \begin{pmatrix} Y_2(\tau) \\ Z_2(\tau) \end{pmatrix}. \tag{34}$$

Using this expression for $\vec{R}_2$ in Eq.(32), multiplying both sides of the equation by $\overline{s}$ from the left side and rearranging, we get an equation for B:

$$\frac{d}{d\tau}B = \frac{e^{-i\omega\tau}}{\overline{s}}\vec{r}\overline{s}\vec{N}_1 \tag{35}$$

which can be integrated to get the value of B in terms of A, A^* and $\omega\tau$'s.

Returning back to Eq.(29) once again, $\mathcal{O}(\epsilon^{\frac{3}{2}})$ terms give a matrix equation

$$\frac{d}{d\tau}\vec{R}_3 + \frac{d}{d\tilde{\tau}}\vec{R}_1 = L_0\vec{R}_3 + L_1\vec{R}_1 + \vec{N}_2, \qquad (36)$$

where

$$\vec{N}_2 \equiv \begin{pmatrix} \xi_3(2Y_1Y_2 + Y_1^3) \\ \zeta_3(2Y_1Y_2 + Y_1^3) \end{pmatrix}. \qquad (37)$$

Expressing $\vec{R}_3$ as

$$\vec{R}_3 = Ce^{i\omega\tau}\vec{r} + C^* e^{-i\omega\tau}\vec{r^*}, \qquad (38)$$

and following the same procedure as we did to obatin Eq.(35), we get

$$\frac{d}{d\tau}C + \frac{d}{d\tilde{\tau}}A = \frac{e^{-i\omega\tau}s}{s\vec{r}}\left(L_1\left(Ae^{i\omega\tau}\vec{r} + A^*e^{-i\omega\tau}\vec{r^*}\right) + \vec{N}_2\right). \qquad (39)$$

This Eq.(39) brings out the slowly varying behavior of the coefficient A (and A^*). Extracting the equation of the slow varying modes by using compatibility condition, we have

$$\frac{d}{d\tilde{\tau}}A = \mu A + \eta|A|^2 A, \qquad (40)$$

and a similar equation for A^*. Here μ and η are complex coefficients that are functions of a, b_0, and c.

Once again, considering the solution for $\vec{R}$ and remembering that it grows continuously out of the critical eigenmodes, we can express it as

$$\vec{R}(\tau) = \psi(\tau)e^{i\omega\tau}\vec{r} + \psi^*(\tau)e^{-i\omega\tau}\vec{r^*}, \qquad (41)$$

where

$$\psi(\tau) \equiv \epsilon^{\frac{1}{2}}A + \epsilon B + \epsilon^{\frac{3}{2}}C + ... \qquad (42)$$

Using Eqs.(41,42) in Eq.(29) and rearranging after multiplying by $\bar{s}$, we get an equation for ψ of $\mathcal{O}(\epsilon^{\frac{1}{2}})$ given by

$$\frac{d}{d\tau}\psi = \epsilon\mu\psi + \eta|\psi|^2\psi. \tag{43}$$

Eq.(43) is a TDGL equation where ψ is the complex order parameter variable. Its steady state solution gives the amplitude of the order parameter as

$$|\psi| = \sqrt{-\epsilon\frac{\mu_1}{\eta_1}} \tag{44}$$

and its frequency, Ω, (with $\psi = |\psi|e^{\imath\Omega\tau}$) as

$$\Omega = \epsilon\left(\mu_2 - \frac{\eta_2}{\eta_1}\mu_1\right). \tag{45}$$

Note that ψ describes the limit cycle oscillation. In order that the solution should exist η_1 should be negative since μ_1 is positive.

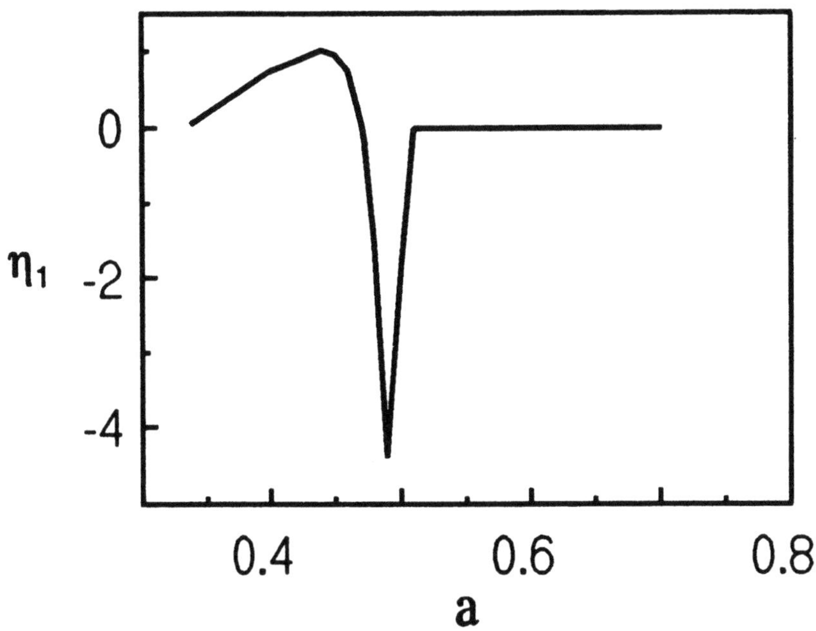

Fig. 2 : Plot of η_1 versus a for $b = 10^{-4}$. Note that the flat region from $a > 0.5$ corresponds to positive η_1 values.

Indeed, there is such a range of a for which η_1 is negative. In this case, since the amplitude of the order parameter grows continuously in proportion to $\epsilon^{\frac{1}{2}}$ from the fixed point to the limit cycle solution, the transition is said to be continuous (or second order transition). The range over which this holds is narrow and is found around $a = \frac{1}{2}$. The value of η_1 as a function of a is shown in Fig.2.

For other values of a (within the instability region),, η_1 is positive. In these cases, the transition is dicontinuous (first order type). One has to then go to higher orders in ϵ to get quintic or even higher terms in the TDGL equation for obtaining an expression for the order parameter. We have carried out the calculation upto $\mathcal{O}(\epsilon^{\frac{3}{2}})$ in the expression for the order parameter and have derived the quintic TDGL equation:

$$\frac{d}{d\tau}\psi = \epsilon\mu\psi + \eta|\psi|^2\psi + \nu|\psi|^4\psi. \tag{46}$$

The steady state solution of this equation, when η_1 is positive, has an order parameter whose amplitude is

$$|\psi| = \sqrt{-\frac{\eta_1}{\nu_1} + \epsilon\frac{\mu_1}{\eta_1}}, \tag{47}$$

and frequency is

$$\Omega = \frac{\eta_1}{\nu_1}\left(\nu_2\frac{\eta_1}{\nu_1} - \eta_2\right) + \epsilon\left[\mu_2 + \mu_1\left(\frac{\eta_2}{\eta_1} - 2\frac{\nu_2}{\nu_1}\right)\right]. \tag{48}$$

In order to obatin an expression for the order parameter from this equation, ν_1 must be negative. Outside the continuous transition range, we found ν_1 to have both negative and positive values. However, in both cases as long as η_1 is positive the order of transition remains the same namely first order.

4. Summary and Discussion

We have carried out the reductive perturbative approach to the problem of steps on creep curve regarding it as a formation of ordered state when the system is driven away from equilibrium. This helps us to obtain the 'potential function' near the point of bifurcation and is obtained from the time dependent Ginzburg-Landau equation for the order parameter ψ. Note that the order parameter represents both the amplitude and the phase of the limit cycle solution when $\epsilon > 0$. Since the above derivation is valid only in the neighbhourhood of the critical value, the expression for the order parameter ψ is valid only for small ϵ. We have shown that both the first order transition as well as the second order transition are seen in the entire range of the parameter a. *While the first order transition seen over most of the values of a is consistent with our earlier calculation [2], we get a second order transition in a narrow mid range of a. The existence of the second order transition was missed by us in our earlier calculation.* Due to the fact that we have a first order transition, we were forced to go to the quintic order in order to obtain the magnitude and the frequency of the order parameter. Thus, this method helps us to study the nature of the transition for the whole range of values of the parameters. From this study, we also see that the model has two tricritical points. The present calculation helps us to map the theoretically introduced parameters to the macroscopic experimentally measured quantities. *For instance, in experiments one measures, the amplitude of the strain jumps on the creep curve as well as their frequency as a function of stress and temperature. The dependence of the amplitude of the strain jumps and its frequency can be evaluated by using the order parameter equations in the Orowan's equation by noting that $|\psi|$ corresponds* *to the amplitude of the strain jumps while the frequency* Ω *gives a measure of the frequency of the steps.* Thus all the three parameters a, b and c can be related to the experimentally measured quantities. This excersize also helps us to demonstrate the complicated dependence of the order parameter variable on the original modes. This will serve as a warning to those using hand waiving arguments for declaring certain modes as fast modes and slow modes in modelling of such of these problems.

Acknowledgement One of us (M.B.) would like to thank the International Program in Physical Sciences, Uppsala University (Sweden) for offering me a fellowship to study at Indian Institute of Science.

References :

[1] G. Ananthakrishna and D. Sahoo, J. Phys.,**D14** 2081(1981).

[2] M. C. Valsakumar and G. Ananthakrishna, J. Phys.,**D16**,1055 (1983).

[3] R. J. Deissler and H. R. Brand, Phys. Rev. Lett.**72**,478 (1994).

[4] E. O. Hall,*Yield Point Phenomena in Metals and Alloys*, (London: Macmillan 1970): J. D. Lubahn and R. P. Felgar, *Plasticity and Creep of Metals* (New York: John Wiley 1961).

[5] T. L. Da Silveria and S. N. Monteiro, Met. Trans., **A10**,1795 (1979): L. N. Zagorukuyko, A. I. Osetskiy and U. P. Soldatov, Phys. Met. Metallogr., **43**, 156 (1977).

[6] B. J. Brindley and P. J. Worthington, Metall. Reviews., **145**, 101 (1970).

[7] K. Chihab, Y. Estrin, L. P. Kubin and J. Vergnol, Scripta. Met., **21**, 203 (1987).

[8] S. R. Bodner and A. Rosen, J. Mech. Phys. Solids.,**15**,63 (1967); P. Penning, Acta Metall.,**20**, 1169 (1972).

[9] A. H. Cottrell, Phil. Mag., **44**, 829 (1953).

[10] See for instance several articles in *Non Linear Phenomena in Materials Science* Ed. L. P. Kubin and G. Martin, (Switzerland:Trans Tech 1988) and other reference therein; *Non Linear Phenomena in Materials Science, Vol II* Ed. G. Martin and L.P. Kubin, L. P. and Martin, G (Switzerland:Trans Tech 1992) and other reference therein; See also the articles in the same volume as Ref. 12.

[11] G. Ananthakrishna and M. C. Valsakumar, J. Phys.,**D15**, L171 (1982) G. Ananthakrishna and M. C. Valsakumar, Phys. Lett., **A95**, 69 (1983) and G. Ananthakrishna in Ref. 10.

[12] G. Ananthakrishna, Scripta. Met. **29**,1185 (1993).

[13] Y. Estrin and L. P. Kubin, J. Mech. Behaviour of Materials, **2**, 255 (1989); L. P. Kubin and Y. Estrin, Acta Metall, **38**, 697 (1990).

[14] See for example E. Aifantis, in Ref. [10]. See also P. Hahaner and L. P. Kubin in Ref. [10].

[15] P. Hahner, Mater. Sci. and Eng., **164**,23, (1993).

[16] Y. Kuramoto and T. Tsuzuki, Prog. Theor. Phys., **54**, 687 (1975).

[17] H. Mashiyama, A. Ito and T. Ohta, Prog. Theor. Phys., **54**, 105 (1975).

[18] P. H. Richter, I. Procaccia and J. Ross, in *Advances in Chemical Physics*, Vol. 43, Eds. I. Prigogine and S. Rice (Wiley-Interscience, New York 1981).

Laboratory Simulation of Dielectric Breakdown

L. Benguigui and P. Ron

Solid State Institute and Physics Department, Technion-Israel Institute of Technology, 32000 Haifa, Israel.

Abstract We realize an experimental analysis of the dielectric breakdown by means of a lattice of resistors and light emitting diodes (LED). The LEDs mimic the behavior of an insulator which cannot stand a voltage higher than a threshold and is transformed in a conductor. We show that the breakdown voltage behaves as the minimum insulation gap. We determined the pattern of the breakdown path made by all the broken insulators, at the breakdown voltage.

1. Introduction

Electrical failures in solids can be classified in two classes: one is the phenomenon of dielectric breakdown and the other is the burning. In the first case, the solid is an insulator and an electric field is created on it. We consider a simple situation of a solid between two parallel electrodes under tension V. If the field $E = V/L$ (L is the thickness of the sample) is low there is no current flowing from one electrode to the other. However, if the field becomes larger than a critical field $E_c = V_c/L$ there is a breakdown: the insulator becomes a conductor and there is a current through the sample. The microscopic mechanism of the breakdown is very important to elucidate, in view of the technological importance of the problem. However, in this paper, we shall only assume the existence of such a critical field regardless of the exact mechanism.

The second problem of the burning concerns a conductor in which a current flows. When one increases the current, one reaches a situation in which the conductor is melted and becomes an insulator. The basic effect for the burning is the heating of the conductor by the Joule effect. In such a problem, the temperature plays an important role and the temperature increase is dependent on several parameters such as the specific heat, the cooling conditions of the sample, etc.

It is clear that in both cases the failure can be influenced by defects which reduce the critical field or the critical current. In the dielectric breakdown, the presence of conductive regions inside the insulator solid has a decisive influence on the voltage that can be applied to the sample until the breakdown occurs. Because the conductive regions (we suppose that the distribution of these inclusions is such that there is no conductive path from one electrode to the other, i.e., the whole sample is insulator at low fields), it is possible that the local field in some point of this inhomogeneous material reaches the critical field before the mean field V/L is equal to E_c. In such a case, there is a local breakdown which may lead to a general failure of the sample. This is because after the first breaking, the local field will increase in another region which will be broken and so on. The breakdown is reduced by the conductive inclusions.

In the case of the burning problem, the defects are insulating regions. For analogous reasons as above (increase of the local current in the conductive channels), it is easy to understand that the total current needed for a complete failure is also reduced by the presence of the insulating regions. There are several studies of the burning problem [1] and I refer the reader to these works. One can find also several tentatives to describe the structure of the sample during and after the breakdown [2] and the connection with fractals.

In this paper, we present the results of an experimental simulation of the breakdown problem in presence of conductive defects. Our simulation represents an original situation which has not been theoretically studied. Preliminary results have already been published in a short paper [3].

2. The Problem

2.1 Description

One considers a lattice of insulating elements located between two neighboring sites. One side of the lattice (of size L) is connected to a voltage V, when the other parallel side is to the ground. Each insulator element can stand a field E_c or a voltage $V_c = E_c a$, when a is the lattice constant, i.e. the distance between two neighboring sites. This lattice is a discrete version of a solid. Since only isotropic solids are considered at the present stage, the chosen lattice has a high symmetry like the square or triangular lattice in two dimensions or the cubic lattices in three dimensions. The breakdown voltage of this lattice is merely MV_c for the square or cubic lattices where $M = \frac{L}{a}$. Once one element is broken, one has to choose how it behaves for $E > E_c$. The simplest choice is that it becomes a conductor – either a superconductor with zero resistance or a regular conductor with a finite resistance. Now, one replaces randomly the insulating elements by resistors with probability p, and one looks for $V_b(p)$, the voltage to be applied between the two electrodes in order to get the breakdown of the whole lattice.

The problem was solved in 2D by means of computer simulations [4,5]. One of the most interesting results[4] is that V_b depends not only on p but also on the size of the sample L. V_b decreases if L increases because the "most dangerous defects" have a larger probability of appearing when the sample becomes larger. If now L is fixed, V_b decreases with p and this decrease is given by $V_b \sim |p - p_c|^\nu$ near the percolating threshold p_c, when ν is the correlation length exponent [4,5]. For $p > p_c$, the sample is a conductor and the breakdown problem is irrelevant. Manna and Chakrabarti [5] calculate $V_b(p)$ for the whole range of $p \geq p_c$. They found that for $p \rightarrow 0$, $(dV_b/dp) \rightarrow \infty$.

Another geometrical approach (to be distinguished from V_b calculated in refs. 4 and 5 where the electrical problem is solved) was proposed for the dielectric breakdown [6]. In consists of looking for the "shortest path" between the two electrodes (minimum insulation gap). This path is defined as follows: one supposes that a walker is able to jump from site to site going from one electrode to another. The shortest path is that for which the walker jumps across the smallest number of insulating elements. This number g can be seen as a measure of the geometrical breakdown. The "shortest path" is not necessarily unique but it is possible that several paths have the property of being the shortest. Examples of a shortest path are shown in Fig. 1. The number g is a function of p and goes to zero for $p \rightarrow p_c$. In refs. 5 and 6 the variation of g with p has been calculated. Near p_c, g goes to zero with the same exponent as V_b. Near $p = 0$, the results of the two groups are different. In particular the slope $(dg/dp)_{p=0}$ is found to have different values in ref. 5 and in ref. 6. In fact the problem studied by Manna and Chakrabarty [6] is not the same as that of ref. 5 and this can explain the difference between the behavior of $g(p)$. In the case of Manna and Chakrabarty the insulating elements are in the bond of the lattice but now the sites are occupied by a conductor (probability p) or not occupied (probability $1 - p$).

2.2 Experimental Realization

The breakdown phenomenon is an irreversible process and in order to measure the breakdown voltage, V_b, it is necessary to increase V beyond V_b. This means that after each experimental realization and determination of V_b, a new sample must be prepared. In such a case the experiments are time consuming and very

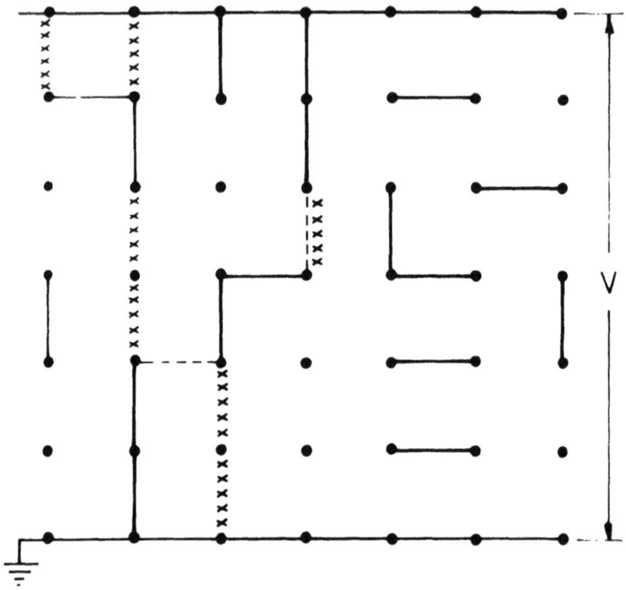

Figure 1: Example of the geometrical shortest path in a lattice of insulating elements (non indicated) and resistors (thick bonds) (- - -). The paths indicated by crosses (× × ×) are shortest paths when horizontal insulating elements are excluded. In the first case $g = 2$ and in the second, $g = 3$.

tedious. Furthermore, a perfect insulating element does not exist and the best we can do is to take an element with a high resistance.

To make the experiments tractable, we need an element which has the following properties: a) it has a high resistance at low voltage, b) for a well defined voltage, its resistance begins to decrease strongly, c) it has a reversible behavior i.e. if the voltage decreases, the element comes back to its initial high resistance. Such an element exists: it is light emitting diode (LED). If one plots the curve $I(V)$ (current versus the applied voltage) in the direction for which the diode emits light, one gets the curve shown in Fig. 2. For V smaller than $V_t \sim 1.15$ volt, the LED is a linear resistor with a resistance of about $10^7\Omega$. For $V \geq V_t$, there is a rapid increase in the current by several orders of magnitude. Afterward, the current increases with V like V^m (($m \sim 3-4$). During the steep increase of the current, the diode begins to emit light. For V equal V_t or slightly larger, the light intensity is low and it is possible to see it only in complete darkness. With the further increase of V, the light intensity also increases.

In Fig. 3, we show the variation of the equivalent resistor (V/I) with the applied voltage. One can see that for $V > V_t$ there is a rapid drop in the equivalent resistance, from $10^7\Omega$ to $10^2\Omega$ for a variation of about 0.4 volt.

Thus the LED can mimic an insulating element which can be broken but with an important feature: the behaviour is reversible with the applied voltage. We have the element we look for and we can do experiments in a very efficient way.

We built a square lattice of 20×20 sites and the bond between two neighboring sites can be made of a LED or a regular resistor (500Ω). The LEDs are put in the same direction: the vertical diodes are perpendicular to the electrodes and their direction gives the direction of the current of the system. In the horizontal direction, they are either to the left or to the right. However, as we discovered in the course of the experiments, the horizontal diodes do not play a role (since they never reach the state of emitting light) and it is possible to omit them without changing the results. We put at random in the bonds of a lattice either a resistor (probability p) or a LED (probability, $1-p$).

We measured the current flowing through the sample as a function of the applied voltage to the whole lattice (Fig. 4). For $p = 0$, (all the bonds are LEDs) $I(V)$ is similar to the curve $I(V)$ for a single LED, except that the current is multiplied by 20 as the threshold voltage, $V_b = 20V_t \approx 23$volts. For $p < p_c$, the curve $I(V)$ exhibits always a well defined threshold voltage V_b, that we identify with the breakdown voltage of the sample. We observe also the strong increase

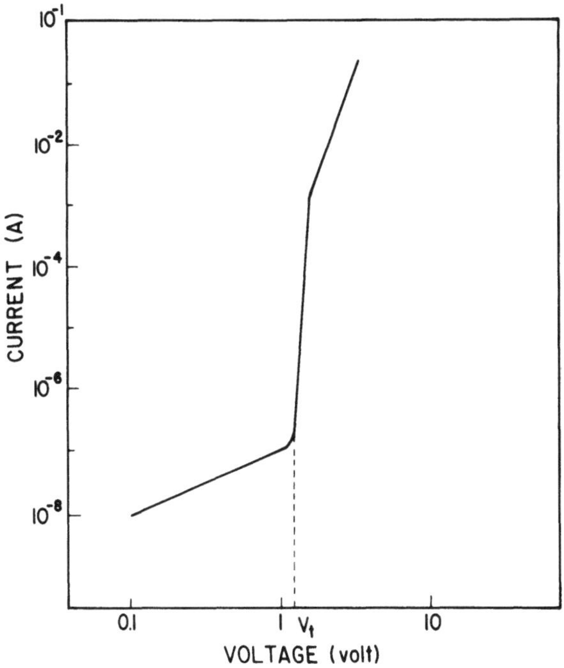

Figure 2: The characteristic current-voltage of a light emitting diode

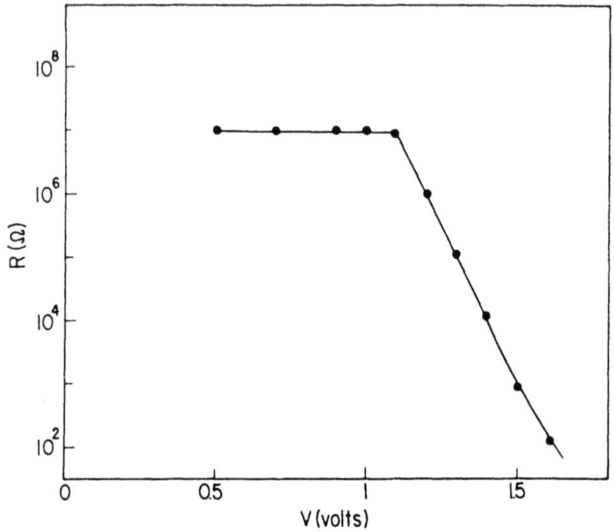

Figure 3: Variation of the equivalent-resistance of a LED with the applied voltage (log-log plot)

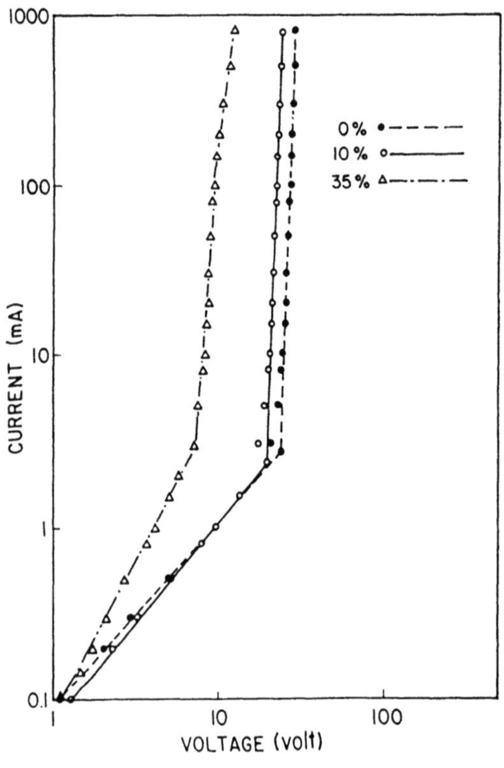

Figure 4: Curve current-tension of the lattice for different values of *p*

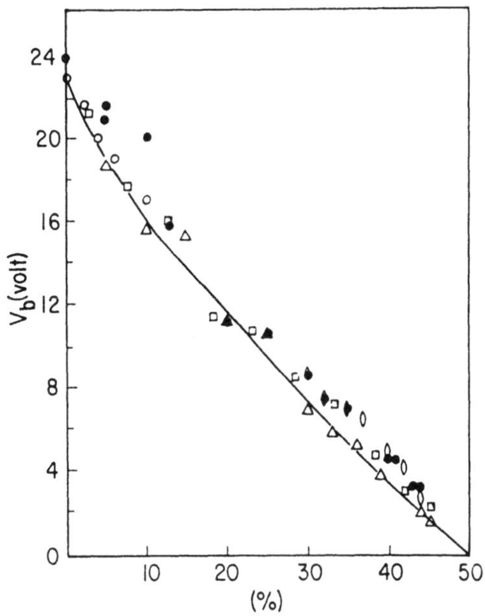

Figure 5: V_b versus p. The different symbols indicate different realizations of the experiment. Thick line: results of ref. 6, normalized to 23 volts for $p = 0$.

of the current for $V > V_b$. The only marked difference with the case $p = 0$ is that the current in the high resistance region is no longer linear with V.

V_b is a decreasing function of p and as expected, it goes to zero when p goes to threshold. In the present case of bond percolation of a 2D square lattice $p_c = 0.5$. In Fig. 5, we plotted the variation of V_b with p.

3. Interpretation of the Breakdown Voltage

We realized a system which mimics very well the breakdown of an insulator. For a well defined voltage V_b, the whole lattice has an important change in the current flowing through it. This shows that the system has two possible states: an almost insulating state below V_b and a conductive one above.

The question we want to give an answer to is: what exactly is V_b? V_b cannot be the breakdown voltage defined in the Section 2. It is because the behavior of a LED is not linear in its emitting state which is the equivalent of the broken state. When a LED is in its emitting state the voltage is approximately constant for important variations in the current. When $V = V_b$, it exists at least one path from electrode to electrode where all the LEDs belonging to this path are in their emitting or conductive state (of course, this path includes also resistors) i.e. the voltage across each LED is near V_t.

We have the possibility of verifying experimentally this assertion. It is possible to observe visually what are the LEDs which begin to emit light when the total voltage is increased. At the same time, we measure the voltage V_b. We always found that $V_b = N_b V_t$, where N_b is the number of emitting LEDs in a conductive path from electrode to electrode. Thus, one could think that V_b is proportional to g, the number of LEDs in the geometrical shortest path, $V_b \sim g$. However, it is not completely correct. It is because we never observed that a horizontal diode emits light. This means that the voltage drop in a horizontal LED never reaches the threshold value V_t. Consequently a conductive path never includes horizontal diodes but only resistors (horizontal or vertical) and vertical LEDs. However, the geometrical shortest path may include horizontal or vertical insulating elements. In Fig. 1 we give an example of the difference between g and $N_b = (V_b/V_t)$, when in the last case the transversal diodes are not taken into account. In general, $(V_b/V_t) \geq g$. In spite of this difference, we shall compare our results with the theoretical results concerning g.

In Fig. 5, we plotted with our experimental results, and the results of the computer simulation from Ref. 6. The results of ref. 6 are normalized to $g = 23$ for $p = 0$. We recall that $V_b = 23$volts is the experimental breakdown voltage for $p = 0$. It is possible to say that there is a fair agreement between the two sets of data. One remarks also that the experimental points are slightly above the theoretical curve, may be because, as explained above, g and (V_b/V_t) are not completely equivalent.

Near p_c we analyzed the results, comparing them with the following expression $V_b \sim |p - p_c|^a$. Since $(V_b/V_t) \geq g$, we expect $a \leq \nu = 4/3$. In Fig. 6, we show the fit of the experimental results. Each point is the average of four realizations and the continuous line is given by $V_b \sim (p - p_c)^a$, with $a = 1.1 \pm 0.05$. a is smaller than ν, but it is very likely that this is due to the small sample we used. To verify this point we measured the resistance of a lattice for which p bonds are resistors and $1 - p$ bonds are without any element. p is taken larger than p_c in order that the sample will be conductive. In this way we realized the conductivity problem near p_c. R is expected to behave like $|p - p_c|^t$ with $t \sim 1.3$. We found again $t = 1.1 \pm 0.05$, smaller than the theoretical value. We believe that the exponent a in our problem is very near ν, but because of the small sample, the values we found are below the theoretical exponent.

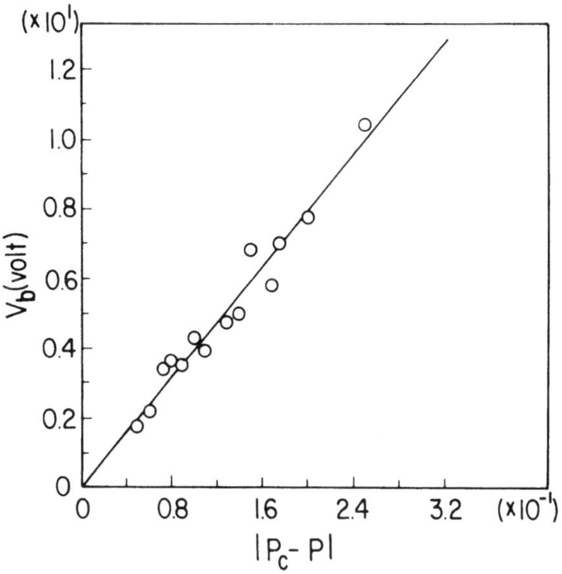

Figure 6: V_b versus $|p_c - p|$ and fit with $V_b \sim |p - p_c|^a$. We found $a = 1.10 \pm 0.05$

4. The Breakdown Path

We define the breakdown path as the path between the two electrodes, which includes continous lines of emitting diodes and resistors. The path may have a complicated pattern and it is a compound of all the ways to pass from one electrode to an another, crossing resistors or emitting diodes.

We proceeded in complete darkness and increased slowly the voltage between the electrodes. We observed visually the emission of light from the diodes and stopped the increase of the voltage when a breakdown path is created. At this state, we can confirm that V_b is very near to $N_b V_t$ when N_b is the number of emitting diodes when one follows a way from one electrode to another, crossing only emitting diodes. From this visual determination of the breakdown path we found that a horizontal diode never emits light and it is the basis of the distinction we made above between $N_b = (V_b/V_t)$ and g. In Fig. 7a-d, we show the breakdown path for different values of p. The non emitting diodes are not represented in these figures, but only the resistors and the emitting diodes. For $p = 0$ (no resistor), it is evident that all the vertical LEDs form the breakdown path. If now, we have only one vertical diode replaced by a resistor, the breakdown path is made by the LEDs belonging to the row where a LED is replaced by a resistor. It is why, at low p (= 5%) the breakdown path consists of some lines which are almost straight.

For $p \leq p_c$ (see Fig. 7d, $p = 45\%$), there are large clusters of resistors, and the breakdown path is made by the LEDs between these large clusters. This explained why the number of LEDs in the breakdown path is small and goes to zero for $p \rightarrow p_c$.

In the intermediate range of p, the breakdown path has a complicated pattern. To the best of our knowledge, there is no theoretical analysis of the breakdown path and the results we bring here are the first determination of the breakdown path.

It is interesting to plot the total number N of LEDs in a breakdown path as a function of p. This number N is larger than N_b, which is the number of LEDs when one crosses the sample passing only on emitting diodes. From the Figs. 7, it is possible to see that N is small at $p \leq p_c$, but relatively large for intermediate values of p. It is clear that we have to exclude to the curve $N(p)$, the portion near $p \sim 0$, since for $p = 0$, the breakdown path is made of all the vertical diodes.

In Fig. 8, we show the variation of N with p, for $p > 5\%$. The important point is the maximum in N, around $p \sim 25\%$. This maximum is related to the fact that for $p \sim 20 - 30\%$, the breakdown path has a branched pattern. For $p \rightarrow p_c$, the breakdown path consists of isolated LED between large clusters of resistors and for $p \sim 0$, it is made of almost straight lines.

Finally we made (with the help of M. Murat) a computer determination of the breakdown path. The insulating elements are chosen with the following property: if the potential drop V_1 is less than 1 volt, this element has a linear resistance of $10^9\Omega$. When the current increases such that V_1 is equal to 1 volt, this voltage is kept constant whatever the current which flows in this element. The resistors have a resistance of 100Ω.

A given voltage V is applied to the lattice and the currents and voltages in the high and low resistors are calculated. V is increased by step until one insulating element has a voltage drop equal to 1 V and it is kept constant in the following calculations. The calculations are made until there is a way to go

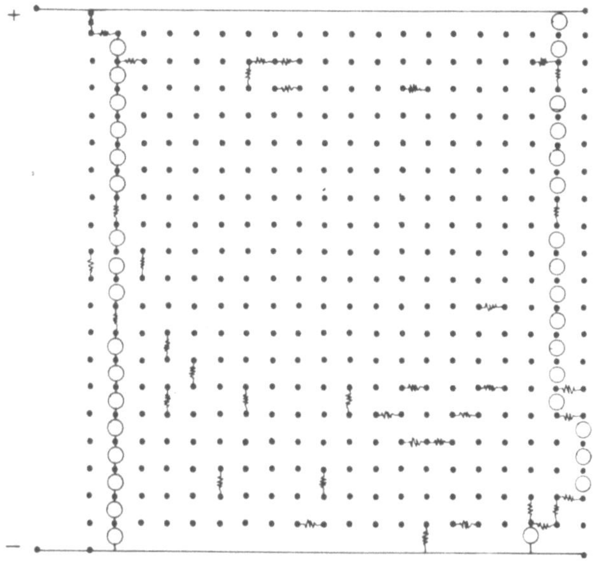

Figure 7a: Breakdown path for $p = 5\%$. The circles represent the emitting diodes.

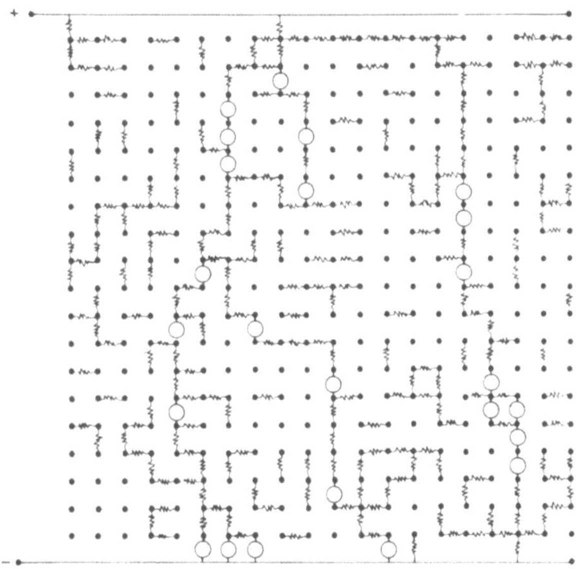

Figure 7b: Breakdown path for $p = 25\%$.

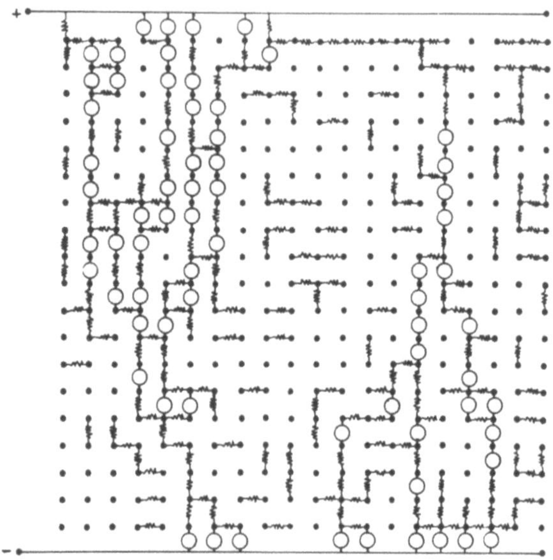

Figure 7c: Breakdown path for $p = 30\%$.

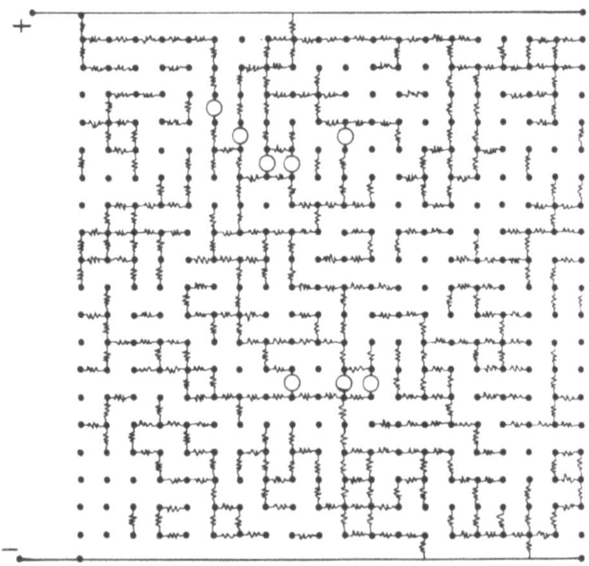

Figure 7d: Breakdown path for $p = 45\%$.

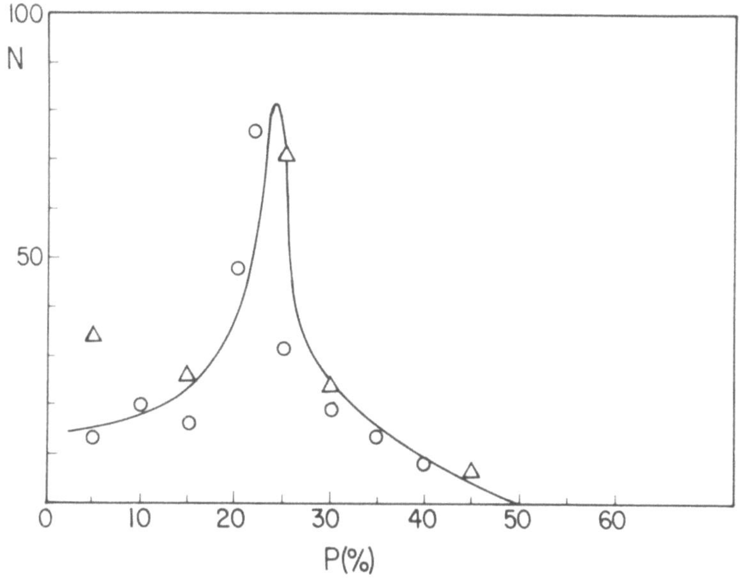

Figure 8: Number of LEDs in the breakdown path versus p.

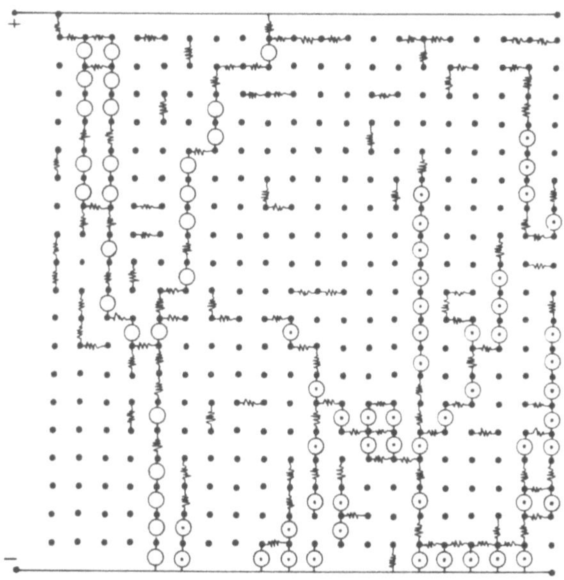

Figure 9: Breakdown path determined by the computer simulation (0). The simulation indicates that there are numerous other broken elements (θ) which do not yet belong to the breakdown path. $V = V_b$, $p = 15\%$.

from one electrode to the other, crossing insulating elements with voltage drops equal to one (broken elements).

We choose a given realization of insulating elements and resistors and we made altogether the measurements and the calculations with this distribution of LEDs-insulating elements and resistors.

In Fig. 9 we show the results. The broken elements determined visually are indicated by circles. The results of the computer simulation give the same breakdown path as that experimentally determined. But there are numerous broken elements which do not belong to the breakdown path. They are indicated in the figure by circles with a point inside. The voltage across all these broken elements is equal to their threshold voltage. We think that the fact that they are not visible in the experiment is due to the very small current flowing through them. However, there are experimental indications of the existence of these broken elements (i.e. elements with voltage equal to the threshold but not visible). Since a LED is a nonlinear element, the fact that the relationship between I and V is not linear below V_b (see Fig. 4 for $p = 35\%$) indicates that some LEDs are already in the broken state. We can conclude then the computer simualtions are in complete agreement with the experiments. At the same time, we see also that there is no cascade effect as in other models of electric failure: the voltage on the sample for which only one element is broken is smaller than V_b, for which the whole sample is broken.

Acknowledgement: The authors thank M. Murat for his help in the computer simulation.

References

1. L. de Arcangelis, S. Redner and H.J. Herrman, J. Phys. (Paris) Lett. **46**, L 585 (1985).
   M. Söderberg, Phys. Rev. B **35**, 352 (1987).
   P.M. Duxbury, P.L. Leath and P.d. Beale, Phys. Rev. B **36**, 367 (1987).
2. L. Niemeyer, L. Pietronero, and H.J. Wiesmann, Phys. Rev. Lett. **52**, 1033 (1984).
   H.J. Wiesmann and L. Pietronero, in "Fractals in Physics", L. Pietronero and Tosatti Ed. North Holland, Amsterdam (1986).
3. L. Benguigui, Phys. Rev. **B 38**, 7211 (1988).
4. P.D. Beale and P.M.. Duxbury, Phys. Rev. **37**, 2785 (1988).
5. S.S. Manna and B.K. Chakrabarti, Phys. Rev. **36**, 4078 (1987)
6. P.M. Duxbury, P. Shukla, R.B. Stinchcombe and J.M. Yeomans, Preprint (1987).

Fracture roughness and physical implications

Stéphane Roux

Laboratoire de Physique et Mécanique des Milieux Hétérogènes, Ecole Supérieure de Physique et Chimie Industrielles, 10 rue Vauquelin, F-75231 Paris Cedex 05, France

Abstract: A large number of studies of fracture surface topography have revealed that for a wide class of materials, and fracture conditions, the surfaces exhibit a self-affine character. Moreover, the roughness exponent which characterizes this self-affinity appears to be universal. In this paper, we analyse some physical consequences of this property. The examples we consider concern the permeability of a crack, a percolation transition through long-range correlated surfaces, the elastic contact between two rough surfaces, and a simple model of dynamic friction.

1. Introduction

We all know from daily experience that fracture surfaces are seldomly perfect planes, or even smooth surfaces. However, a proper description of the statistical properties of the roughness of crack surfaces has only emerged in the past ten years, through the use of self-affinity concepts, and extensive experimental studies. The first pioneering work in this field is due to Mandelbrot, Passoja and Paullay [1].

Since then, a wide corpus of experiments has been published. A recent review covering the statistical modelling of rough surfaces has been written by Meakin [2], with a section concerning crack surfaces. Other recent reviews more focused on fracture have been published in Refs.[3-5].

The primary concern of people interested in this field was to extract from roughness measurement some information on material properties such as the toughness, or on the fracture process itself. However, after some initial encouraging results [1] reporting a correlation between the fractal dimension of the surface and the toughness of the material, further works have reported an inverse correlations, or no correlation at all [6].

Today, after a long list of published results, it appears that some early works have been carried out without paying enough attention to all sorts of bias resulting from the use of different methods of analysis. Therefore reconsidering previously published works with more reasonable estimates of error bars, it appears that the range of variation of the measured roughness exponent is extremely small for most materials and loading conditions which have led to fracture. Moreover this range of variation is well within error bars when they are estimated carefully. Therefore, it has been suggested, first by Bouchaud *et al* [7], and then by others [8] that the roughness exponent might be universal, and only dependent on the space dimensionality of the system. At present, the roughness exponent ζ — defined precisely in the following section — amounts to $\zeta = 0.66 \pm 0.05$ in two-dimensional systems [9-12], and $\zeta = 0.85 \pm 0.1$ in three dimensions [7,8].

Today, two main directions for research in this field can be distinguished: the first one aims at answering the question of the origin of this property. At the present stage, no satisfactory modelling has been proposed in order to justify the value of the roughness exponent, nor even to explain the occurence of self-affinity in a qualitative manner. The second direction consists in trying to explore all physical properties which may result from such a property. The point being to try to go beyong the purely academic description of the statistical properties of the surface and to investigate up to what point, the roughness may controll some physical properties of a fracture. This contribution deals with the second subject.

2. Self-affinity

2.1. Definition

Self-affinity is a particular symmetry which must be considered here in a statistical sense. In the particular case of a fracture surface which is on average parallel to the (x, y) plane, the surface will be said to be self-affine, if it is (statistically) invariant under the following scale transformation [13]

$$\begin{cases} x \to \lambda x \\ y \to \lambda y \\ z \to \lambda^{\zeta} z \end{cases} \tag{1}$$

where λ is any real scaling factor. The above transformation may be more general, introducing different scaling factor in each space direction. However, most studies which have investigated this problem have found that the surface could be considered as isotropic in its mean plane. Let us note that this property is far from being obvious, and that we may have expected that it is possible to "read" from the surface the direction of propagation of the crack front. The fact that the scaling factor in the z direction is a power-law of λ, can be seen as a result of a group structure of the transformation (1). Figure 1 shows an example of an artificially generated self-affine surface with an exponent $\zeta = 0.85$ close to the measured value for real fracture surfaces. The algorithm used to generate the surface is presented in Ref. [14]

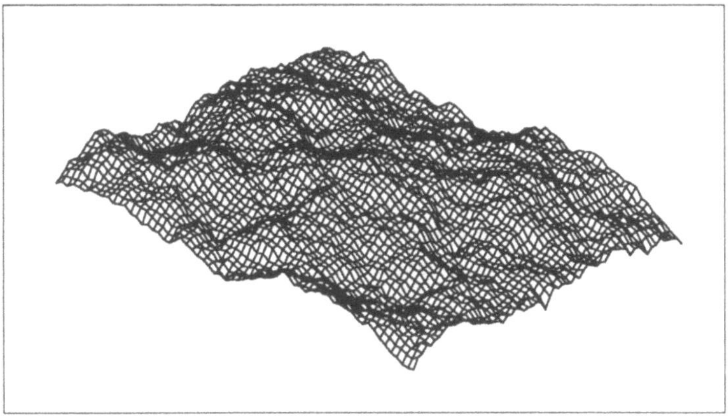

Fig. 1. An example of an artificially generated self-affine surface, with a roughness exponent $\zeta = 0.85$.

One important point to be noted, is that there is no intrinsic length scale attached to the surface. In particular, the roughness — defined as the standard deviation of the height — depends on the size of the window over which it is computed. More quantitatively, the roughness w estimated over a window of size δ scales as $w \propto \delta^\zeta$ as a result of the scaling transformation (1). Thus the concept of a scale independent roughness does not exist !

We skip in this text all details concerning the method to use in order to analyse the self-affinity of a surface. Details can be found elsewhere (see e.g. Ref.[8]). However, we would like to emphasize that any statistical analysis in this domain should be done with a lot of care in particular in connection with the final estimate of the roughness exponent and of error bars associated to it. In particular, different methods should be used simultaneously in order to test the realibility of the result. Some caution has also to be taken for the sample size. A series of data which is too short may bias the result in a drastic way. An easy way to check this effect is to undersample a given data series.

2.2. Geometry of contact between self-affine surfaces

Before getting to the heart of the subject, let us investigate the geometry of contact between two self-affine surfaces. Let us consider two surfaces $z_1(x,y)$ and $z_2(x,y)$. These two surfaces are supposed to be the boundary of two solids, one below the first surface $z < z_1(x,y)$ and the second above $z > z_2(x,y)$. If the two solids are rigid, in order to set a contact between the two surfaces, the second should be translated by a distance h along the z direction in such a way that

$$h = \max_{(x,y)} (z_1(x,y) - z_2(x,y)) \tag{2}$$

The local aperture $a(x,y)$ is then defined as $a(x,y) = h + z_2(x,y) - z_1(x,y)$. Let us note that we consider here only translational degrees of freedom. If the two surfaces are free to rotate, more than one contact point will be necessary in order to have a stable contact (3 contact points are necessary if rotations around all axes in the (x,y) plane are allowed). We will not consider this case further, although the following discussion could be extended to include rotations.

Two frequently encountered cases are to be distinguished. The first case is when the two surfaces are statistically independent. If we assume that both surfaces have nevertheless the same roughness exponent, then the aperture a is also self-affine with the same roughness exponent. However, if we consider the aperture relative to the contact point, translational invariance will be broken. The mean aperture at a distance d from the contact point $\langle a(d) \rangle$ varies with d as a power-law

$$\langle a(d) \rangle \propto d^{\zeta} \tag{3}$$

as shown in Figure 2 [15].

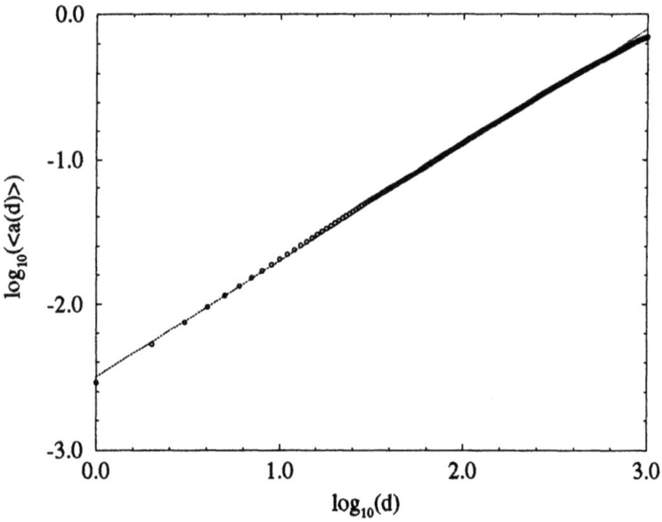

Fig. 2. Mean aperture $a(d)$ at a distance d from the contact point between two independent self-affine surfaces. The data shown are the result of numerical simulations over three dimensional surfaces using $\zeta = 0.8$. The dotted line is a power-law with exponent ζ.

The second case concerns two conjugated surfaces, i.e. the two sides of the same crack, which may have been displaced one from the other, i.e. $z_2(x,y) = z_1(x + d_x, y + d_y)$. Then the two surfaces will be strongly correlated and thus the aperture is no longer a self-affine function. However, at a small scale, (for distances smaller than $||d|| = (d_x^2 + d_y^2)^{1/2}$), one does not yet feel the identity between the two surfaces, and thus the aperture appears to be self-affine with the same roughness exponent as the original surface, z_1. For distances larger than $||d||$, then, the mean aperture becomes independent of the scale of analysis. The standard deviation of the aperture $\sigma_a(\delta)$ over a window of size δ can be cast into the following scaling form [16]

$$\sigma_a(\delta) = \delta^\zeta \varphi(\delta/||d||) \tag{4}$$

where $\varphi(x)$ is a non-zero constant for $x \ll 1$, and $\varphi(x) \propto x^{-\zeta}$ for $x \gg 1$. Above the scale $||d||$, the aperture function is homogeneous, and no longer scale dependent.

3. Permeability

One particularly important geophysical application of the analysis of fracture roughness concerns the problem of fluid flow in cracks [17,18]. The permeability is directly linked to the aperture of a crack which itself is scale dependent.

We consider the situation presented in the last section of a crack which has been subjected to a slip $||d||$. The material is assumed to be undeformable — note however that we will present in Section 5 the problem of the elastic contact which allows to relax this hypothesis. We have seen that the slip $||d||$ can be seen as a correlation length above which the aperture is homogeneous. Therefore it suffices to estimate the permeability over a scale $||d||$. The hydraulic conductance relating the flow to the pressure drop is well-known to be proportional to the cube of the aperture for a simple Hele-Shaw geometry (flow through parallel plane plates). In the case of a self-affine aperture, the same dependence is expected to hold, because of the fact that the relative roughness (roughness divided by the scale over which it is estimated) vanishes for large distances. Therefore, the hydraulic conductance G scales as $\langle a \rangle^3$ or using the result (4), we obtain the following non trivial scaling [16]

$$G \propto ||d||^{3\zeta} \tag{5}$$

The permeability K is usually defined as the hydraulic conductance divided by the mean aperture (and other non-geometric factors such as the viscosity of the fluid). Although this is unpractical to use in practice because the mean aperture is difficult to estimate, we obtain $K \propto ||d||^{2\zeta}$. Therefore, the self-affinity of the crack rules the scaling of the permeability as a function of the slip. This relation has recently been confirmed experimentally on a granite sample [19].

Some other effects may be worth being underlined. Apart from the mean value of the permeability, its statistical variance is of importance. The variance $\sigma^2(L)$ of K over a sample of size $L \times L$ is related to the variance at the scale $||d||$ through

$$\sigma(L) \propto (||d||/L)^{1/2} \sigma(||d||) \tag{6}$$

At the scale $||d||$, the self-affinity of the aperture gives rise to a different behavior: $\sigma(||d||) \propto ||d||^{2\zeta}$ [20]. Thus the variance scales as

$$\sigma(L) \propto L^{-1/2}||d||^{2\zeta+1/2} \tag{7}$$

Again, this is a non trivial manifestation of the self-affine geometry of the crack.

Another important experimental feature to be noted is the following. A self-affine surface is non-differentiable. This means in practice that there exist very sharp kinks on the surface at a micro-scale. Those kinks have a drastic influence on the validity of a linear Darcy's law (i.e. a linear dependence of the flow on the pressure gradient). Indeed, the transition to a turbulent flow will happen for very low pressure (much lower than those necessary to reach this regime in a Hele-Shaw cell). This point has not yet been investigated quantitatively and certainly it deserves a careful study.

4. Percolation transition

Also related to the problem of flow through cracks, the case of two immiscible fluids is extremely important for oil industry. Two-phase flow is a very wide and complex topic partly addressed in three communications of these proceedings [21-23].

One particular case is the capillary driven (slow) invasion of a non-wetting fluid. This case is well described by "invasion percolation" [24]. It consists in successively invading the "pore" with the largest aperture among those available in the neighborhood of the interface. The invaded domain in a homogeneous porous media is identical to the incipient infinite cluster of percolation theory. The latter theory justifies some universal features in this process such as the fractal dimension of the invaded domain, which are unsensitive to the microscopic details of the porous media, when it is spatially uncorrelated.

When the invasion process happens in a crack, the spatial correlation in the aperture of the crack may induce a change of universality class (i.e. the critical exponents may no longer be independent of the microstructure of the medium) [25].

The problem had been considered initially using an ϵ-expansion technique [26]. It has been recently analysed numerically [27]. We have solved exactly the problem on a hierarchical structure [28] and found that the conjectured relation from Refs. [25,26] holds.

The result can easily be stated considering the case of the correlation length exponent ν. It is one of the fundamental critical exponent which gives the divergence of the correlation length ξ as a function of the distance to the percolation threshold $(p-p_c)$: $\xi \propto |p-p_c|^{-\nu}$. In two dimensions, ν is equal to 4/3 [29]. At the percolation threshold, the statistical fluctuations of the effective percolation threshold vanishes with the system size L as $L^{-1/\nu}$. The parameter p is directly related to the aperture of the crack. Therefore, the fluctuation in threshold is to be compared with the roughness estimated over the system size, which has been seen to vary as L^{ζ}. Thus if $\zeta < -1/\nu$, the correlation in the height fluctuations decays faster than the correlation created by the connectedness. In this case, we may neglect the initial surface correlations and thus the problem belongs to the

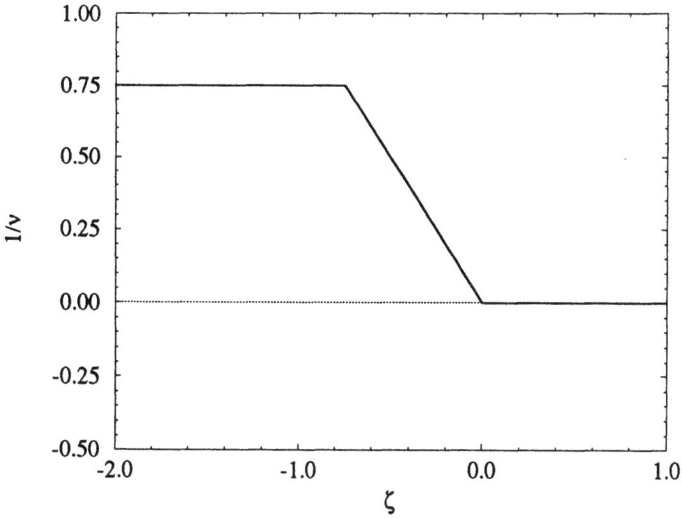

Fig. 3. Inverse correlation length exponent $1/\nu$ as a function of the roughness exponent ζ. For small values, $\zeta < -0.75$, ν assumes its usual value $\nu = 4/3$. In the intermediate range $-0.75 < \zeta < 0$, the correlation length exponent depends explicitly on the roughness exponent $\nu = -1/\zeta$. Finally, in the last regime $\zeta > 0$, there is no longer a genuine transition, and even in the thermodynamic limit, the system behaves as if it was of finite size. Conventionally, $\nu = \infty$.

same class of universality as the standard uncorrelated case. On the contrary, if $\zeta > -1/\nu$, the self-affinity of the surface imposes the long range correlations and an effective ν exponent appears $\nu = -1/\zeta$. This holds as long as the roughness decreases with the system size $\zeta < 0$. In this case, the percolation transition is still a genuine critical point.

For positive roughness exponent ζ, it appears that only the longest wavelength modes are important, and basically, the smaller length scale simply "decorates" the incipient infinite cluster. In this last case, even going to an infinite system size (i.e. with a small length scale going to zero), the statistical properties of the system remain those of a coarser description, only preserving the longest wavelength modes. The system thus behaves as if it was finite, and thus there is no genuine phase transition, and the system is not self-averaging. Figure 3 summarizes the evolution of the inverse correlation length exponent as a function of ζ.

The discussion concerning the correlation length exponent can be extended to other critical features. For instance the fractal dimension of the invaded domain is that of standard percolation (i.e. $D = 91/48$) when $\zeta < -3/4$, it then increases with ζ. For $\zeta > 0$, the invaded domain is compact ($D = 2$).

5. Elastic contact

Let us now consider a different problem, namely the elastic contact between self-affine rough surfaces. This problem has been solved for smooth surfaces by Hertz last century [30]. However, here again the existence of long range correlations in the aperture prior to elastic deformation will play a determining role.

We report here the simple dimensional analysis given in Ref. [16]. Let us mention that a similar analysis has been proposed independently in Ref. [31]. We consider now two independent surfaces with the same roughness exponent ζ. In fact only the aperture matters. These surfaces are the boundary of two linearly elastic solids which may have different elastic properties. We assume no friction at the interface, and we will analyse the elastic penetration δ along the z direction as a function of the normal force F. The area of contact is denoted S.

Let us imagine two geometrical transformations and check the effect of their effect on δ, F and S. The first transformation is simply a homogeneous dilation in all directions (x, y, z) by a factor λ: $x \rightarrow \lambda x$, ... The elastic properties are kept unchanged. The transformation results in

$$\begin{cases} \delta \rightarrow \lambda \delta \\ S \rightarrow \lambda^2 S \\ F \rightarrow \lambda^2 F \end{cases} \tag{8}$$

The first two transformations on geometrical quantities are obvious, and the last one comes from the fact that the deformations are unchanged (dimensionless), and so are the stress, and thus F scales as the area S.

The second transformation is an affinity along the z axis by a factor μ : $z \rightarrow \mu z$, whereas x and y are unchanged. The equivalent to (8) is now

$$\begin{cases} \delta \rightarrow \mu \delta \\ S \rightarrow S \\ F \rightarrow \mu F \end{cases} \tag{9}$$

The latter scaling of F comes from the fact that the contact area is unchanged and that the problem is thus linear. Since the displacement δ is multiplied by μ, so is the response F.

None of these two transformations preserve the self-affinity of the aperture. However, if μ is set equal to $\lambda^{\zeta-1}$, and both transformations are combined, then the surfaces are statistically unchanged. Combining (8) and (9) and eliminating λ gives the following scaling :

$$\begin{cases} \delta \propto F^{\zeta/(\zeta+1)} \\ S \propto F^{2/(\zeta+1)} \end{cases} \tag{10}$$

Numerically, using $\zeta = 0.85$, we obtain $\delta \propto F^{0.46}$ and $S \propto F^{1.1}$. The previous approach permits also to recover Hertz' law, by formally setting the ζ exponent to 2 in order to describe a parabolic aperture for smooth surfaces.

The above results have motivated an experimental study of rough fracture surfaces onto a rubber block limited by a plane surface [32]. The analysis of the force displacement relation first revealed for very small penetration an exponent much closer to the result of Hertz, than the expected result. This in fact can be understood as follows. For small δ, the contact is made through a tiny area, and the material cannot yet feel long-range correlations. Instead, the contacting part of the solid may be described roughly as a paraboloid. However, as the penetration increases, the force-displacement relation could be accurately described by the power-law given in (10).

6. Dynamic friction

As a last example of the importance of fracture roughness, we now address the problem of friction between two rough surfaces. It is obvious that the micro-scale roughness will play a determining role in the way two surfaces can slip onto each other. A recent review on friction is given in Ref.[33]. For static friction, Barton [34] has shown the influence of joint roughness on the resulting friction behavior, however, a proper relation with the self-affine geometry of the surfaces remains to be developped. Part of the analysis should include the elastic punch problem of Section 5, but it should be complemented by the anlysis of erosion and fracture processes of contacting asperities.

We will rather focus on a very simple minded model [35] which address the velocity dependence of friction taking into account the occurence of ballistic flights where the two solids are no longer in contact during some period of time. A similar model has been considered in [36] for a simple periodic geometry, and using a renormalization approach.

We consider two rigid solids limited by independent self-affine surfaces. The lower solid is fixed whereas the upper one is driven at fixed horizontal velocity V along the x axis and subjected to gravity (mass m and gravity g). Moreover, as in section 2.2, we consider that rotations are blocked, and thus only translations are considered. In fact, for large system sizes, rotations can be neglected. They will only introduce correction terms in the scaling form written below. For this reason, we consider the simpler case where rotations are forbidden, and where one single point of contact suffices to determine the geometry. When the two blocks are in contact, the friction at the contact point is locally given by Coulomb's law. Due to the imposed velocity, the upper block may loose contact and undergo a ballistic flight until it hits again the surface. The collision between the two blocks is assumed to be perfectly inelastic so that the two blocks remain in contact and do not bounce.

¿From the construction of section 2.2, we can compute the height $h(x)$ of the upper block necessary to maintain a contact as a function of the displacement along the x axis. The original problem for two-dimensional surfaces can thus be seen as being equivalent to the motion of a point-like mass along a *one-dimensional* fictituous profile $h(x)$. The roughness of $h(x)$ estimated over the sample scale L (i.e. the standard deviation of $h(x)$) is noted A. Furthermore the surfaces and thus the function $h(x)$ is discretized at a scale ℓ. $h(x)$ is piecewise linear over intervals of length ℓ, and continuous.

It is easy to realise that a ballistic flight will start as soon as the slope of $h(x)$ decreases. Depending on the imposed horizontal velocity, the typical flight distance can cover any range from 0 to the system size. It seems reasonnable however to expect a different behavior when the typical flight distance is smaller or larger than the small length scale that appears in our model: ℓ. This allows to introduce a "natural" local scale for the horizontal velocity, U_x, such that the typical ballistic flight distance will be comparable to ℓ. A simple computation leads to $U_x = \sqrt{\ell g/s}$ where $s \approx (A/\ell)(\ell/L)^\zeta$ is a typical slope estimated a scale ℓ. (s is computed by dividing the typical height difference at a distance ℓ, by ℓ). We can now scale the velocity with respect to U_x to obtain the dimensionless reduced velocity:

$$v = V/U_x = V\sqrt{\frac{A}{\ell^2 g}}\left(\frac{\ell}{L}\right)^{\zeta/2} \tag{11}$$

For a given velocity, we can compute the temporal average of the friction coefficient $\varphi(v) = F(v)/mg$. Figure 4 gives the result of numerical simulations for system size $L/\ell = 8192$ and $\zeta = 0.8$. The result is averaged over 100 different surfaces.

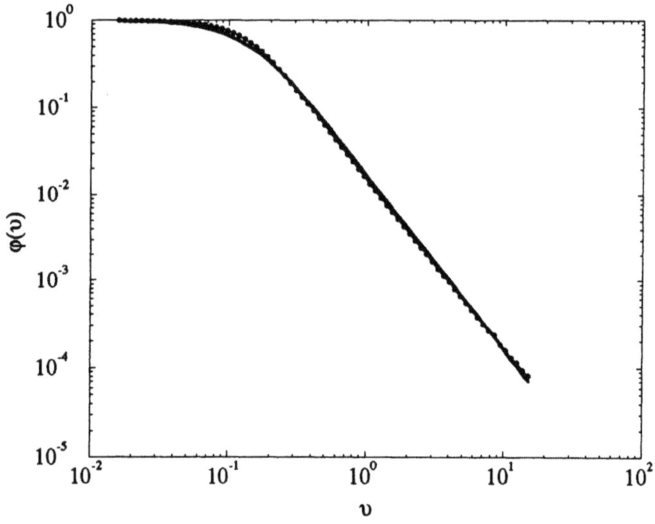

Fig. 4. Evolution of the friction coefficient as a function of the reduced sliding velocity. The solid curve corresponds to the approximation given in Eq. (13).

We clearly see two regimes. For low velocities, $v \ll 1$, the friction coefficient is roughly constant, whereas for large velocities, $v \gg 1$, it decreases algebraically with $\varphi(v) \propto v^{-2}$.

In order to understand this velocity weakening regime, we note that when the blocks are in contact, the friction they feel is velocity independent. Therefore, the effective friction coefficient should simply be proportional to the fraction of the time where the two blocks are in real contact,

$$\varphi(v) = \varphi(0)\frac{t_{contact}}{(t_{contact} + t_{flight})} \tag{12}$$

In fact $t_{contact}$ is roughly equal to ℓ/V (up to a constant factor) since it corresponds to the occurence of the next kink where the slope $h(x)$ decreases. On the other hand, the typical time of flight can be estimated to be proportional to $t_{flight} \approx sV/g$. Inserting those two dependence in Eq. (12) gives the observed v^{-2} dependence at large velocities. It is important to note that the self-affine nature of the surfaces only enters in the definition of the reduced velocity but not in the form of the velocity weakening.

It is possible to generalize the problem to include a different type of loading, such that the upper block is pressed onto the lower one by a spring of constant stiffness. We refer the reader to Ref.[35] for details about this case. Let us simply mention that we then observe a new high velocity regime where φ crosses over to a v^{-1} dependence.

This model is extremely simplistic. More than being an accurate description of a given experiment, it provides us with a limit case and shows the effect of partial temporal contact due to the roughness of the facing surfaces under an imposed velocity.

7. Conclusions

We have tried in the few examples discussed in the above sections, to show that the self-affinity of cracks may give rise to unexpected scaling laws some of which have been observed experimentally but did not receive a satisfactory explanation.

We are far from having exhausted the list of different applications where roughness of the interface may play a determining role. We strongly believe that exploring those properties in the spirit of a simple dimensional analysis is a very rich field, with many practical applications, in particular for geophysics.

Acknowledgements

Most of the material presented in this contribution has been obtained in collaboration with T. Engøy, A. Hansen, J. P. Hulin, P. Kurowski, K. J. Måløy, F. Plouraboué, J. Schmittbuhl and J. P. Vilotte. We acknowledge the support of the GdR "Physique des Milieux Hétérogènes Complexes" of the CNRS.

References

1. B. B. Mandelbrot, D. E. Passoja and A. J. Paullay, Nature **308**, 721, (1984)
2. P. Meakin, Phys. Rep. **235**, 189, (1993)
3. B. L. Cox and J. S. Y. Wang, Fractals **1**, 87, (1993)
4. E. Bouchaud, *in Proceedings of "Dislocations'93"*, meeting held in Aussois, Fr. (April 1993)
5. R. H. Dauskardt, F. Haubensak and R. O. Ritchie, Acta Met. Mat. **38**, 143, (1990)
6. D. L. Davidson, J. Mat. Sci. **24**, 681, (1989)
7. E. Bouchaud, G. Lapasset and J. Planés, Europhys. Lett. **13**, 73, (1990)
8. K. J. Måløy, A. Hansen, E. L. Hinrichsen and S. Roux, Phys. Rev. Lett. **68**, 213, (1992)
9. A. Hansen, E. L. Hinrichsen and S. Roux, Phys. Rev. Lett. **66**, 2476, (1991);
10. C. Poirier, M. Ammi, D. Bideau and J.-P. Troadec, Phys. Rev. Lett. **68**, 216-219 (1992)
11. J. Kertész, V. K. Horváth, and F. Weber, Fractals, **1**, 67 (1993)
12. T. Engøy, K. J. Måløy, A. Hansen, and S. Roux, preprint
13. J. Feder, *"Fractals"*, Plenum Press, (New York, 1988)
14. R. F. Voss, in *Fundamental algorithms in computer graphics*, ed. R. A. Earnshaw, Springer-Verlag, Berlin, pp 805-835, (1985)
15. S. Roux, J. Schmittbuhl and J. P. Vilotte, preprint
16. S. Roux, J. Schmittbuhl, J. P. Vilotte and A. Hansen, Europhys. Lett. **23**, 277, (1993)
17. K. J. Evans, T. Kohl, R. J. Hopkirk and L. Rybach, Report to Swiss National Energy Research Fund, NEFF Project N° 359, (April 1992)
18. S. R. Brown, J. Geophys. Res. **92**, 337, (1987); *ibid.* J. Geophys. Res. **94**,429, (1989); *ibid.* Geophys. Res. Lett. **13**, 1430, (1986)
19. P. Kurowski, I. Ippolito, S. Roux, M. Rosen, R. Chertcoff and J. P. Hulin, *in Proceedings of "Porous Media"*, D. Houi ed., meeting held in Toulouse, Fr., (June 1993)
20. S. Roux and A. Hansen, preprint
21. D. Salin, these proceedings
22. K. J. Måløy, these proceedings
23. H. Van Damme, these proceedings
24. R. Chandler, J. Koplik, K. Lerman and J. F. Willemsen, J. Fluid Mech. **119**, 249, (1982)
25. M. B. Isichenko, Rev. Mod. Phys. **64**, 961, (1992)
26. A. Weinrib and B.I. Halperin, Phys. Rev. B, **29**, 387, (1984)
27. S. Prakash, S. Havlin, M. Schwartz and H. E. Stanley, Phys. Rev. A, **46**, 1724, (1992)
28. J. Schmittbuhl, J. P. Vilotte and S. Roux, J. Phys. A **26**, 6115, (1993)
29. D. Stauffer, *"Introduction to percolation theory"*, (Taylor and Francis, London, 1985)
30. K. L. Johnson, *Contact Mechanics*, Cambridge Univ. Press, (New York,1985)
31. F. M. Borodich, C. R. Acad. Sci. **316**, 281, (1993)
32. K. J. Måløy, X.-l. Wu, A. Hansen and S. Roux, Europhys. Lett. **24**, 35, (1993)
33. J. T. Oden and J. A. C. Martins, Comp. Meth. in Appl. Mech. Eng. **52**, 527, (1985)

34. N. Barton and V. Choubey, Rock Mech., **10**, 1, (1977)
35. J. Schmittbuhl, J. P. Vilotte and S. Roux, to appear in J. Physique (Feb. 1994)
36. T. Pöschel and H. J. Herrmann, Physica A **198**, 441, (1993)

Rock fracturing by gas loading for well stimulation

L. Petitjean and B. Couët

Schlumberger-Doll Research, Old Quarry Rd, Ridgefield, Connecticut 06877, U.S.A.

Abstract. Wellbore gas loading has been used since the 80's to generate small fractures to overcome near wellbore damage. Two techniques are currently used. The first uses a propellant tool to dynamically pressurize the formation whereas the second store the fracturing energy through the well pressurization before the casing perforation and the fracture initiation. These two techniques are reviewed and laboratory and field results are presented as well as modeling results.

1 Introduction

Since its introduction in 1949, hydraulic fracturing has allowed the petroleum industry to enhance oil and gas producing rates. About 35 to 40 % of all currently drilled wells are hydraulically fractured. To date, the only proved development method for low permeability reservoir has been from Massive Hydraulic Fracturing (MHF) [1]. The fracturing process is a major undertaking involving a large amount of equipment and quantities of fluids and proppants that are pumped into the targeted formation. The desired fracture extension is a function of the reservoir permeability and will impact the amount of material to be injected and obviously the cost and the duration of the treatment. For formations with permeability higher than 1 mD, fracture extension requirements are typically less than 300 m. A typical fracture propagation speed is 0.1 m/s. For lower permeability formations, longer fractures are targeted and the cost of a MHF treatment of 500000 gal can approach half the total well cost [1].

On the other hand a technique that can only bypass the near wellbore damage, for instance by the generation of small fractures, very often can be the optimal economic treatment . Two cost effective methods that have been recently developed use a gas loading technique to store the fracturing energy and to deliver it in a very short period of time compared to the hydro-fracturing technique. The first method that uses a propellant tool to dynamically pressurize the formation [2] can be applied in both open holes and cased wells. Experimental and theoretical studies of this method are reviewed in Section 2 and generally concern open holes unless otherwise stated. The second method can be used only in cased wells as a large portion of the well is filled with gas at very high pressure before the shooting of the perforations [3]. This technique is called overbalanced perforating as the well pressure is well above the in-situ stresses and the formation pressure, and will be reviewed in Section 3.

2 Formation stimulation with propellant tools

In this stimulation method, the wellbore is pressurized with a controlled realease of fluid energy provided by the combustion of propellant tools. These tools use propellants that burn or deflagrate rather than detonate, to tailor the pressure loading of the formation to be fractured. By a proper choice of propellants, multiple radial fractures are created emanating from the wellbore. These fractures are extended by the crack internal pressurization. An extensive amount of theoretical works and experimental efforts, both field tests and laboratory studies, have been devoted to this stimulation technique.

2.1 Sandia National Laboratories well tests

Sandia's well tests were conducted from 1979 to 1988 in the DOE Nevada Test Site (NTS), where mineback permitted direct observation of the amount of fracturing obtained [2,4-6]. Tests were done in a tunnel drilled in a volcanic tuff formation. The average in-situ stress was 10 MPa. Several kinds of propellant were used to vary the well pressure loading rate to investigate its impact on the fracture regime. Measurements of borehole pressure were made during these tests. Post-job measurements include the determination of the fracture extension and the study of the well transmissivity improvement. These field tests were very useful to study the dynamic fracturing process. Indeed, models were not capable of handling the overall problem of both the crack initiation and the crack propagation due to gas pressurization. Similarly laboratory experiments can be inconclusive as realistic boundary conditions are difficult to reproduce or because large rocks are necessary to study the effect of gas pressurization.

The first three well tests (GF1-3) were conducted in 7.6 cm holes drilled in ash-fall tuff and spaced 4.5 m apart. Three different propellants with different burn rates, and therefore different pressure-time profiles, were used, and represent the main difference in the conditions of these three experiments [2]. As shown by Table 1 and by other series of tests, the pressure loading rate has a dominant impact on the fracturing regime. The increase in the loading rate changes the fracturing regime from hydraulic (low rate) to multiple and finally to explosive (high rate). Given the loading rate, either the mass load or the peak pressure have little impact.

Table 1 : Propellant stimulation of Open Hole : tests characteristics and resulting behavior

	Mass load (kg/m)	Peak Pressure (MPa)	Loading rate (GPa/s)	Transmissivity improvement	Fracture Behavior
GF 1	12.0	43	0.6	N/A	Hydraulic
GF 2	12.0	95	138	N/A	Multiple
GF 3	12.0	138	10300	N/A	Explosive
GF 4	9.9	248	428	833	Multiple
GF 7	0.74	76	1030	0.35	Explosive
GF 8	0.89	29	4.1	3.8 - 5.6	Hydraulic (mainly)

Figure 1 summarizes schematically the results of the mineback for tests GF1, 2 and 3. GF2 shows the typical behavior of multiple fractures generated with a gas fracturing technique, the fracture wings in the hydraulic direction being the most developed. A typical dimension of a fracture wing in the most favorable direction is 3 m. In open holes, Cuderman has observed that the major fracture was always parallel to the maximum in-situ stress (the hydraulic direction) [5].

The created fracture network obviously impacts the flow-coupling communication between the well and the formation as was shown by tests GF4,7,8 (see Table 1). Creation of a bi-wing fracture or of multiple fractures can significantly enhance the well transmissivity. On the contrary, an explosive behavior of the tool creates a crushed and compacted zone around the wellbore that reduces the near wellbore permeability and therefore decreases the transmissivity as shown by GF7 in Table 1.

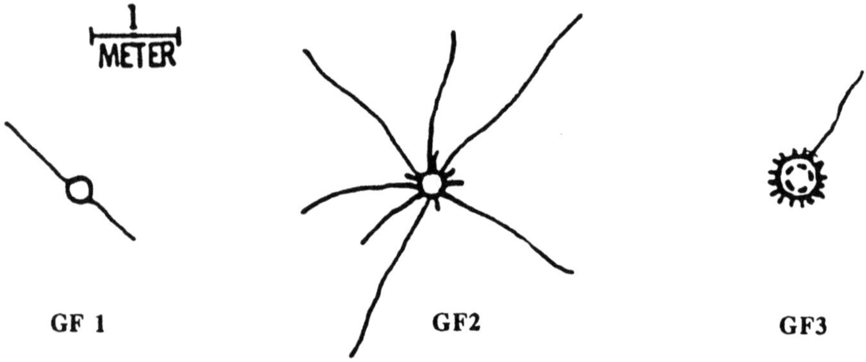

Figure 1 : Schematic of mineback observation [2].

Figure 2 reports the pressure loading and the schematic of the mineback for tests GFSE-2C1 and GFSE-2D1. The modification of the risetime was primarily due to the nature of the propellant : faster propellant was used in 2C1 test. From these tests a typical risetime to obtain either multiple or hydraulic fracturing can be estimated. For

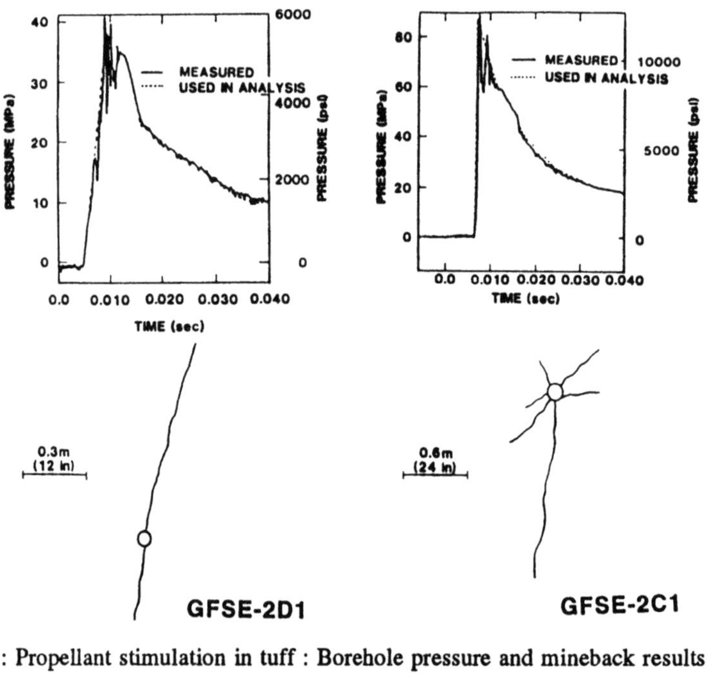

Figure 2 : Propellant stimulation in tuff : Borehole pressure and mineback results [5].

these experimental conditions (i.e. rock type, well diameter), the risetime was 0.11 and 4.0 ms for multiple (2C1) and hydraulic fracturing (2D1) respectively. Note that some ringing effects are superimposed to the pressure curves [5]. These effects have nothing to do with the fracturing process itself. They have been also observed in closed pipe tests of propellant tools [6]. Pressure wave effects for air-filled, long cylindrical vessels are well documented in ballistic litterature. The time scale is of the order of l/c, l being a characteristic length of the system and c the sound velocity. Using l = 0.1 m and c = 1000 m/s, this time scale is of the order of 0.1 ms.

Well tests done in cased and perforated wellbores have shown that the initial direction of the fractures is given by the orientation of the perforations, the fracture surfaces gradually turning toward the hydraulic direction [7]. Figure 3 reports the pressure record and a schematic of the mineback obtained during the test G where the angle between two lines of perforations was 120°. Use of propellant tools in a cased well sets a new challenge to design a safe operation. To obtain the maximum possible fracture extension, tool with high mass of propellant must be used. This mass is such that if no flow -or insufficient flow- occurs through the perforations and within the fractures, the casing could be severely damaged. The primary parameters studied in [7] include the burn rate and the mass of propellant, the presence or absence of liquid in the wellbore, the in-situ stress orientation and the perforation parameters. Among the 8 well tests, 6 have shown casing damage. It points out the importance of having a reliable model to keep the peak pressure below the casing burst pressure while extending the longest possible fractures.

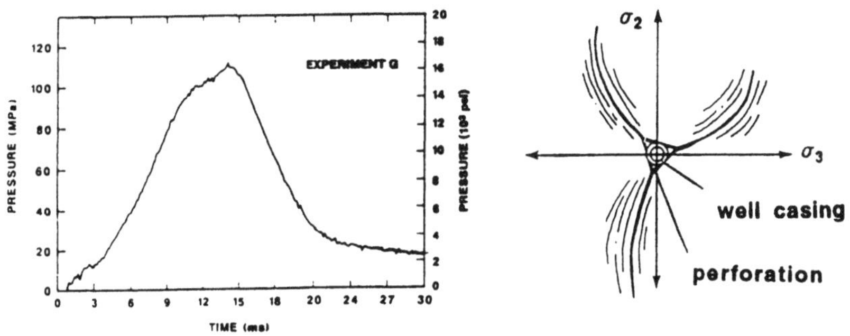

Figure 3 : Test in a cased and perforated well (σ_2 represents the minimum in-situ stress).

2.2 Laboratory investigations

A laboratory experiment was devised by Swift and Kurosov [8] to provide controlled pressurization of the borehole of rock samples at loading in the range of 10 to 10^5 GPa/s. Observations on Nugget sandstone showed that the degree of multiple fracturing depends on the loading rate as previously demonstrated by Sandia : the number of fractures increases with an increase in the loading rate, but is also dependent on transient pore-pressure effects. Although some scaling criteria that can determine the boundaries of the three fracturing regimes can be applied for tuff [4], for instance they are not usable for Nugget sandstone.

Using models made of plexiglas, Fourney *et al* [9] studied both the crack initiation and propagation by high-speed photography and photoelasticity. The camera took 16 photographs during one test at rates between 30,000 and 850,000 frames per second. It showed the instantaneous location of the stress waves, the fracture front and the fluid front; in particular the fluid front continuing to lag behind the fracture front 0.7 ms following the ignition of the propellant. It was reported that the distance between the crack tip and the fluid front increases in absolute terms but decreases in term of percent of total fracture length. Fracture speeds were in the 200-250 m/s range. Propellant charges with a loading rate in the 30-140 GPa/s range produced very little damage was done to the borehole and propogated only 2-4 fractures.

Schatz *et al* [10] have made extensive small scale laboratory experiments with 3/8" borehole in blocks loaded to a true 3D stress state with miniature propellant tools. As reported by previous studies [2,4-8], with an increase of the loading rate, the regime evolves from hydraulic to multiple and finally to explosive. Effects of the confining stresses, of the material (Young modulus, permeability and porosity), of the propellant characteristics on the wellbore pressurization, and of the fracture number and extension, were investigated. As shown by Figure 4, and with constant tool parameters, total fracture length is related inversely to the confining stress. Variation in the material type has also an impact on fracture extension. As described below, with numerical modeling, it was shown that the lower the permeability or the porosity or the higher the Young modulus, the longer the fracture length.

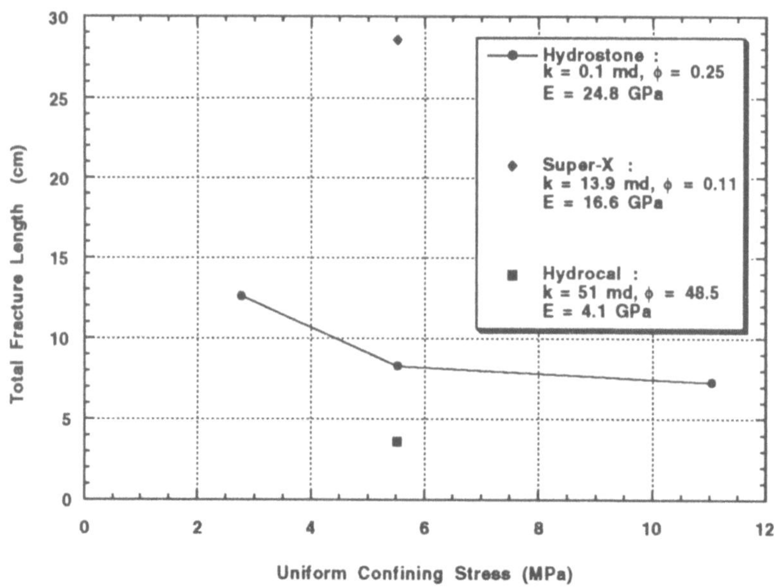

Figure 4 : Laboratory studies of propellant stimulation in 0.95 cm diameter wellbore [10].

2.3 Modeling approaches

The modeling of gas fracturing is a complex problem involving the coupled physics of the non-linear solid deformation, fracture mechanics and fluid flow. In addition, attention is required to consider the in-situ stress state, the compressible nature of the fracturing fluid, the heat exchange between the propellant gas and the surrounding rock, and the seepage losses through the fracture walls. All these phenomena could contribute to the overall technique efficiency.

Most theoretical approaches have focused only on one or two aspects of the process. In several models for instance, the number of fractures to propagate is an input to the model, while the pressure distribution within the fracture is prescribed in others. To determine the fracture extension, these models generally consider the process as a quasi-static one. The problem of fracture process in gas fracturing operations can be viewed as an intermediate situation between rock blasting and hydraulic fracturing. During rock blasting, when the explosive releases its energy, stress waves propagate into the material, crushing the zone around the wellbore and generating a number of radial cracks [11-13]. This phase is a *dynamic phase*, its duration being of the order of hundreds of microseconds [11]. When the pressure pulse dies out, the cracks are submitted to the expansion of the wellbore gas pressure. If the cracks are open to the gas, the internal pressurization will propagate them. This second phase is called the *quasi-static phase* because the fracture speed is small compared to the wave speed in the solid. The importance of this second phase when the major part (80 to 98%) of the explosive energy is being dissipated, has been pointed out by many experimental and theoretical studies [11-13], which concluded that the fracture extension is essentially due to the gas expansion. During the *dynamic phase*, only the immediate vicinity of the wellbore will be affected: crushing can occur but also initiation of cracks from micro-fractures inherent in the rock. This *dynamic phase* can create 10 to 20 fracture wings, but only 4 to 6 will grow during the *quasi-static phase* that determines the fracture extension [11-13]. In contrast, during hydraulic fracturing of vertical wells, a single bi-wing fracture is propagated generally and quasi-static analysis applies. Although a quasi-static analysis is not sufficient to explain the detailed course of the events, it often captures the essential information, like the fracture extension, as long as the number of fractures to be propagated is known. On the other hand this approximation does not allow a prediction of the number of initiated cracks.

2.3.1 Analytical modeling

Many analytical approaches have been applied to simulate multiple fracturing [14-16]. Ouchterlony [14,15] has studied the radial growth of multiple cracks and their interplay. He derived stability growth conditions under quasi-static behavior. He has shown that the uniform growth of all cracks in a radial system is virtually impossible, the non-uniform growth becoming more accentuated with an increasing number of cracks. Hanson *et al* [16] have used a mixed numerical/analytical approach to calculate the stress intensity factor for an arbitrary pressure profile within the cracks. They showed fairly good agreement with Ouchterlony's results. They have considered an even number of equally-spaced radial cracks of alternating lengths and have shown that the propagation of the longer cracks goes faster, these cracks "clampling" the shorter. This can qualitatively explain the results of Sandia well tests where the hydraulic fractures were the longer ones. The stress acting on the crack is lower in the hydraulic direction and then the early propagation of this crack is favored while its extension tends to clamp the others later. Hanson *et al* have also shown that multiple fractures (as opposed to a simple hydraulic fracture) can be propagated if an appropriate pressure is applied without the need for stress wave effects.

2.3.2 Continuum modeling

A continuum model for dynamic ductile and brittle fractures has been developed by Seaman *et al* [17]. The model is based on tensile stress-wave induced activation, and growth and coalescence of inherent flaws in rock. It treats crack densities and orientation in a statistical manner, and therefore does not describe the behavior of an individual crack but gives the average behavior of a crack population. This model has however fourteen independent parameters including nucleation size, nucleation rate, nucleation threshold stress and strain, spoliation criterion, etc., which are difficult or impossible to measure directly. When this model is "calibrated" by fitting the model parameters to the experimental values [18], it gives a qualitative agreement with the experimental data obtained by Sandia in the NTS. Despite the number of parameters used to describe crack nucleation and growth, this type of model is useful to address the initiation phase of the process.

Several simulations have been done by McHugh [13] to compare the crack length increase caused by tensile stress acting alone to the one induced by internal gas pressurization. The crack nucleation and growth model described above [17,18] was used to model the fracturing caused solely by dynamically applied tensile stress. Laboratory tests were also performed with Plexiglas block to verify the computational results. A confining pressure of 6.9 MPa was used. These experiments have shown that the crack length is approximately in agreement with the predictions of the model where the length increase is caused by internal pressurization. The crack length generated by tensile stress acting alone is one order of magnitude lower than the experimental results.

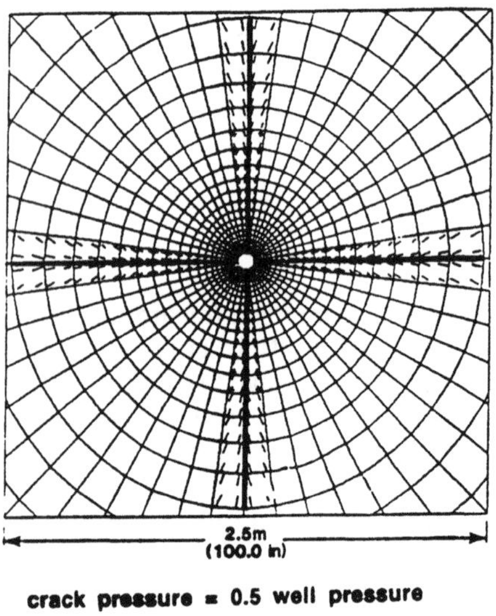

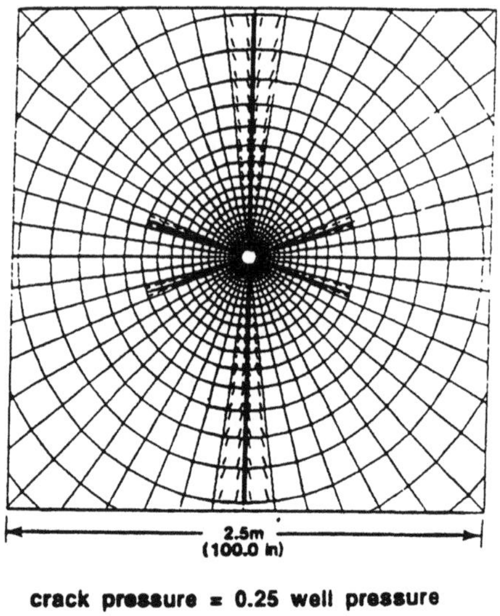

Figure 5 : Predicted fracture pattern for GFSE 2-C1 test. In-situ stress components were 8.6 and 5.4 MPa in the y and x direction respectively [19].

A 2-D model has been derived by Swenson and Taylor that assumes that crack initiation and growth depend on tensile failure criteria [19,20]. Tensile failure is assumed to occur when the maximum principal stress is higher than a specified tensile strength, the crack plane being normal to the maximum principal stress. Elastic approximation is used to adjust the stress components to the presence of cracks. This study demonstrates that the borehole loading rate is a primary parameter determining the fracture pattern, the peak pressure being of less importance. It was also shown that in open hole when a significant stress difference exists (> 3 MPa), the dominant crack is perpendicular to the minimum stress. Another conclusion is that the fracture pressurization plays a major role, both by inhibiting initiation of new fractures and by extending the existing ones. Comparisons with well test data were done. As shown by Figure 2 (dotted line) the well pressurization was set to the experimental data. A qualitative agreement was observed with an ad-hoc fracture pressurization, equal to one-half or one-fourth of the wellbore pressure. Once again it points out the importance of having a good description of the gas flow to determine the pressure profile within the propagating cracks and the wellbore pressure.

2.3.3 Discrete fracture propagation and fluid flow modeling

A 1-D model able of handling both the problem of the fracture propagation and of the well pressurization has been developped by Nilson *et al* [21,22]. The loading rate of the wellbore is controlled by a propellant burn model and by the rate of gas flow into the fractures. Only the quasi-static phase of the process is taken into account, and the number of fractures is therefore an input to the model. Opening displacements are calculated using quasi-static integral representations from linear elastic mechanics. Fracture propagation is described as a mobile-equilibrium process in which the stress intensity factor is maintained to a critical level given by the fracture toughness. Only mode I fracture is considered. To account for multiple fractures the stress intensity is modified as suggested by [14,15]. The formation is supposed to be isotropic, i.e., the cracks have equal length. The flow velocity and the pressure profile are determined by solving the mass, momentum and energy conservation equations. Seepage losses through the fracture walls are taken into account. The momentum equation leaves a balance between the gradient of the pressure and the laminar and/or turbulent frictional forces. The fluid pressure and temperature distribution, aperture profile and borehole pressure are computed in a fully time-marching fashion such that all the transport equations are satisfied. The model was applied to the the test GSFE-2C1 and 2D1 (see 2.1 and Fig 2). Calculations of peak pressure and of pressure decay are well predicted by the model. Fracture extents are within a factor of two with the experimental data as shown by Figure 6 for the GSFE-2D1 test.

Schatz *et al* have developed a model very close to the Nilson *et al* ' s approach [10]. This model was able to reproduce observed laboratory crack lengths fairly well. It was also able to reproduce the parameter sensitivities shown in Figure 4. The most significant role was found to be played by the permeability. Seepage losses can reduce substantially the fracture extension. They have also developed a 2-D model which was carried out with the same fluid flow approach and a displacement discontinuity technique developed by Cornet [23]. The fracture model used by Cornet is based on strain energy rate release to control direction. This 2-D approach is able to account for stress anisotropy and crack curvature. It was shown than the gas pressurizartion can drive crosscutting fractures a substantial distance (typically 15-25 borehole diameters) before curvature effects become significant.

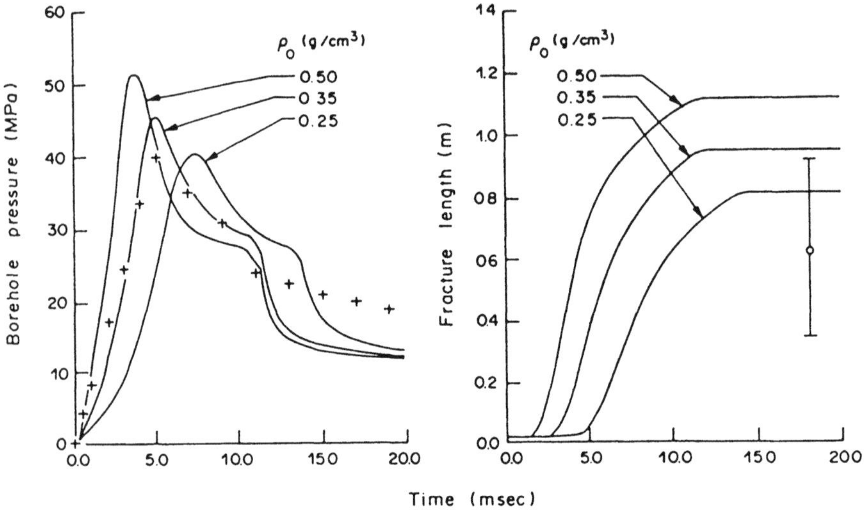

Figure 6 : Predicted borehole pressure and fracture extension for GFSE 2- C1 test [21]. The largest loading density was the one suggested by the experimentalist [5]. Some experimental pressure points are reported with the minimum,mean and maximum fracture wing lengths.

A coupled discrete finite-element fracture model and fluid flow model was developed by Heuze *et al* [24] . The solid and fracture mechanics are solved by an implicit finite-element approach which provides for mixed (I and II) fracture propagation in arbitrary stress field. Scaled hydrofracture tests were done in blocks of gypsum cement with sandstone lenses in a triaxial state of stress [25]. The fractures were tracked and the well pressure was monitored allowing the validation of this coupled model. This model was used also to simulate gas-driven fracturing of open holes. Propagation of cracks of unequal length were studied. It was shown that shorter cracks that are 60 % of the length of the longer cracks propagate at a speed which is about 1/3 of that of longer cracks. This is in qualitative aggreement with previous theoretical investigations [14-16,19,20].

2.3.4 Modeling of solid and fluid dynamics for explosively driven fractures

A study has been recently published by Nilson *et al* [26] that combines stress wave dynamics with fluid mechanics to achieve a fully interactive model of explosively driven hydrofracture propagation. It is a combination of a previous model of fluid dynamics [21,22] and of a dynamic nonlinear deformation model that uses a Lagrangian formulation of the equations of motion. This method was applied to containment of underground nuclear tests [26]. Three different fracture mechanisms were distinguished. In stress-limited cracking, the hoop stresses around the cavity are always sufficient to constrain the growth of a crack. In speed-limited cracking the crack speed is controlled by both solid and fluid dynamic constraints. In the last regime, depletion-limited cracking, stress wave dynamics become less important and the propagation is controlled by the decay of cavity pressure and fluid dynamic considerations. This last regime is characteristics of propellant stimulation when the propellant deflagrates, as the loading rate is "low" compared to the explosive regime. This result validates once again the use of a quasi-static approach for the estimation of the fracture extension in the case of formation stimulation with slow-burning propellant tools.

3 Overbalanced perforating

In this method, a high pressure is applied to the cased wellbore before the shooting of the perforations [3,27,28]. The purpose of perforations is to provide effective flow communication between the wellbore and the reservoir. This involves shooting a hole through casing and cement and providing a perforation tunnel into the formation. Lower wellbore pressure than reservoir pressure are generally applied to flush the tunnel just after the perforating. However the perforation efficiency can be unpredictable with underbalanced technique, the perforation efficiency being altered by perforation damage (crushed zone aroung the tunnel) and/or the required underbalance to clean the perforation tunnel being unachievable. A recent review revealed inconsistent results achieved with underbalanced perforating, with an average perforation efficiency of less than 25% [27]. Another reason to use an overbalance treatment comes from the problems of fracture initiation and fracture link-up which are dependent on the near wellbore stresses. It has been shown that the fracture opening surface is not generally a smooth surface and that there are a large amount of rock masses entrapped by connecting cracks [29]. This situation could be greatly improved with overbalanced perforating which shall be able to favor fracture link-up from each individual perforations [29]. Overbalance-generated fracture may also initially align with the perforation axis.

3.1 Field data

Figure 7 illustrates a typical well configuration during an overbalanced perforating job [27]. The height of the liquid column can be as high as 1800 m whereas the height of the gas column is 900 m, the surface and bottomhole pressures being 55 and 76 MPa respectively [28]. Typical thickness of the pay zone is in the 3-60 m range and tubing diameter is 3.8 cm. 50 wells have been perforated with this technique and higher production rates were reported compared to wells perforated underbalanced [27]. For instance, a negative completion skin factor was achieved with a median value of -2.0. The pressure decay is very rapid, the wellbore pressurization lasting only several seconds as reported by field user [28].

3.2 Modeling approach

A model of fluid-driven fracture propagation has been recently developed by Petitjean and Couët that can be applied to this fracturing technique [30,31]. An integral/numerical approach was employed, the fluid dynamic model being similar to the approach of Nilson *et al* [21, 22]. Numerical simulations were done using typical conditions that could be found in the field. A 15.24 cm O.D. well was placed in an isotropic formation made of carbon tan sandstone (fracture toughness = 0.561 MPa$\sqrt{m}$, shear modulus = 4.64 GPa, Poisson's ratio = 0.37). The rock permeability and porosity were 17.8 mD and 15.3 p.u respectively. The 3-meter thick pay zone is taken to be at a depth of 1500 m where the in-situ stress and the ambient pore pressure were 24.1 and 16 MPa respectively. Water was used as the fracturing fluid and a 40-meter high gas column to pressurize it at 50 MPa is applied before the shooting of the perforations which we take first to be at 180° phasing. Friction losses in the wellbore or the perforations are neglected. The normalized pressure distribution (P/P_0) is reported in Figure 8 for various times as the normalized fracture

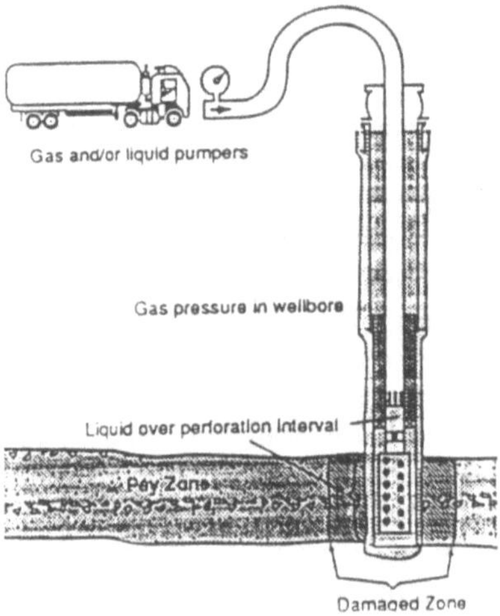

Figure 7 : Overbalanced perforating well configuration [27]

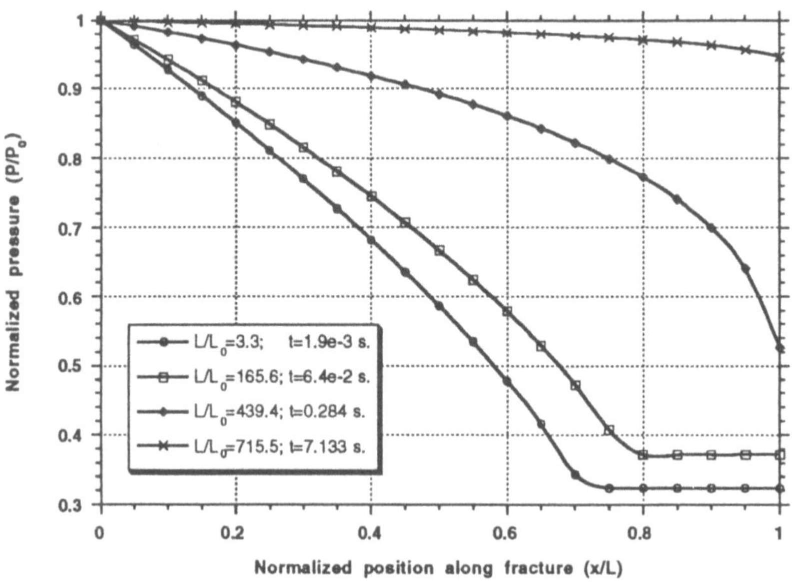

Figure 8 : Evolution as length increases of the pressure profile along the fracture during overbalance perforating. Initial crack length L_0 = 2 cm [31]

length L/L_0 increases. At the onset of the simulation, as the large pressure is applied, the effect on the fracture is similar to applying a wedge to open the rock and the fracturing fluid does not reach completely the end of the fracture. As the fracture increases in length, the fracturing fluid wets the whole fracture, finally reaching a level of almost constant pressure across the complete length of the fracture. Figure 9 reports the behavior of the fracture length and of the well pressure with time. The pressure decay is very rapid as observed in the field [28]. The fracture speed is high compared to hydrofracture operations allowing to minimize seepage losses through the fracture walls [31]. During the fracture propagation the fluid dynamics within the crack is fully turbulent, in sharp contrast with an hydrofracture one where the fluid flow is generally considered to be laminar.

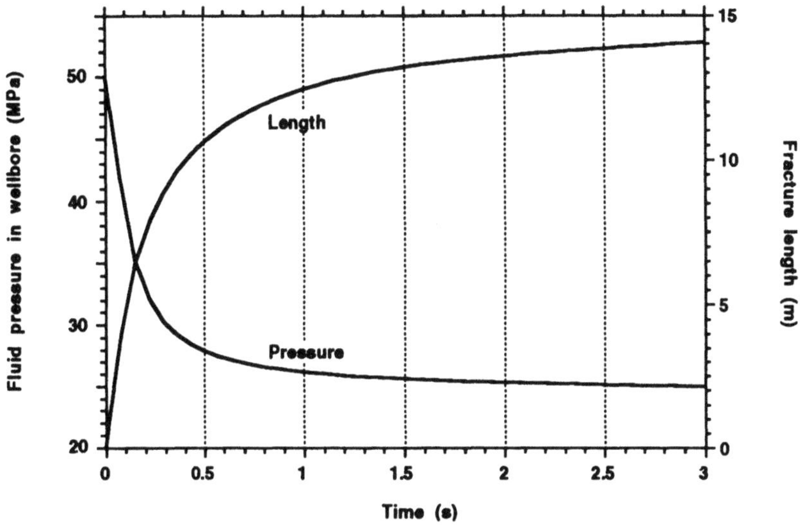

Figure 9 : Fracture extension and wellbore pressure during overbalance perforating [31]

4 Conclusion

Fracturing by gas loading the well offer new opportunities to the petroleum industry But it also offers new challenges to the scientific community. The application, sometimes dynamically, of a high pressure to the well compared to the formation pressure and the in-situ stresses modifies the near wellbore stress of state. It therefore changes the fracture initiation process compared to current hydrofracture techniques. Although some studies have been done this last decade, a firm description of the initiation of the fracturing process is missing. It is very often of vital importance as it can govern the flow communication between the well and the near wellbore formation. On the other hand, study of the fracture propagation process is more advanced. Several models that give reasonable comparison with experimenmt exist, although improvements are needed to take into account more sophisticated, but real, problems like heterogenity or anisotropic stress.

REFERENCES

1. R.W. Veach, J. Pet. Tech. **35** **(4)**, 677 (Oct 1983).
2. N.R. Warpinski, R.A.Schmidt , P.W. Cooper, H.C. Walling and D.A. Northrop, Proc. of the 20th U.S. Symp. on Rock Mech., Austin, Texas, U.S.A. (1979).
3. P.T. Branagan and R.H. Wilmer, SPE 17716, SPE Gas Tech. Symp. ,Dallas, Texas, U.S.A. (1988).
4. J.F. Cuderman, SPE/DOE 10845, Proc. of the Unconventional Gas Recovery Symp., Pittsburgh, Pennsylvania, U.S.A. (1982).
5. J.F. Cuderman, GRI-84/0090, Gas Research Institute, Chicago, Illinois, U.S.A. (1984)
6. J.F. Cuderman, T.Y. Chu, J. Jung and R.D. Jacobson SAND85-2809, Sandia National Laboratories, Albuquerque, New Mexico, U.S.A. (1986).
7. J.F. Cuderman , SAND86-1133, Sandia National Laboratories, Albuquerque, New Mexico, U.S.A. (1986).
8. R.P. Swift and A.S. Kurosov, Soc. Pet. Eng J. **22**, 923 (1982).
9. W.L.Fourney, D.B. Barker and D.C. Holloway, Int. J. Rock. Mech. Min. Sci. **20**, 91 (1983).
10. J.F. Schatz, B.J. Zeigler, R.A. Bellman, J. Hanson, M. Christianson, R.D. Hart, GRI-87/0199, Gas Research Institute, Chicago, Illinois, U.S.A. (1987), see also SPE 18866 (1989).
11. H.K. Kutter and C. Fairhurst, Int. J. Rock. Mech. Min. Sci. **8**, 181 (1971).
12. S. Bhandari, Acta Astronautica **6**, 365 (1979).
13. S. McHugh, Int. J. of Fracture **21**, 163 (1983).
14. F. Ouchterlony, 3rd Congress of Int. Soc. for Rock Mech., **IIB**, 1377 (1974).
15. F. Ouchterlony, Report DS1982:4, Swedish Detonic Research Foundation, Stockholm, Sweden (1982).
16. J.M. Hanson, R.A. Schmidt, C.H. Cooley and J.F. Schatz, SPE 12839, SPE/DOE/GRI Unconventional Gas Recovery Symp., Pittsburgh, Pennsylvania, U.S.A. (1984).
17. L. Seaman, D.R. Curran and D.A. Shockey, J. App. Phys. **47**, 4814 (1976).
18. S.L. McHugh and D. Keough, SPE/DOE 10844, SPE/DOE Unconventional Gas Recovery Symp., Pittsburg, Pennsylvania, U.S.A. (1982).
19. D.W. Swenson and L.M. Taylor, SAND82-0633, Sandia National Laboratories, Albuquerque, New Mexico, U.S.A. (1982).
20. D.W. Swenson and L.M. Taylor, Int. J. Num and Anal. Meth. Geomech., **7**, 469 (1983).
21. R.H. Nilson, W.J. Propper, R.E. Duff, Int. J. Rock Mech. Min. Sc. & Geomech. Abs **22**, 3 (1985).
22. R.H. Nilson, Intl. J. Num. Anal. Meth. Geomech. **48**, 191 (1986).
23. F.H. Cornet, J. Appl. Mech. **46**, 349 (1979).
24. F.E. Heuze, R.J. Shaffer, A.R. Ingraffea and R.H. Nilson, Int. J. Rock Mech. Min. Sc. & Geomech. Abs **27**, 243 (1990).
25. S.C. Blair, R.K. Thorpe and F.E. Heuze, Int. J. Rock Mech. Min. Sc. & Geomech. Abs **27**, 255 (1990).
26. R.H. Nilson,N Rimer and E. Halda, J. Geophys. Res. **96**, 18.081 (1991).
27. P. J. Handren, T. B. Jupp and J. M. Dees, SPE 26515, SPE Annual Technical Conf.,Houston,
Texas,USA (1993).
28. J. Schmidt, Private communication (1993)
29. C. H. Yew, M. E. Mear, C. C. Chang and X. C. Zhang, Annual Technical Conf. SPE 26514, SPE Annual Technical Conf., Houston, Texas, USA (1993).
30. L. Petitjean and B. Couët, SPE 28084, Eurock' 94, Delft, Nederlands (1994)
31. L. Petitjean and B. Couët, SPE Annual Technical Conf., New Orleans, Louisiana, USA (1994)

Physics of Random Nonlinear Composites

P. M. Hui

Department of Physics, The Chinese University of Hong Kong
Shatin, New Territories, Hong Kong

1. Introduction

Many of the talks in this Workshop are concerned with nonlinearity due to some "breakdown" mechanisms, e.g., nonlinear response in carbon-wax composites due to possible tunneling between deadends of the cluster formed by the better conductor as applied voltage is increased.[1] In this article, we discuss another class of nonlinear composites in which one (or many) of the constituents possesses intrinsic nonlinear response. The topic forms a part of the larger subject of physics of macroscopically inhomogeneous media, which has been reviewed recently.[2] Experimentally, random composites consisting of dots with nonlinear optical properties embedded in glass have been studied with the aim to control the effective nonlinear response of the composites by tuning parameters such as the concentration and size of the dots.[3] On the theoretical side, much effort have been concentrated on deriving calculation schemes for linear and nonlinear response in these random mixtures.

The study of effective response in macroscopically inhomogeneous systems has benefitted much from technique developed in other areas of condensed matter physics. Percolation theory[4] and the idea of fractal objects, for example, have been widely used in describing effective response in random mixture consisting of components with very different physical properties, e.g., conductor mixed with insulator. Theories developed in treating the electronic problems in alloys[5] can also be applied. For example, effective medium type approximations[6] for the effective response stem from the coherent potential approximation in the alloy problem. Here, we review some of the approximation schemes for estimating the linear and nonlinear response in random mixtures. We shall focus on the discssion of the effective transport properties in this article.

The plan of this article is as follows. In Sec.2, we set up the problem and define the effective response that we want to calculate. The relationships between response in nonlinear mixture and that in linear random mixture are discussed in Sec.3. In Sec.4, such relations are used to derive approximate expression for the case of a dilute concentration of nonlinear component embedded in a linear host medium. Section 5 reviews an effective medium type approximation, which is applicable to the whole concentration range. Effects of the restricted geometry near the percolation threshold may lead to enhanced nonlinear response. Such effects are discussed in Sec.6 within the context of response in a nonlinear conductor - perfect conductor mixture. Results are summarized in Sec.7.

2. Effective response

We consider a two-component composite in which each component can be described by a weakly nonlinear relation between the current density $\mathbf{J}$ and the electric field $\mathbf{E}$ of the form

$$\mathbf{J} = \sigma_i^0 \mathbf{E} + \chi_i |\mathbf{E}|^2 \mathbf{E}, \tag{1}$$

where $i = A, B$ runs over the two components A and B, and σ_i^0 and χ_i are the linear conductivity and nonlinear susceptibility of the i-th component, respectively. By weak nonlinearity we mean that the nonlinear response is assumed to be small compared to the linear response. Our aim is to study the effective response of such random nonlinear composites. For simplicity, we will restrict our discussion to the case of only one component in the composite exhibits nonlinear response, though our approaches can be readily generalized to more complicated cases.

The equations satisfied by $\mathbf{J}(\mathbf{x})$ and $\mathbf{E}(\mathbf{x})$ in the inhomogeneous media follow from the Maxwell's equations. They are $\nabla \cdot \mathbf{J} = 0$ and $\nabla \times \mathbf{E} = 0$ for steady state plus some appropriate boundary conditions. Let Ω be the volume of the random mixture and S be the surface enclosing Ω. A convenient way is to impose the boundary condition $\phi = -\mathbf{E}_0 \cdot \mathbf{x}$ for $\mathbf{x}$ on the surface S, where ϕ is the scalar potential. If the system were uniform, such boundary condition on ϕ would give rise to a uniform electric field $\mathbf{E}_0$ everywhere in Ω. Even in an inhomogeneous medium, with the imposed boundary condition, the space-averaged electric field in Ω is given by $\mathbf{E}_0$, i.e.[7]

$$< \mathbf{E} >= \frac{1}{\Omega} \int \mathbf{E}(\mathbf{x}) d^3x = \mathbf{E}_0. \tag{2}$$

There are, at least, two natural ways to define the effective response of the random mixture. One way is to relate the space-averaged current density $< \mathbf{J} >$ to the space-averaged field $< \mathbf{E} >= \mathbf{E}_0$ using

$$< \mathbf{J} >= \sigma_e^0 \mathbf{E}_0 + \chi_e |\mathbf{E}_0|^2 \mathbf{E}_0, \tag{3}$$

where σ_e^0 and χ_e are the effective linear conductivity and nonlinear susceptibility, respectively. In fact, higher order nonlinear terms may be generated but we will restrict ourselves to discussion on the leading nonlinear response.[8] Another way to define σ_e^0 and χ_e is to use the expression for power dissipation

$$W \equiv \int \mathbf{J} \cdot \mathbf{E} d^3x = \Omega(\sigma_e^0 |\mathbf{E}_0|^2 + \chi_e |\mathbf{E}_0|^4). \tag{4}$$

It has been shown that the definitions (3) and (4) are equivalent and one may choose whichever is the more convenient for any given calculation.[7]

3. Nonlinear response and linear random system

Equation (4) gives a convenient way to relate the problem of effective nonlinear response to that of electric field distribution in random *linear* problem, i.e. in system in which all components have only linear response. We write Eq.(4) as

$$W = \int \sigma(\mathbf{x})|\mathbf{E}(\mathbf{x})|^2 d^3x + \int \chi(\mathbf{x})|\mathbf{E}(\mathbf{x})|^4 d^3x = W_2 + W_4. \tag{5}$$

To first order in $\chi(\mathbf{x})$,

$$W_4 = \int \chi(\mathbf{x})|\mathbf{E}_{lin}(\mathbf{x})|^4 d^3x, \tag{6}$$

where $\mathbf{E}_{lin}(\mathbf{x})$ is the electric field which could exist at position $\mathbf{x}$ if $\chi(\mathbf{x})$ were identically zero. Since the correction to $\mathbf{E}(\mathbf{x})$ due to the presence of nonlinear response $\chi(\mathbf{x})$ is at least first order in $\chi(\mathbf{x})$, such correction to W_4 will only be a second order effect. Comparing with the definition (Eq.(4)), the nonlinear response χ_e is given by[2,7]

$$\chi_e = \frac{1}{\Omega|\mathbf{E}_0|^4} \int \chi(\mathbf{x})|\mathbf{E}_{lin}(\mathbf{x})|^4 d^3x. \tag{7}$$

Similarly, the effective linear conductivity can be expressed as

$$\sigma_e^0 = \frac{1}{\Omega|\mathbf{E}_0|^2} \int \sigma(\mathbf{x})|\mathbf{E}_{lin}(\mathbf{x})|^2 d^3x. \tag{8}$$

Equation (7) gives a relation between the effective nonlinear response and the electric field distribution in a random *linear* mixture.[7,9] If only one component has nonlinear response, χ_e is related to the space average of $|\mathbf{E}_{lin}(\mathbf{x})|^4$ within the volume occupied by that component. Such a dependence is also found in the so-called noise problem in linear random composites in which one or more constituents of the composite exhibits fluctuating physical response. Equation (7) forms a convenient starting point for developing different approximation schemes for evaluating the effective response.

4. Dilute limit approximation

Consider the case that only component A is nonlinear. The concentration of A is p_A and $p_A \ll 1$. In this case, Eq.(7) gives

$$\chi_e = \frac{1}{\Omega|\mathbf{E}_0|^4} \chi_A \int_{\Omega_A} |\mathbf{E}_{lin}(\mathbf{x})|^4 d^3x, \tag{9}$$

where the integration is over the portion Ω_A of Ω occupied by material A. Since $p_A \ll 1$, the grains of component A in the random mixture are well separated and $\mathbf{E}_{lin}(\mathbf{x})$ can well be approximated by the local field inside a spherical grain of material A embedded in the host material B with an external field $\mathbf{E}_0$.[10] Thus,

$$\mathbf{E}_{lin} = \frac{3\sigma_B^0}{\sigma_A^0 + 2\sigma_B^0} \mathbf{E}_0. \tag{10}$$

Since $p_A = \Omega_A/\Omega$, Eq.(9) for χ_e can be rewritten as

$$\chi_e = p_A \chi_A \left(\frac{3\sigma_B^0}{\sigma_A^0 + 2\sigma_B^0}\right)^4, \tag{11}$$

to lowest order in χ_A. The expression can be easily generalized to finite frequencies for the case of nonlinear **D-E** response, where **D** is the displacement field.[7,11] The result (Eq.(11)) corresponds to the case that the effective nonlinear response is driven by the local field in the linear limit. Identical result can be obtained using the Maxwell-Garnett approximation, which is good for dilute linear problems, with a local field dependent conductivity for the nonlinear component.[12] For the case of arbitrary nonlinearity such that material A has a response of the form $\mathbf{J} = \sigma_A^0 \mathbf{E} + \chi_A |\mathbf{E}|^\beta \mathbf{E}$, the corresponding result in the dilute limit is

$$\chi_e = p_A \chi_A \left(\frac{3\sigma_B^0}{\sigma_A^0 + 2\sigma_B^0}\right)^{\beta+2}. \tag{12}$$

If the host medium is also nonlinear, the resulting expression for χ_e is more complicated and has an additional term involving χ_B.[13]

5. Effective-medium-type approximations

An effective-medium-type approximation has been proposed for estimating the effective response in nonlinear random mixture.[14] We briefly review the approach. For linear composites consisting of two materials with only linear response characterized by linear conductivities σ_A^0 and σ_B^0, the effective response is linear and can be represented by

$$\sigma_e^0 = F(\sigma_A^0, \sigma_B^0, p_A), \tag{13}$$

where F is some function which will, in general, depend on the geometry of the composite. The effective medium approximation amounts to the assumption that even in the presence of nonlinearity, the effective response of the mixture can be approximated by

$$\sigma_e = F(\sigma_A, \sigma_B, p_A) \tag{14}$$

with the same function F and

$$\sigma_i = \sigma_i^0 + \chi_i < |\mathbf{E}_i|^2 > \qquad (i = A, B) \tag{15}$$

and $< |\mathbf{E}_i|^2 >$ is the mean-square of the electric field in the i-th component in the linear limit. This amounts to the approximation that the electric field is uniform in the nonlinear component. In the case of only material A is nonlinear and that the nonlinearity is weak, an approximation of σ_e can be obtained by expanding the function F in a Taylor series about the linear σ_e^0,

$$\sigma_e \approx F(\sigma_A^0, \sigma_B^0, p_A) + F'_A(\sigma_A^0, \sigma_B^0, p_A)\chi_A < |\mathbf{E}_A|^2 >, \tag{16}$$

where the first term is σ_e^0 and $F'_A = \partial\sigma_e^0/\partial\sigma_A^0$. This partial derivative can be expressed exactly in terms of the average squared electric field in component A in the linear limit;[15] the relation is

$$F'_A(\sigma_A^0, \sigma_B^0, p_A) = \frac{p_A}{E_0^2} < |\mathbf{E}_A|^2 > . \tag{17}$$

Combining Eqs.(16) and (17), the effective nonlinear susceptibility χ_e is given by

$$\chi_e = \frac{\chi_A}{p_A}\left(\frac{\partial\sigma_e^0}{\partial\sigma_A^0}\right)^2. \tag{18}$$

Thus, by invoking an approximation for the linear effective conductivity σ_e^0, one also gets an approximation for the nonlinear susceptibility. When both components A and B are nonlinear, the result Eq.(18) can be generalized to get

$$\chi_e = \frac{\chi_A}{p_A}(F'_A)^2 + \frac{\chi_B}{p_B}(F'_B)^2, \tag{19}$$

where p_A and p_B are the volume fractions of the components A and B, respectively. An alternative derivation of the results has been proposed by Stroud and Wood[16] using a decoupling scheme to relate approximately the space average of $|\mathbf{E}_{lin}|^4$ to the average of $|\mathbf{E}_{lin}|^2$. Besides the one discussed here, other effective medium approaches has also been proposed.[13,17]

We can obtain the dilute limit approximation in the previous section using the effective medium approximation. For the dilute linear random mixture problem, a useful expression for the effective linear response is[2]

$$\sigma_e^0 = \sigma_B^0 + 3p_A\sigma_B^0\frac{\sigma_A^0 - \sigma_B^0}{\sigma_A^0 + 2\sigma_B^0}, \tag{20}$$

for a dilute concentration p_A of material A embedded in a material B. Equation (20) is simply the lowest order correction in p_A to the conductivity taking local field effect into account. Using this dilute mixing rule and evaluating F', one recovers the result Eq.(11) in the previous section.

For general volume fractions, one can invoke the effective medium approximation (EMA) for linear problem[2]

$$p_A\frac{\sigma_A^0 - \sigma_e^0}{\sigma_e^0 + g(\sigma_A^0 - \sigma_e^0)} + p_B\frac{\sigma_B^0 - \sigma_e^0}{\sigma_e^0 + g(\sigma_B^0 - \sigma_e^0)} = 0, \tag{21}$$

where g is a depolarization factor, as the starting point to evaluate the effective nonlinear response χ_e using the result Eq.(18). In the linear problem, EMA represents a good approximation to the effective behaviour, at least for problems away from the percolation transition. In the nonlinear problem, due to the factor $|\mathbf{E}_{lin}(\mathbf{x})|^4$ in the expression of χ_e (Eq.(7)), the effective response is more sensitive to the local field and thus one expects that EMA may not be as useful as in the case of linear problem. Numerical simulations on weakly nonlinear random resistor networks have been performed and results have been compared with EMA.[18] Qualitatively, results of simulations can be described reasonably well by EMA, except near the percolation threshold. It is typical of any effective medium type approximation that the predictions become less reliable as the critical region is approached.

6. Percolating effects

There are interesting percolating effects due to the restricted geometry of the percolating cluster near the percolation threshold.[19] One has a percolation problem when a nonlinear component is randomly mixed with either an insulator or a perfect conductor. It has been pointed out that nonlinear response is enhanced as the percolation threshold is approached in a nonlinear component - perfect conductor (N/S) mixture.[20,21] We illustrate the effect by considering the crossover electric field $|\mathbf{E}_c|$, which is defined as the magnitude of the applied field at which the linear and nonlinear response of the random mixture become comparable.[22] For a N/S composite in which the normal component is nonlinear, the effective response of the mixture is nonlinear below p_c, where p_c is the percolation threshold of the perfectly conducting component, and is given by Eq.(3). It follows from the definition of $|\mathbf{E}_c|$ that

$$|\mathbf{E}_c| = (\frac{\sigma_e^0}{\chi_e})^{1/2}. \tag{22}$$

Hence, by using established methods to obtain σ_e^0 and χ_e for p near p_c, the behaviour of $|\mathbf{E}_c|$ can be extracted.

For the N/S system under consideration, a simple approximation for the linear effective conductivity is the effective medium approximation.[6] This can be used in conjunction with Eq.(18) to estimate χ_e. In fact, this is the simplest approximation that predicts a percolation transition, although the exponent s describing the divergence of the conductivity in a N/S composite is numerically incorrect. Within this approximation,

$$\sigma_e^0 = \frac{\sigma_A}{3}(p_c^{EMA} - p)^{-1}. \tag{23}$$

Here, $p_c^{EMA} = 1/3$ is the percolation threshold in three dimensions within EMA and $s = 1$. Substituting Eq.(23) into Eq.(18) with $p_A = 1 - p$, we have

$$\chi_e = \frac{\chi_A}{9(1-p)}(p_c^{EMA} - p)^{-2}. \tag{24}$$

The crossover field follows from Eq.(22) and gives

$$|\mathbf{E}_c| = (\frac{\sigma_e^0}{\chi_e})^{1/2} = (\frac{3(1-p)\sigma_A^0}{\chi_A})^{1/2}(p_c^{EMA} - p)^{1/2}. \tag{25}$$

Thus, within EMA, the crossover field behaves as

$$|\mathbf{E}_c| \sim (p_c - p)^M \tag{26}$$

with $M = 1/2$ for all spatial dimensions d. It is well known that EMA may give the correct behavior near p_c qualitatively but usually predicts the incorrect exponents.

One can also relate the exponent M in Eq.(26) to the noise exponent κ' in a linear N/S mixture. A formal relation between the problem of nonlinear effective response in random nonlinear composites and the problem of relative fluctuations (noise) in *linear* random composites has been demonstrated by Stroud and Hui.[7] For our N/S system, only the normal component is nonlinear. It follows from Eq.(9) that the effective nonlinear susceptibility χ_e is proportional to the fourth moment of the field distribution in a linear random composite. Next, consider a linear N/S composite in which the normal component has a fluctuating conductivity given by $\sigma'(\mathbf{x}) = \sigma^0(\mathbf{x}) + \delta\sigma^0(\mathbf{x})$, where $\delta\sigma^0(\mathbf{x})$ represents the fluctuation. Since $\sigma'(\mathbf{x})$ is fluctuating, the effective conductivity of the composite is also fluctuating. It has been shown that the mean square fluctuation in the effective linear response is also related to the fourth moment of the local field.[23] Therefore, χ_e is proportional to the mean square fluctuation in linear composite. The *relative* fluctuation, which is the ratio of the mean square fluctuation to the square of the effective conductivity, behaves as $(p_c - p)^{-\kappa'}$ in a S/N mixture for p approaches p_c from below. The exponent κ' is referred to as the noise exponent.[24,25]

Using the established relation, it follows that

$$\frac{\chi_e^2}{(\sigma_e^0)} \sim (p_c - p)^{-\kappa'}. \tag{27}$$

Combining Eq.(22) with Eq.(27), we have

$$|\mathbf{E}_c| = (\frac{\sigma_e^0}{\chi_e})^{1/2} = \frac{1}{(\sigma_e^0)^{1/2}}(\frac{(\sigma_e^0)^2}{\chi_e})^{1/2} \sim (p_c - p)^{(\kappa'+s)/2}. \tag{28}$$

The crossover field exponent M, which is an exponent defined in a nonlinear random composite problem, is related to the exponents κ' and s, which are exponents defined in a *linear* problem, by

$$M = \frac{\kappa' + s}{2}. \tag{29}$$

Using the bounds[24] derived for the noise exponent κ' and with the known values for the s-exponent, we have $1.18 \leq M \leq 1.33$ in two dimensions and $0.66 \leq M \leq 0.88$ in three dimensions. It is noted that M is bounded from above by the exponent ν describing the divergence of the correlation length in the percolation problem.

Due to the restricted geometry of the path through which current flows in the vicinity of the peroclation threshold, the crossover electric field $|\mathbf{E}_c|$ behaves as $|\mathbf{E}_c| \sim (p_c - p)^M$ with $M > 0$ for p approaches p_c from below. It corresponds to an enhancement in the nonlinear response near the percolation threshold relative to a system consisting only of the nonlinear material A. It implies that the region of linear response shrinks and a small electric field is sufficient to lead to an appreciable nonlinear response near p_c.

7. Discussion

In this article, we reviewed some approximation schemes for estimating effective linear and nonlinear response in random mixture of conductors with some intrinsic nonlinear behaviour. Relations of the nonlinear response problem with the field distribution and the relative fluctuation problems in linear random mixtures are discussed. Approximations for estimating the effective nonlinear response in the dilute limit and for general concentration are derived.

Percolating effect will, in some cases, enhance the nonlinear response of a random mixture near the percolation threshold. We discussed this effect within the context of the crossover electric field strength in random nonlinear conductor - perfect conductor mixture. Although our discussion is made within the framework of **J**-**E** relation, it can be readily generalized to systems with nonlinear **D**-**E** relation, where **D** is the displacement field; and to problems on the effective elastic properties in random mixtures.[26] In the latter problem, a "superconductor" corresponds to a hard material and the nonlinear material corresponds to a medium with a stress-strain relation exhibiting nonlinear response for large strain.[27] Measureable linear response of random elastic composites include such quantities as the bulk modulus, sheer modulus and Poisson ratio. However, the linear response regime corresponding to such measureable physical quantities shrinks as p approaches p_c. With minor modifications, our discussion here can be applied to this elastic percolation problem.

Acknowledgments

The author would like to thank the organizers of the workshop for their hospitality. Supports from the Indo-French Centre for the Promotion of Advanced Research and from the Department of Physics at the Chinese University of Hong Kong are gratefully acknowledged. He also thanks D. Stroud, D.J. Bergman, X.C. Zeng and K.W. Yu for fruitful collaborations that lead to some of the work reported here. This work was supported in part by a grant through the Research Grants Council (RGC) of the University and Polytechnic Grant Committee (UPGC) of the Hong Kong Government.

References

1 See, for example, the articles by K.K. Bardhan and A.K. Sen in this volume; R.K. Chakrabarty, K.K. Bardhan and A. Basu, Phys. Rev. B **44**, 6773 (1991).

2 D.J. Bergman and D. Stroud, in *Solid State Physics: Advances in Research and Applications*, edited by H. Ehrenreich and D. Turnbull (Academic Press, New York, 1992), Vol. 46, p.147; J.P. Clerc, G. Giraud, J.M. Laugier, and J.M. Luck, Advances in Phys. **39**, 191 (1990); and the

articles in the *Proceedings of the Second International Conference on Electrical Transport and Optical Properties of Inhomogeneous Media* in Physica A **157**, No.1 (1989).

3 See the collection of articles on this subject in J. Opt. Soc. Am. B **6**, No.6 (1989).

4 D. Stauffer and A. Aharony, *Introduction to Percolation Theory*, 2nd edition (Taylor and Francis, London, 1992).

5 H. Ehrenreich and L. M. Schwartz, in *Solid State Physics: Advances in Research and Applications*, edited by H. Ehrenreich, F. Seitz and D. Turnbull (Academic, New York, 1976), Vol. 31, p.149.

6 S. Kirkpatrick, Rev. Mod. Phys. **43**, 377 (1978); D. Stroud, Phys. Rev. B **12**, 3368 (1975).

7 D. Stroud and P.M. Hui, Phys. Rev. B **37**, 8719 (1988).

8 J.W. Haus, R. Inguva and C.M. Bowden, Phys. Rev. A **40**, 5729 (1989); K.W. Yu, P.M. Hui and D. Stroud, Phys. Rev. B **47**, 14150 (1993); P.M. Hui, J. Appl. Phys. **73**, 4072 (1993).

9 A. Aharony, Phys. Rev. Lett. **58**, 2726 (1987).

10 See, for example, J. D. Jackson, *Classical Electrodynamics*, 2nd edition (John Wiley & Sons, New York 1975), p.151.

11 G.O. Olbright, N. Peyghambarian, S.W. Koch and L. Banyai, Opt. Lett. **12**, 413 (1987).

12 P.M. Hui, J. Appl. Phys. **68**, 3009 (1990).

13 D.J. Bergman, Phys. Rev. B **39**, 4598 (1989).

14 X. C. Zeng, D. J. Bergman, P. M. Hui, and D. Stroud, Phys. Rev. B **38**, 10970 (1988).

15 D.J. Bergman, Phys. Rep. **43**, 377 (1978).

16 D. Stroud and Van E. Wood, J. Opt. Soc. Am B **6**, 778 (1989).

17 J.W. Haus, M. Bloemer and C.M. Bowden, J. Opt. Soc. Am B **6**, 797 (1989); G.Q. Gu and K.W. Yu, Phys. Rev. B **46**, 4502 (1993).

18 C.S. Yang and P.M. Hui, Phys. Rev. B **44**, 12559 (1991); P.M. Hui and K.H. Chung, (unpublished).

19 R. Blumenfeld and D. J. Bergman, Phys. Rev. B. **43**, 13682 (1991).

20 T. Ohtsuki and T. Keyes, J. Phys. A **17**, L559 (1984).

21 P.M. Hui, Phys. Rev. B **41**, 1673 (1990).

22 P.M. Hui, Bull. Am. Phys. Soc. **36**, 462 (1991); and unpublished.

23 R. Rammal, C. Tannous and A.-M.S. Tremblay, Phys. Rev. A **31**, 2662 (1985).

24 D.C. Wright, D.J. Bergman, and Y. Kantor, Phys. Rev. B **33**, 396 (1986).

25 P.M. Hui and D. Stroud, Phys. Rev. B **34**, 8101 (1986).

26 D.J. Bergman, Phys. Rev. B **33**, 2013 (1986); S. Feng, M. F. Thorpe, and E. Garboczi, *ibid* **31**, 276 (1985); *ibid* **35**, 8579 (1987).

27 L. Benguigui and P. Ron, Phys. Rev. Lett. **70**, 2423 (1993); and references therein.

Scaling Behavior of Electric Response in a Non-linear Composite

Asok K. Sen and *Abhijit Kar Gupta*

Saha Institute of Nuclear Physics
1/AF, Bidhannagar, Calcutta 700 064, INDIA

Abstract: A short, general overview of non-linearity in electrical (current) response in two different regimes and their origins are discussed. Special attention is paid to some recent experiments on non-linear response in inhomogeneous (composite) materials. A model percolating structure to study this behavior has been introduced. Results for this model and some of its predictions for both dc and ac non-linear response have been discussed.

1. Introduction

In the well-studied linear response theory, one assumes the response to be proportional to the driving field, at least when the perturbation due to the applied field is small enough. For electrical transport properties, this means that the current (I) is proportional to the external applied voltage (V), where the constant of proportionality is called the conductance (G). This is OHMIC or DIFFUSIVE behavior for the charge carriers and may be alright for (dirty) metals or semiconductors at low voltage and/or high enough temperature, where scattering due to phonons is the chief mechanism for dissipation. But at low enough temperature, and in the presence of disorder or other scattering mechanisms, quantum mechanical *tunneling* (or, *hopping*) between sites may become important and the above linearity breaks down, to give for the voltage-dependent conductance:

$$G(V_0) = I/V_0 \qquad (\textit{chord rule}), \tag{1.1}$$

or

$$= (dI/dV)|_{V=V_0} \qquad (\textit{tangent or derivative rule}), \tag{1.2}$$

where I is the current through the sample at a voltage $V = V_0$. Since these two rules may give rise to quite different values one should state clearly which is being used. For example, if the $I-V$ curve is sloping down in some region, the differential conductance is negative whereas the first rule always gives a positive conductance. Also if the $I-V$ curve becomes asymptotically linear for large voltages, the two rules give different functions for the asymptotic approach to constant conductance. In all our discussions below, we will be talking about the differential conductance unless otherwise stated. Some early observations on this non-linearity were in varistors [1] as well as semiconductors in high fields [2]. This non-linearity is manifested somewhat differently in two different regimes of disorder as discussed below, and there seems to be no simple unifying picture at present of a crossover from one to the other. Attempts in this direction in recent literature may be found in some recent books and proceedings [2]. Since these lecture notes are meant to be short reviews in the field, we have tried to give brief overviews of both the regimes in Secs. 2.1 and 2.2 (while pointing out more detailed reviews) for the benefit of newcomers to the field. We have also presented some new results for non-linear composites, by describing first some key experiments in Sec. 2.2.1 and then our model and the relevant scaling behavior in Sec. 3 (occasionally making contact with experiments). Readers already acquainted with the general problem may straight go to Sec. 2.2.1 from here.

2.1 Electric Field and Quantum Transport

When the disorder is truly microscopic (of atomic dimensions), one needs to consider the effects of wave-function localization. Interestingly, in an ordered, or a weakly disordered (dirty metal) system, a strong electric field tends to produce localization, in the form of quantized Stark-Wannier ladders with the ladder (or, level) spacing proportional to the external electric field (see [3] for a recent review). In a strongly disordered (Anderson-localized) system, on the other hand, several theoretical studies [4] indicate that there is a crossover from exponential localization (insulator) to power-law localization (metal?). In this context, it may be noted that while typical metals show a very specific form of power-law localization (i.e., Ohm's law with $G \propto L^{-1}$, L being the length in the direction of the current), other non-integer negative exponents of L generally imply critical (instead of metallic) states. Even if one accepts that a power-law behavior should imply an insulator to metal transition (i.e., presence of quantum diffusion), the transition seems to be taking place at arbitrarily small V [4]. Thus *'quantum breakdown' does not yet seem to be a well-posed problem* (for a different, somewhat speculative view see [5]).

But, whether a insulator-metal transition takes place or not under the action of an electric field, non-linearity of transport is intrinsic in a mesoscopic quantum disordered system. By the word '*mesoscopic*', one means systems with sizes less than the localization length, as well as the inelastic scattering length (or phase coherence length) due to phonons because the experiments are done at a finite temperature, however low. Recent experiments [6] by Webb et al on metal wires and rings and by de Vegvar et al on GaAs wires clearly demonstrate the existence of a strong non-linear conductance (somewhat oscillatory as a function of the field, as well as asymmetric under field reversal). The explanations put forth are as follows:

In a disordered quantum system, the energy levels are randomly spaced. The energy range within which this random distribution is correlated is called the Thouless energy (E_T) and is a measure of the degree of disorder in the system [7]. In passing note that for ordered systems (extended states), this energy covers the entire energy domain, which in this case are the energy band/s. According to a theory due to Altshuler, Khmelnitskii and Larkin [8], the non-linearity shows up for $V > V_c = E_T/e$, where e is the charge of the carrier. According to a more recent work by Tang and Fu [9], this non-linearity becomes observable as soon as δE, the change in electronic energy due to the imposed finite voltage, becomes equal to ϵ, the level spacing of the random spectrum. Since the levels are not symmetrically placed around the Fermi energy (for $V = 0$), asymmetry under $V \to -V$ is also explained.

Before ending this section, we remark that resistance (conductance) is a rather delicate concept in a truly quantum system at zero (or very low) temperature. In such a system, all the scatterings are elastic even in the presence of impurities and thus there is no dissipation within the sample. Hence no constitutive relation like eq. (1.1) or (1.2) holds for small parts of the sample. One visualizes two reservoirs (two electrodes of a battery) connected by semi-infinite perfect leads at the two ends of the sample. Resistance appears only because of backscattering of waves incident on the sample, but all the non-equilibrium processes (like dissipation) are assumed to occur in the reservoirs [10]. Thus the connection between the truly quantum regime and the regime where microscopic parts of the sample themselves have dissipative mechanism built-in (section 2.2 below) involves the difficult issue of how macroscopic laws develop dissipation (time irreversibility) from microscopic laws which are reversible. In this work, we will not discuss about this kind of systems any further.

2.2 Electric Field and Non-linear Composites; General Considerations

For systems away from the quantum regime, the microscopic disorder is not of atomic dimension (e.g., metal granules in insulating host or vice versa), but not completely macroscopic (of lab dimensions) either. If one takes the charge carrier (electron or hole) as a random walker, a metal corresponds to diffusive motion giving rise to Ohm's law. This diffusive behavior changes in the presence of tunneling across a barrier, like M-I-M, S-M-S, M-s-M, s-I-s, etc. junctions (M stands for a metal, S a superconductor, and s a semiconductor). So, these systems are *semi-classical* in the sense that one does not solve for the wave function but takes the quantum mechanical effects indirectly by including only the tunneling currents through barriers [11]. A very good, though somewhat dated, sourcebook for the tunneling behavior of different types of junctions, is by Burstein and Lundqvist [12]. We discuss a few cases from that book. For example, a potential barrier called Schottky barrier arises within a semiconductor in intimate contact with a metal and gives rise to $i \propto \exp(v/E_0)$, where E_0 is a function of temperature. In the case of a p-n junction heavily doped with Sb donors, the current seems to go from one linear region to another in steps, or the conductance smoothly connects two plateaus. Qualitatively similar curves are observed for InP tunnel diode, Au doped Ge tunnel diode, etc. [13]. Further there could be inelastic processes by which an electron loses energy to the vibrational modes of species within the barrier. These processes thus provide extra paths or channels for current flow (thus increasing the conductance), and hence one may observe breaks in the $i - v$ curves joining two piece-wise linear (Ohmic) regions.

It is clear that if these tunneling ('non-linear') bonds, present in the system, become linear as discussed above beyond some characteristic microscopic voltage, then the non-linear conductance of the whole system would change over to Ohmic behavior (constant differential conductance according to Eq. (1.2)) beyond some system-specific applied voltage. The point which may not be apparently obvious is that even if the non-linear bonds with slope greater than unity never become linear, their resistances become zero ('super'-conductor) at infinite voltage or frequency (see Secs. 3.3.1 and 3.4), and hence the system saturates at high voltage or frequency to another linear region or Ohmic behavior (since the tunneling bonds do not percolate by themselves). *In either case, the resistive behavior of the original percolating structure (no voltage) guides the resistive behavior of the shorted percolating structure* (see Sec. 3.3.1). Thus, for vanishing driving fields, the system is in some state providing a fixed number of channels for the response, but asymptotically for infinitely large driving fields, the system is in another state representing a much higher number of channels. The generalized susceptibility (which could be thermal or electrical conductivity, any elastic moduli, or permeability, etc.) of this two (or, multi)-level system crosses over from one asymptotic linear region (value) to another (of higher saturation value) through some non-linear region [14]. It may be noted here that in breakdown (e.g., electrical fuse) phenomena, essentially the opposite thing happens, namely the system crosses over from a higher susceptibility value (highly efficient and stable network) to a lower one where the system cannot hold itself anymore. In electrical terminology, the system as a whole fuses to give rise to an insulator, or in mechanical terminology, the structure becomes mechanically unstable. *Thus this enhancement or (de-enhancement) of the number of system-spanning channels and their eventual saturation as a function of some driving field is at the heart of all the non-linear phenomena* discussed, for example, in the present volume.

2.2.1 Experimental Facts

There have been many experiments [15] on non-Ohmic behavior, including an early one by van Beek and van Pul on carbon-black loaded rubbers and a very recent one by Bhattacharyya et al on a lipid bilayer doped with charge-transfer complexes. References [15-18] are meant to be representative and not exhaustive. Key features collected from various of these experiments are summarised below.

(i) Low percolation threshold: For example, in the experiment on carbon-wax system, there is a percolation transition at $p_c = 0.0076$ [16]. Scanning electron micrograph [17] shows a filamented and highly tenuous structure for the carbon phase above p_c. Similar low value for p_c is also reported for carbon-black-polymer composites by van der Putten et al [15].

(ii) Qualitatively identical I-V (as well as dI/dV vs. V) response for volume fractions both above and below the threshold: In the Chen and Johnson experiment [18] on Ag particles in KCl matrix similar non-linear transport was reported both below and above $p_c = 0.213$, even though the exponents for these two regions were grossly different. This behavior along with low p_c (which we call p_{ct}, see Sec. 3.2) indicates identical origin (namely, *tunneling/ hopping between disconnected conducting regions*) for this non-linearity and we predict that the exponents should be same both below and above p_c.

(iii) Power-law growth of conductance: For small V, the excess conductance (above the constant/ Ohmic part) is claimed to grow as a non-integer power law with exponent $\alpha \cong 1.36$ in 3D [16b], whereas previous works took α to be integer $= 2$.

(iv) Cross-over current for non-linearity: The current (or voltage) at which the zero-voltage conductance G_0 deviates by some chosen fraction ϵ is seen to scale as $I_c \propto G_0{}^x$ ($V_c \propto G_0{}^{x-1}$), where $x \cong 1.4$ [16, 17] in 3D, and $x \cong 1.5$ in 2D as found by Gefen et al [15] (Actually, the latter experiment is also consistent with 3D; see Sec. 3.3.2). One can easily check that this exponent is related to α above by $\alpha = 1/(x-1)$. We would like to address the discrepancy in this regard in ref. [16].

(v) Non-linear ac response: An excellent review of both theoretical and experimental works devoted to non-linear ac response of two-phase inhomogeneous materials with percolating networks, their scaling properties etc. has been given in [19]. Experiments on carbon-wax [16, 17], alongwith many previous ones in disordered/ amorphous systems, find a non-integer power-law behavior [2] even for $p > p_c$. More interestingly it has been claimed [16b] that for a fixed voltage, $[G(\omega) - G(0)] \propto \omega^\alpha$, where $\alpha = 1.36$ is the same as in (iii) above.

3.1 A Model

For model we take a bond-percolating structure in the presence of tunneling bonds. In some sense, this may be thought of as a new model for highly correlated percolation, as explained below, on a lattice [20]. For simplicity, we consider here the situation in two dimension only, and take the lattice as a square lattice. One first throws Ohmic bonds (o-bonds) of conductance g randomly on this lattice at a volume fraction p. One knows that geometrical bond-percolation occurs for this lattice at a critical value of $p_c = 0.5$. Additionally now one imagines tunneling to occur wherever two o-bonds are geometrically separated by the nearest-neighbor distance (lattice constant) and no further. We represent this tunneling by t-bonds. The conductance of the t-bonds are explicitly non-linear (or even complex for the study of frequency dependence of conductivity). Again for simplicity we have chosen for the static (dc) case, the non-linearity in the t-bonds to be piecewise linear (as has been discussed above to be the case for tunnel junction diodes), in the sense that the conductance has been chosen to be zero upto some threshold voltage v_g and to be some fixed finite value above this voltage. This kind of non-linearity was also chosen by Roux and Herrmann [21], except that the disorder in their model did not originate from an underlying percolating structure, but rather from the fact that v_g for each bond in their model was chosen to be random.

3.2 A New Percolation Threshold and its Universality Class

Let us, for simplicity, assume in this section that the t-bonds, i.e., the bridge-bonds, conduct at all voltages (i.e., $v_g = 0$). Obviously in the presence of the fully correlated t-bonds at positions determined by the formations of single gap

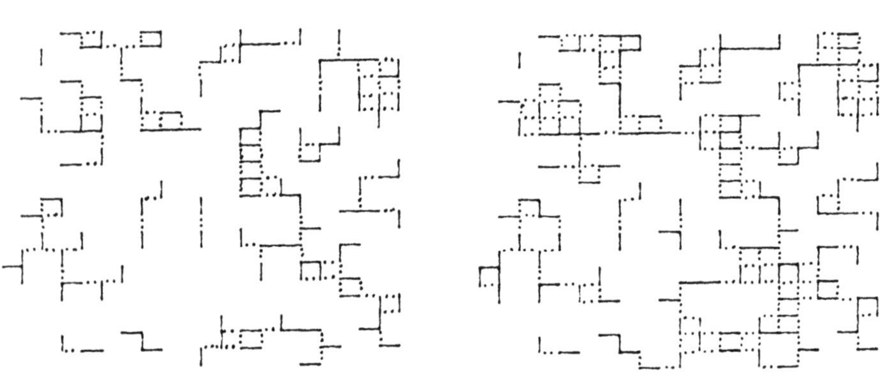

(a) (b)

Fig. 1. Two realizations of a square lattice with L=20. The solid bonds with concentration p are ohmic and dotted bonds between two nearest-neighbor solid bonds are tunneling bonds. (a) $p = 0.16$, the lattice does not percolate, and (b) $p = 0.2$, where it does.

between the randomly occupied o-bonds, the system should electrically percolate even when the o-bonds are not geometrically connected, i.e., close to but below $p = 0.5$. To show this effect, we show in Fig. 1 two different realizations of the square lattice for $p = 0.16$ and 0.2 respectively for a system size $L = 20$ (we avoid bigger L for clarity in the figure). The o-bonds are shown by full lines and the the t-bonds are shown by dotted lines in the figure. Clearly, the o- and t-bonds together span the system for $p = 0.2$ but do not do so for the case of $p = 0.16$. The question of whether this newly proposed model of bond percolation has a new threshold or not, or more precisely, whether there is a true phase transition at a very different threshold or not has been considered by us in detail elsewhere [20]. We find by using small cell renormalization [22] and finite size scaling that for the square lattice, the transition (electric percolation) occurs at $p_{ct} \cong 0.18$ which is significantly smaller than the well-known geometrical bond-percolation threshold, $p_c = 0.5$. The *universality class* for this new percolation, though, seems to be the same as that for simple geometrical percolation, because the correlation length exponent ν near p_{ct} still comes out to be about 1.35 and 1.4 (by using two different methods), which is close to the value of 4/3, the ν exponent for geometrical bond-percolation in 2D.

The results above indicate a few interesting features: (i) The effective percolation threshold may become very small in the presence of t-bonds. Clearly, the threshold becomes progressively lower if tunneling occurs between farher and farther neighbors and may well explain why p_c in the carbon-wax composite [16] in 3D is so small (0.0076)! Of course, there could be other reasons for this smallness, like very elongated shapes of the carbon inclusions, but still the effect of the t-bonds cannot be ruled out, since it seems quite obvious that tunneling through dangling bonds must play an important role in that system. (ii) When all the bridge bonds in the system are connected, i.e., in the piece-wise linear case, when all of them have voltages greater than v_g across them, the singularity behavior of the lattice is not different from that when all these bonds are disconnected (since the correlated lattice was shown to fall in the same universality class). Thus it is unlikely that the singularity behavior will be different in-between where all the t-bonds are not connected, i.e., in the non-linear regime where the system crosses over from one linear region (asymtotically zero voltage regime) to the other (asymptotically infinite voltage regime).

3.3 The Model Random Resistor Network and Non-linear I-V Response

Next we proceed to solve for the response of the dilute square lattice (as in Sec. 3.1) under the action of a dc voltage V across its ends by solving for the Kirchoff's laws at each node of the random resistor network. Numerically this is done for several lattice sizes from $L = 20$ to 80, by using the standard

Gauss-Seidel relaxation technique. We start from some initial arbitrary voltage distribution at each node and iterate through the lattice many times until a certain desired convergence criterion for the final voltage distribution is satisfied. The Ohmic bonds are assumed to have a conductance $g = 1$ (to set the conductance scale) and the t-bonds are piecewise linear as described above (see Fig. 2) with a conductance $t = 0.9$ above a fixed threshold voltage $v_g = 0.5$. We found this simple model to be able to qualitatively explain the hopping-dominated non-linear transport.

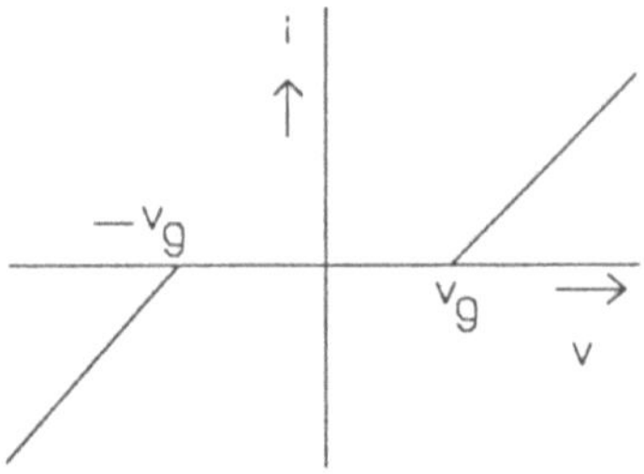

Fig. 2 The current-voltage characteristic assumed for each tunneling bond.

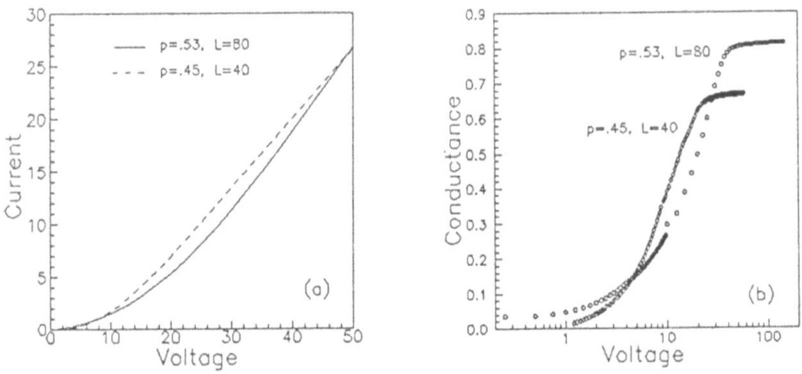

Fig. 3. (a) The non-linear current-voltage characteristic for the whole system (one above and the other below the geometrical percolation threshold, p_c=0.5). (b) Plot of differential conductance against the applied voltage obtained from the direct derivative of I-V curve in fig. 3a.

The response was studied at several p both below and above p_c. As stipulated before we found qualitatively identical behavior both above and below p_c, but always above p_{ct} (otherwise the system never conducts). We show some typical $I-V$ responses in Fig.3 (a) and the resultant dI/dV vs. V in Fig.3 (b) for both $p = 0.53$, $L = 80$, and $p = 0.45$, $L = 40$. The conductances were calculated by first obtaining a smooth fitting of the $I-V$ data and then by taking the derivatives of the fit. It is to be noted that the dI/dV vs. V curve (found numerically from $I-V$ curve) is a distorted S-shaped function between two asymptotic flat regions, the S-shaped non-linear behavior being in the region where electrically straining the system tend to make newer connectivities. It is also to be noted that this non-linear response starts appearing in this model for $V > V_{gs}$ (a configuration-dependent system threshold voltage) since there is a threshold for the t-bonds.

3.3.1 Analysis of data:

To get some hint on how to analyse the non-linear regime, we would first like to analytically solve for a prototype situation, which is something like a minimal representation of our model. This is depicted in Fig. 4, where we have three ohmic

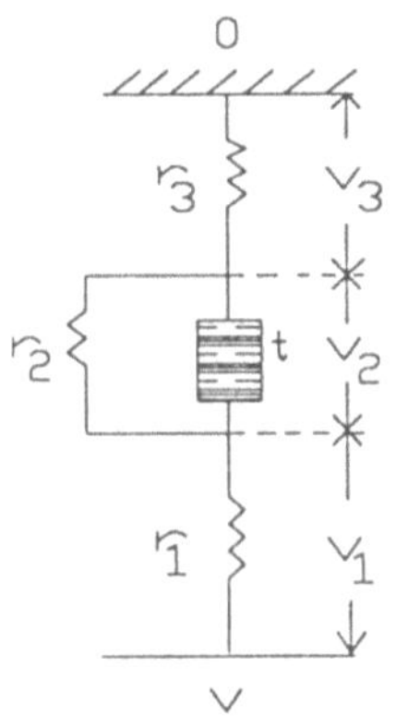

Fig. 4 A model network with the ohmic conductors (resistances r_i) and a nonlinear one (of conductance t) placed between two nearest-neighbor ohmic conductors.

conductors with resistances r_i (conductance $g_i = 1/r_i$) and a non-linear conductor of conductance $t = f(v)$ across the conductor 2 where a bridge bond is assumed to exist. It is to be noted that since there is no randomness in this case, a piecewise linear function $f(v)$ in this case simply gives a piecewise linear response for the whole system. Since that is not interesting enough, we choose first $f(v) = 2qv$. If the voltage applied across the sample is V and if the drop across each resistor is v_i, then one has from Kirchoff's laws:

$$I = v_1 g_1 = v_3 g_3 = v_2 g_2 + q{v_2}^2, \tag{3.1}$$

and

$$V = v_1 + v_3 + v_2 = g_2 v_2/G_0 + q{v_2}^2/G_f, \tag{3.2}$$

where the initial resistance for vanishing V is $R_0 = r_1 + r_2 + r_3 = 1/G_0$ and the asymptotically saturated final resistance (when the t-bond essentially becomes a zero-resistance), is $R_f = r_1 + r_3 = 1/G_f$. Let us call the difference between these asymptotic values $G_d = G_f - G_0$. Then solving for the current from the above two equations (one of which is a quadratic equation for the unknown v_2 in terms of V) and taking derivative, one obtains for the non-linear conductance at any V:

$$G = dI/dV = G_f - G_d[1 + kV]^{-1/2}, \tag{3.3}$$

where the constant $k = 4q{g_s}^2/({g_2}^2 g_f)$. For small V, this function increases from G_0 linearly with V, and for $V \to \infty$, it approaches the higher conductance value G_f as $V^{-1/2}$.

Next we look at the case of a cubic $i - v$ relation for the t-bond, i.e., $f(v) = 3qv^2$. One first obtains a similar set of equations as in (3.1) and (3.2), the latter being a cubic equation for v_2. In this case the final result for the response is:

$$I = VG_f - G_d g_2/G_0[(-b/2 - \sqrt{D})^{1/3} + (-b/2 + \sqrt{D})^{1/3}], \tag{3.4}$$

where $a = g_2 G_f/(qG_0)$, $b = -(G_f/q)V$, and $D = (b^2/4 + a^3/27)$. By taking derivative one may now obtain the conductance, which is some messy looking function. But the important point is that in this case we find that for $V \to 0$, the conductance again increases from G_0 linearly with V, but as $V \to \infty$, conductance saturates to the value G_f as $V^{-2/3}$ (which is faster than in the case of quadratic $i - v$ for t-bonds). One may continue the above exercise with a quartic or higher order function $f(v)$ for the non-linear bond and one finds that while the behavior near $V = 0$ is as before, the approach to the saturaion value of G_f for $V \to \infty$ becomes progressively faster than before.

Guided by the above experience, we try to fit the whole region in our simulation for $V > V_{gs}$ by a function of the form:

$$G = G_0 \qquad (V < V_{gs}) \tag{3.5}$$

and

$$= G_f - G_d[1 + \lambda V'^{\mu}]^{-\gamma} \qquad (V \geq V_{gs}) \tag{3.6}$$

where $V' = V - V_{gs}$ is the shifted voltage from where non-linearity appears. For concreteness, we discuss here the fitting of the data for $L = 40$, $p = 0.6$. For this sample $G_0 = 0.154$, $G_f = 0.881$. The parameter for the best fit in this case as obtained by a simplex search procedure are $\lambda = 3.27 \times 10^{-4}$, $\mu = 1.408$, and $\gamma = 125$. Such large values of γ are obtained for all the cases studied and yet the approach to G_f was slower than in actual data. *This obviously indicates that the approach to saturation is not power-law type.* We tried four possible exponential functions replacing eq. (3.6):

$$G = G_0 + G_d \exp(-\lambda/V') \tag{3.6a}$$

$$= G_0 + G_d \tanh(\lambda V'^{\alpha}) \tag{3.6b}$$

$$= G_f - G_d \exp[-\gamma \exp(\lambda V'^{\alpha}) - 1] \tag{3.6c}$$

$$= G_0 + G_d[1 - \exp(-\lambda V'^{\mu})]^{\gamma} \tag{3.6d}$$

Out of all the optimally fitted functions as described above, (3.6d) is the best in the sense that it gives the minimum mean square deviations (MSD) from the data. (Out of 15 cases studied, only one with $L = 80$ did fit best with eq. (3.6c), the double exponential.) It may be noted that a special case of (3.6d) with $\gamma = 1$

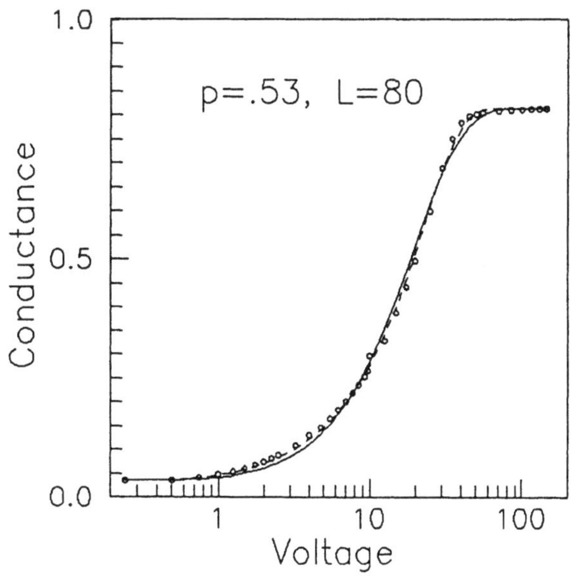

Fig. 5 One example of a typical fit of G (differential conductance) against V (voltage) using the functional form (3.6d). The full line with fixed γ=1 does not fit the data as well as the dashed line with $\gamma = 0.45$.

was the form of non-linearity used by Chen and Johnson in fitting their data [18]. But for all p and L's considered by us, $\gamma = 1$ was found inadequate for fitting the S-shape. For example we have shown in Fig. 5, the fitting with the function (3.6d) for $p = 0.53$, $L = 80$: the restricted case of $\gamma = 1$ shown in full line, and an unrestricted, optimally fit $\gamma \neq 1$ in dashed line. Clearly the unrestricted case fits the data extremely well and the restricted case does not (see also sec. 3.3.2 below). Next important point is that for small V', near the onset of non-linearity, the excess conductance changes with voltage as a power-law:

$$(G - G_0) \sim V'^{\mu\gamma} = V'^{\alpha} \qquad (say). \tag{3.7}$$

Following Gefen et al [15], we find that $\alpha = t/\nu$, where t is the conductivity exponent given by $G_0 \sim (p - p_c)^t$. Thus one expects that $\alpha = 1.3/1.33 \cong 0.97$ in 2D. What we find considering samples both below and above p_c that, on an average $\alpha \cong 0.9 \pm 0.2$ for our model 2D system. Next we found that G_d is dependent on G_0 (as discussed Sec. 2.2), and by plotting data for these two between $(p - p_c) = 0.02$ to 0.22, we find that $G_d = a - bG_0^y$, where a, b are positive constants. For $L = 40$, 50, and 80 (Fig. 6), y is about 1.73, 1.38, and 1.27 respectively. Considering finite size effects, it seems that $y \to 1$, which is consistent with the fact that the correlated lattice (Sec.3.2) is in the same universality class [i.e., $G_f \sim (p - p_c)^t$].

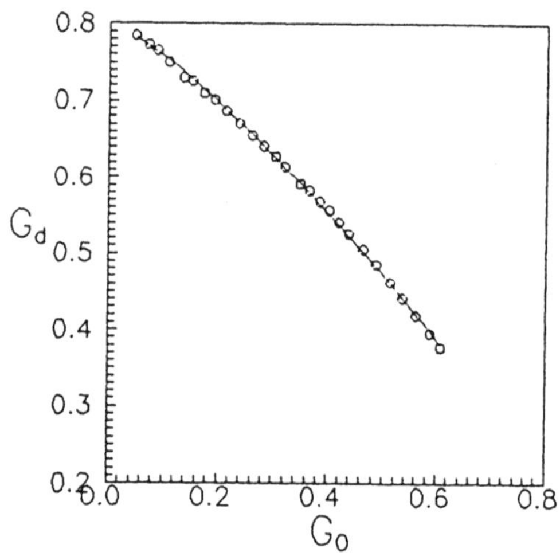

Fig. 6 Best fit of G_d (difference between conductances at zero and infinite voltage) against G_0 (conductance at zero voltage) for a lattice of linear size $L = 80$ with p ranging from 0.52 to 0.72. The best fit line is $G_d = 0.8 - 0.79 G_0^{1.27}$.

3.3.2 Relevance to Carbon-Wax Experimental Data

We have also applied the above analysis on an experimental set of data on carbon-wax [16] and again found that the MSD of the fitted data with the function

(3.6d) with $\gamma \neq 1$ is much smaller than that for the function (3.6d) with $\gamma = 1$. In Fig. 7 we show the best fit of these experimental data with Eq.(3.6d). To further indicate how good this fit is, we used the parameters obtained above with the function (3.6d), to calculate dG/dV and plotted them against their experimental values obtained independently from second harmonic amplitude with a low-frequency ac field [16]. This agreement should be considered rather good given that G and dG/dV are independent measurements, and that higher the harmonic less is its contribution to current and hence it becomes more error-prone. In contrast, one notes that the empirical scaling function for G as given in Eq. (8) of ref. [16b], which has a real part essentially similar to Eq.(3.6) above, seems to fit the low voltage data for G (below the saturation G_f) quite well, but the parameters obtained therefrom give predicted dG/dV which are very different from experimental dG/dV. Next this fitting of the data predicts that the power-law for small V has an exponent $\alpha \cong 1.74$, while a simple power-law fitting in [16b] predicts $\alpha \cong 1.36$. We would simply comment here, as discussed in Sec. 2.2.1(iv), that experiments on crossover current indicate the exponent x to be close to 1.5 and hence $\alpha = 1/(x-1) \cong 2$. Looked at differently $\alpha = t/\nu \cong 1.9/0.88 \cong 2.1$ in 3D. Thus a value of $\alpha \cong 1.74$ for the experimental data [16b] is more consistent than 1.36 reported in [16b].

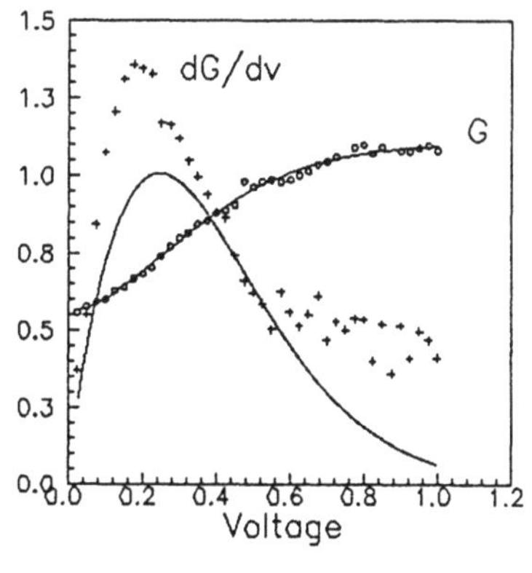

Fig. 7 Fitting of a set of experimental data (from ref. [16]) with Eq. (3.6d). We get $\gamma = 1.19$, $\mu = 1.52$, giving $\alpha \cong 1.74$.

3.4 A Model Random RC Network and Non-linear Frequency Response

If one now considers an ac electric field, the tunnel junctions are expected to behave as leaky capacitors (C) with a high resistance in parallel, because of polarization effects in an insulator in the vicinity of two nearby conductors. Within our correlated RC model (where *capacitors occur only at the positions of the bridge bonds*), the ac conductance is expected to behave non-linearly (for any $p > p_{ct}$) between two saturation regions of $\omega \to 0^+$ and $\omega \to \infty$ (just like the dc response). Indeed one can do an exercise similar to that in Sec. 3.3.1, just by replacing t in Fig. 4 by $j\omega C$, where $j = \sqrt{-1}$ and ω = circular frequency, and one finds a behavior qualitatively similar. In Fig. 8, we have shown the real part of the complex conductance for $p = 0.52$ and $L = 20$. The details have been discussed elsewhere [23] and the main result is that the real part of the ac conductance $G(\omega)$ for a constant $V(= 1)$, is very well described by a function of the form similar to Eq (3.6d), namely:

$$G(\omega) = G(0) + G_d[1 - \exp(-\lambda\omega^{\mu})]^{\gamma} \tag{3.8}$$

But the exponent $\alpha = \mu\gamma$ in this case is about 0.7. Qualitatively similar behavior is obtained for carbon-wax sample [17], where the possibility of saturation at high frequency does not seem to be fully explored and the data is fitted in the intermediate frequency range only by a monotonic power-law increasing behavior ($G \to \infty$ for $\omega \to \infty$). The exponent α obtained in [17] and many other experiments on

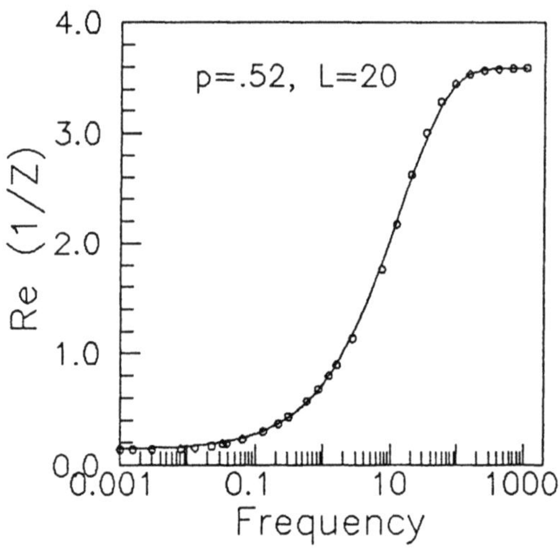

Fig. 8 Real part of the complex-conductance plotted against ω (circular frequency) for our model R-C network, with leaky capacitors placed at the tunnel junctions. The solid line shows the functional form (3.8).

amorphous systems [24] as well as in an 'Extended Pair Approximation' (EPA) theory [25] in the intermediate ω range is close to what we obtain here. Also note

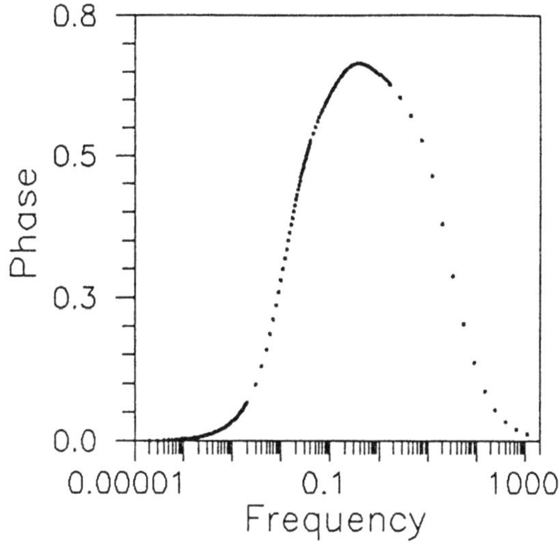

Fig. 9 Plot of phase angle against ω for the same model as in fig. 8. The network is seen to be purely resistive for both $\omega \to 0$ and $\omega \to \infty$.

that for a 2D system exactly at p_c it is 0.5 within a RC model of Clerc et al [19]. Two things to note here: (i) While the experimental samples [16, 17] are above threshold, we predict similar ac behavior for samples both above or below p_c. (ii) Careful analysis of data [23] shows that for very low ω, $\alpha = 1$, which then crosses over first to 2 and then to 0.7 in the intermediate frequency range, i.e., before the effects of saturation show up. Thus there is no evidence of the exponent for the dc and the ac case to be the same as claimed in [16b]. We have also calculated the variation of the phase angle (ϕ) with frequency (Fig.9). It is seen clearly that ϕ is zero (the sample is purely resistive) both for $\omega = 0$ and for $\omega \to \infty$. Interestingly the peak value $\phi_m \cong 0.7$ (radian) for $p = 0.52$ in this case seems to be very close to the universal phase angle at p_c predicted by Clerc et al for 2D [19].

Acknowledgement

We thank B.K. Chakrabarti, K.K. Bardhan and S. Basak for many useful discussions in the early stages of the work. Thanks are also due to K.K. Bardhan and R.K. Chakrabarty for discussion of their experiments and for supplying some data on carbon-wax composites and to B.K. Chakrabarti for a critical reading of the manuscript alongwith very useful criticism. Discussions with S. Roux and H. Herrmann are also gratefully acknowledged.

References

1. G.D. Mahan, L.M. Levinson and H.R. Philipp, J. Appl. Phys. **50**, 2799 (1979)
2. *Hopping and Related Phenomena*, eds. H. Fritzsche and M. Pollak, World Scientific, Singapore (1990); *Hopping Transport in Solids*, M. Pollak and B. Shklovskii, Elsevier, North Holland (1991); *Fourth International Conference on Hopping and Related Phenomena*, Marburg 1991, ed. H. Bottger, Phil. Mag. B **65**, 593-894 (1992)
3. E.E. Mendez and G. Bastard, Phys. Today **46**, 34 (1993) and references therein.
4. C.M. Soukoulis, J.V. Jose, E.N. Economou and Ping Sheng, Phys. Rev. Lett. **50**, 764 (1983); A.M. Jayannavar, Solid State Commun. **69**, 703 (1989)
5. B.K. Chakrabarti, in this Proceeding
6. R.A. Webb, S. Washburn and C. P. Umbach, Phys. Rev.B **37**, 8455 (1988); P.G.N. de Vegvar, G. Timp, P.M. Mankiewich, J.E. Cunningham, R. Beringer and R.E. Howard, Phys. Rev. B **38**, 4326 (1988). See also the excellent review on the subject by S. Washburn and R.A. Webb, Rep. Prog. Phys. **55**, 1311 (1992)
7. B.L. Altshuler and P.A. Lee, Phys. Today **41**, 36 (1988); R.A. Webb and S. Washburn, *ibid* p. 46 (These are two introductory reviews).
8. B.L. Altshuler and D.E. Khmelnitskii, Pis'ma Zh. Eksp. Teor. Fiz. **42**, 291 (1985) [JETP Lett **42**, 359 (1985)]; A.I. Larkin and D.E. Khmelnitskii, Zh. Eksp. Teor. Fiz. **91**, 1815 (1986) [Sov. Phys. JETP **64**, 1075 (1986)]
9. H. Tang and Y. Fu, Phys. Rev. Lett. **67**, 485 (1991)
10. See, e.g., Y. Imry in *Directions in Condensed Matter Physics*, pp. 101-163, eds. G. Grinstein and G. Mazenko, World Scientific, Singapore (1986)
11. Even in this semi-classical sense, these barriers are still fundamentally different from the Ohmic resistors since they contribute to the resistance but not to heat dissipation.
12. E. Burstein and S. Lundqvist eds., *Tunneling Phenomena in Solids*, Plenum, New York, 1969
13. For quite high voltages, other more complicated situations (e.g., negative differential conductance) may arise, but we donot address them in our work discussed here.
14. This behavior is reminiscent of Daniel Bideau's movie on elastic networks under increasing strains (shown in the present Workshop) and discussed in the present volume.

15. L.K. H. van Beek and B.I.C.F. van Pul, Carbon **2**, 121 (1964); G.E. Pike and C.H. Seager, J. Appl. Phys. **48**, 5152 (1977); Y. Gefen, W-H. Shih, R.B. Laibowitz and J.M. Viggiano, Phys. Rev. Lett. **57**, 3097 (1986); E.K. Sichel, J.I. Gittelman, and P. Sheng, Phys. Rev. B **18**, 5712 (1978); D. van der Putten, J.T. Moonen, H.B. Brom, J.C.M. Brokken-Zijp and M.A.J. Michels, Phys. Rev. Lett. **69**, 494 (1992); D. Bhattacharyya, R. Basu, A. Ghosh, A. Manna, A. Nandy and P. Nandy, Biophys. J. **64**, 550 (1993)
16. (a) K.K. Bardhan and R.K. Chakrabarty, Phys. Rev. Lett. **69**, 2559 (1992); (b) Phys. Rev. Lett. **72**, 1068 (1994)
17. R.K. Chakrabarty, K.K. Bardhan and A. Basu, J. Phys.: Condens. Matt. **5**, 2377 (1993); see also J. Robertson, *Amorphous Carbon*, Adv. Phys. **35**, 318-374 (1986) for filamented structure of a-C.
18. I-G. Chen and W.B. Johnson, J. Mat. Sc. **27**, 5497 (1992)
19. J.P. Clerc, G. Giraud, J.M. Laugier and J.M. Luck, *The Electrical Conductivity of Binary Disordered Systems, Percolation Clusters, Fractals and Related Models*, Adv. Phys. **39**, 191-309 (1990)
20. A.K. Sen and A. Kar Gupta, unpublished
21. S. Roux and H.J. Herrmann, Europhys. Lett. **4**, 1227 (1987)
22. H.E. Stanley, P.J. Reynolds, S. Redner and F. Family in *Real-Space Renormalization*, pp. 169-206, eds. T.W. Burkhardt and J.M.J. van Leeuwen, Springer-Verlag, Berlin (1982)
23. A.K. Sen and A. Kar Gupta, unpublished
24. A.R. Long, *Hopping Conductivity in the Intermediate Frequency Range*, pp. 207-231 in *Hopping Transport in Solids* [2]
25. S. Summerfield and P.N. Butcher, J. Phys. C **15**, 7003 (1982)

Non-Linear Effects at the Critical Supercurrent in Josephson Junctions Arrays

P. L. Leath and Y. Cai

Department of Physics and Astronomy Rutgers University, Piscataway, NJ 08855-0849

Abstract: The work of Leath and Xia on the behavior of disordered arrays of Josephson Junctions near the critical current is reviewed with particular emphasis on the non-linear aspects which distinguish it from the breakdown behavior in linear cases such as fuse network failure and fracture with breakable linear springs. In particular we discuss the creations of vortices at breakdown and the self-organized criticality that comes from a sharing of the supercurrent as the entire network fails. Also we give a preliminary review of our recent work on the fluctuation behavior at the transitions on the risers of the Shapiro steps that occur in perfect Josephson junction arrays which are driven by an ac-current.

1 Introduction

Breakdown phenomena in various randomly disordered systems have been the subject of extensive study in recent years as has been made clear from the papers presented at this conference. But in most cases, such as electric and dielectric breakdown, and fracture, the theory and numerical simulations have been carried out almost exclusively on linear systems, such as fuse networks and arrays of breakable, linear springs as discussed in my earlier talk. In these linear systems, at least when the breakdown is brittle, the statistics of extremes can be used to predict the average breaking point and the failure distribution for an ensemble of random samples. The breakdown is caused by the most critical defect or cluster of defects. Such was also found to be the case in a linearized model superconducting breakdown at the critical current[1].

But for many examples found in nature nonlinearity is very important and affects the breakdown process. A major point of this report (besides understanding the superconducting case better) is to show the role of the non-linearity in a very realistic model. We find that [2,3] in the non-linear model studied here that the statistics of breakdown are qualitatively very much like the linear models[1,4] at least for brittle failure, but that the dynamics near breakdown are very different.

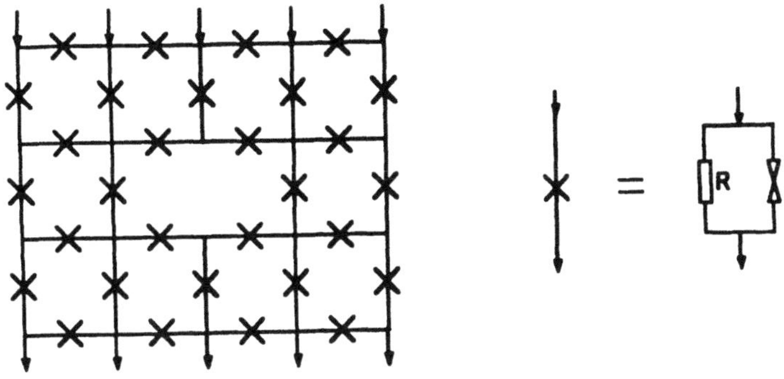

(a) (b)

Fig. 1. a) 5 x 4 junction array with a single defect. b) Each resistively shunted junction can be considered as a microbridge and a resistor in parallel.

Granular superconductors essentially consist of superconducting grains connected by intergrain tunneling through weak links. For overdamped systems, such as thin superconducting films, these links can be modeled by resisitively shunted Josephson-(RSJ-)junction arrays.[5,6] Here we shall consider only the case of two-dimensional arrays of superconducting grains on the sites of a square lattice connected by nearest-neighbor bonds of RSJ junctions. The RSJ junctions (bonds) are considered to be identical and randomness is introduced by randomly removing the bonds and thereby diluting the lattice. An example of a 5 x 4 lattice with a single missing bond is shown in Fig. 1a, where the external current I^{ext} flows in at the top and out at the bottom of the array. Each junction can be considered (Fig. 1b) as a resistor and a microbridge in parallel. For small currents all the current flows through the microbridge as supercurrent and there is no voltage across the bond, but as the current in any junction exceeds the maximum supercurrent J_o, some current must flow through the resistor producing a net voltage and dissipation. The capacitance or charging effect has been ignored; this approximations is valid for small or proximity-effect junctions[6]. The complex superconducting order parameter is of the form $\Delta = \Delta_o e^{i\phi}$, where Δ_o is assumed constant for each grain and ϕ may vary from grain to grain. The Josephson equation for a single junction connecting grains 1, and 2 is

$$J_o sin(\phi_1 - \phi_2) + \frac{h}{2eR}\left[\frac{d\phi_1}{dt} - \frac{d\phi_2}{dt}\right] = I_{12}, \qquad (1)$$

where R is the normal resistance of each junction, and where I_{12} is the total current flowing between the two grains. The first term, on the left-hand side of

(1), is the Josephson supercurrent which can't exceed J_o, and the second term is the normal current due to the voltage, caused by a time-varying phase difference between the two grains. For a network of junctions we sum the total current at site by summing the total current in the surrounding bonds to obtain the equation

$$\sum_{\ell} \left[\frac{d\phi_k}{d\tau} - \frac{d\phi_\ell}{d\tau}\right] = i_k^{ext} - \sum_{\ell} sin(\phi_k - \phi_\ell), \tag{2}$$

where $\tau = 2eRJ_o t/h$ is the dimensionless time, and $i_k^{ext} = I_k^{ext}/J_o$ is the external current inserted into site k (in units of J_o)[i_k^{ext} is zero except at the top (or bottom) of the sample where external current is inserted (or withdrawn)], and where the summations are over the nearest-neighbor sites ℓ. Equation (2) can be written in the generalized matrix form

$$G\frac{d\phi}{dt} = C(\phi), \tag{3}$$

and then solved[7,8] by multiplying this equation by G^{-1} and integrating numerically using a fourth-order Runga-Kutta integration scheme with appropriate initial phases.

When this scheme is used for a perfect array (p = 1.0) it behaves like a single, giant Josephson junction with no voltage across the sample until the bias super current in each level reaches J_o, above which point all junctions begin to pulse in unison with the phase differences across the junctions behaving like a staircase of steps versus time with a regular period

$$T = (i^2 - 1)^{-\frac{1}{2}}, \tag{4}$$

in these dimensionless units, so that the time-averaged voltage is

$$< V >= NR(i^2 - 1)^{\frac{1}{2}}, \tag{5}$$

where N is the number of junctions in the vertical direction. As the current becomes larger, the steps become smaller and more closely spaced, finally merging into a steady ramp corresponding to Ohmic behavior.

2 Disordered Josephson Junction Arrays

The situation is substantially different when defects are present which disturb the supercurrent flow pattern. In order to study this effect we introduce a single, horizontal, linear defect by removing one vertical bond (as in Fig. 1.) The current must flow around this defect which produces a local current enhancement at the ends of the defect. When the applied bias current is small (below the critical current i_c), the phases of each grain are constant in time (reflecting zero voltage) but inhomogeneous in space (representing the current distribution around the defect). But the enhanced current circulating about the defect tips represents an embryonic vortex and anti-vortex attached to the opposite defect ends. When the external current is increased until the enhanced current at the defect tip reaches J_o, there begins a slow pulling of the vortex lines away from the defect tips as the critical bonds at the defect tip refuse to accept additional current

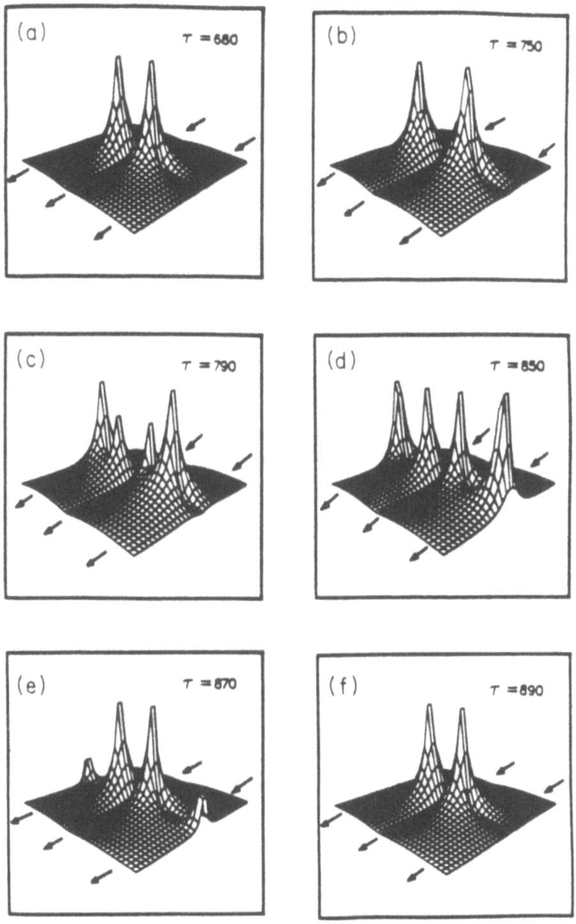

Fig. 2. (a) - (f) Snapshots of vortex movement in a 35 x 34 array with a single linear defect (nine missing bonds in a row) at $i > i_c$ at different times indicated by τ. Picture (f) is identical to (a), and the (a) - (f) process repeats indefinitely. The plot is of the supercurrent in the horizontal bonds vs. positions in the array.

diverting the extra current enhancement further away from the defect tips. The non-linearity thus gives rise to a delay in the onset of voltage while the vortex line is stretching away from the defect before depinning but finally, at $i = i_c$, the critical current, the vortices depin from the defect tips and travel in opposite directions across the the central row of the network perpendicular to the direction of the bias current as is illustrated in the time snapshots of Fig. 2 (which show the supercurrent in each horizontal bond of the array at successive time steps). Immediately when each vortex depins and starts moving across the sample another begins to form at each defect tip (see Fig. 2) so that a procession of vortices is generated which travel across the sample producing the Josephson oscillations (pulses in $d\phi/dt$) in the voltage (Fig. 3) as each vortex line passes

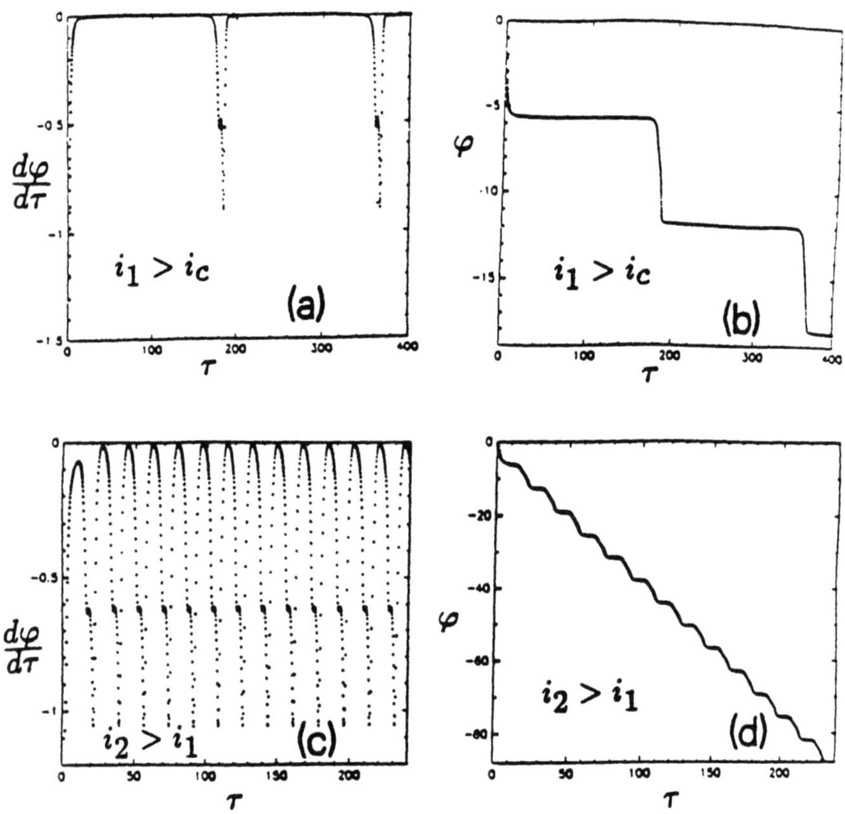

Fig. 3. Phase ϕ (T) vs time τ [(b) and (d)] measured at a single site at the top center of a perfect array with a single vertical missing junction. The rate of change $d\phi$ /d τ vs time τ [(a) and (c)]. Curves (a) and (b) are for an arbitrary current i_1 just above i_c, curves (c) and (d) are for a larger current $i_2 > i_1$.

any particular point of the network. The vertical voltage difference seen across the sample appears entirely across the central row (see Fig. 4) indicating that the central row has gone normal. As in the linear case of the fuse network[4], the critical current, for long defects of length n, obeys the asymptotic power law

$$i_c \sim n^{-\alpha}, \tag{6}$$

with $\alpha = 1/2$, for this case of simple, linear defects. For still larger bias currents i Leath and Xia[2,3] reported the rather complicated I-V curve shown in Fig. 5 which shows another critical current $i_{c3} > i_c$ at which the current in the adjacent rows (just above and below the central row) also goes normal. And, indeed, there must be an infinite series of such critical currents and transitions as, row by row, the entire sample goes normal, but those transitions occur so close together that they are difficult to resolve.

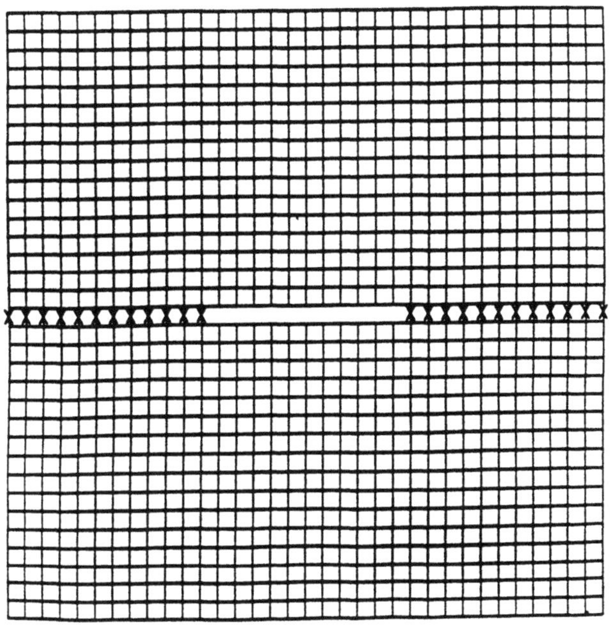

Fig. 4. A perfect 35 x 34 array with a single linear defect. The x's mark those bonds that carry a normal current and that are transversed by the moving vortex lines for current i just above i_c.

Next, Leath and Xia[2,3] reported a numerical study of the funnel defect shown in Fig. 6 which had been purposely constructed to funnel a lot of the applied current into the central region between the points of the two triangular voids and thereby make the central region go normal first, before the sides. But, of course, this scenario doesn't happen because if the central region were to go normal it would develop a resistance which would divert all the current to the superconducting shunts on the sides of the sample. Perhaps, you might think that, then the central region would become superconducting again and an oscillation between the central region and the side channels would develop. Instead,

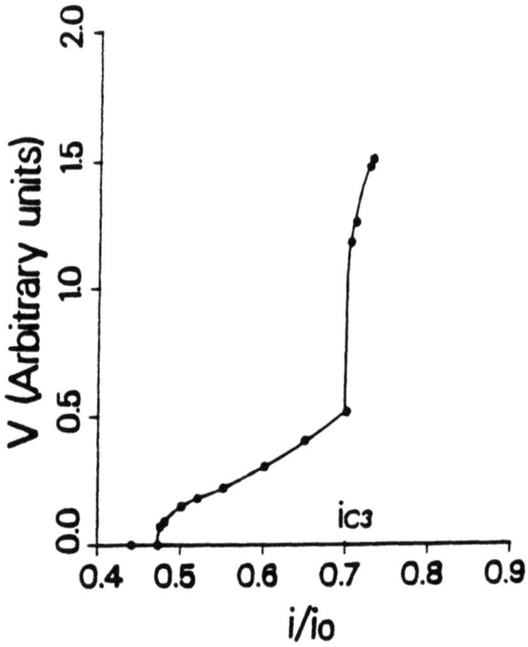

Fig. 5. I-V curve of a perfect array with a simple linear defect of nine missing bonds.

what actually happens is that these oscillations are damped out and the central region just goes up to, but does not exceed, the critical current required to depin the vortices on the interior tips of the triangular defects. Then any *excess* current is shunted around to the side channels until vortices can depin from the outer corners of the triangular defects and an entire path across the sample can go normal at once. In the more general case of a disordered sample, then, one would expect that region after region would go just to the critical current density locally and then accept no further current until a vortex path could be traced across the entire sample. So these self-organized critical regions would grow with increasing bias currents and merge until they could percolate across the sample at i_c blocking all superconducting paths so that the network could develop a potential difference.

Now we consider numerical simulations on randomly diluted, NxN arrays, as illustrated by the 25 x 25 sample at p = 0.90, which is shown in Fig. 7. Leath and Xia [2,3] found several tricks necessary to numerically study such randomly disordered arrays. First, it was necessary to add "bus bars" connecting sites and bonds at the top (and bottom) of each sample to distribute the current across the boundaries of the sample without forcing the current to go through narrow constrictions that randomly can occur at the boundaries of a disordered sample. Second, there are vortices (and anti-vortices) attached to virtually every

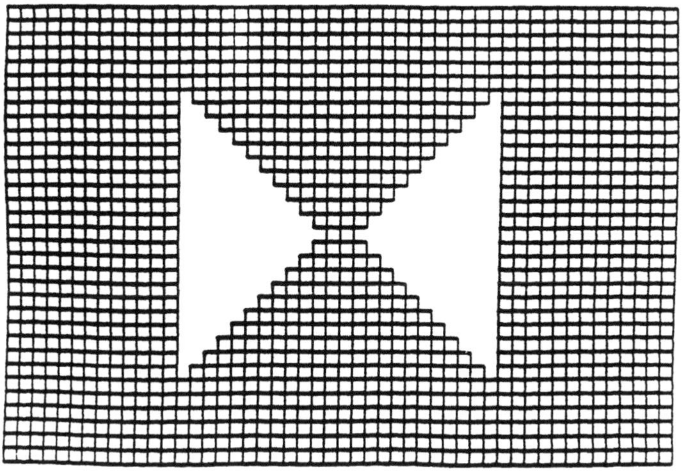

Fig. 6. Perfect array with a funnel defect.

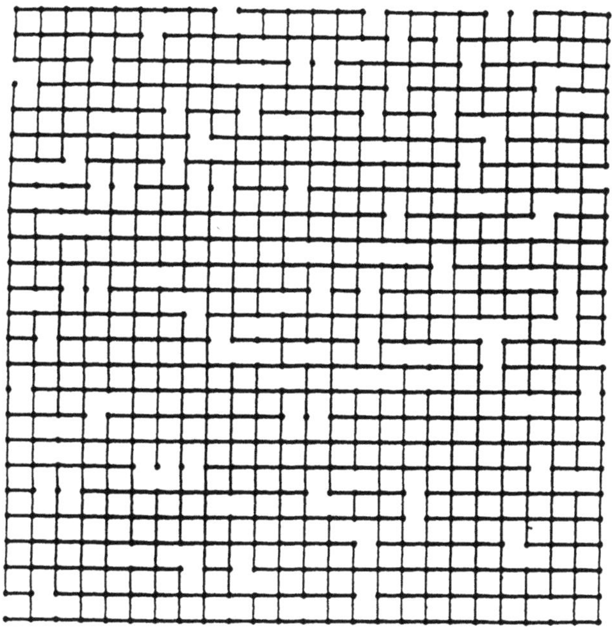

Fig. 7. A configuration of randomly missing bonds in a 25 x 25 array with a concentration p = 0.90 of present bonds.

defect in the array, so any plot of the supercurrent distribution (as in Fig. 2) simply shows a forest of vortex lines, and thus to get a picture of the vortices moving above the critical current it is useful to plot the *normal* current which is non-zero *only* when there are *moving* vortices. Such snapshots of the moving vortices just above the critical current (for the random sample of Fig. 7) are shown in Fig. 8. As the critical current i_c is approached from below, more and more regions become critical (i.e.-locally the bonds carry the Josephson current J_o) until such critical regions which represent a kind of self-organized criticality, span the sample defining the first vortex path. The first vortex path for this disordered sample (i.e.- the path traced by the vortices in Fig. 8) is illustrated in Fig. 9. The solid square near the left edge of Fig. 9 marks the place where the vortices first began to move (once the critical current i_c was exceeded). This vortex motion then spread (at fixed applied current $i \sim i_c$) across the sample

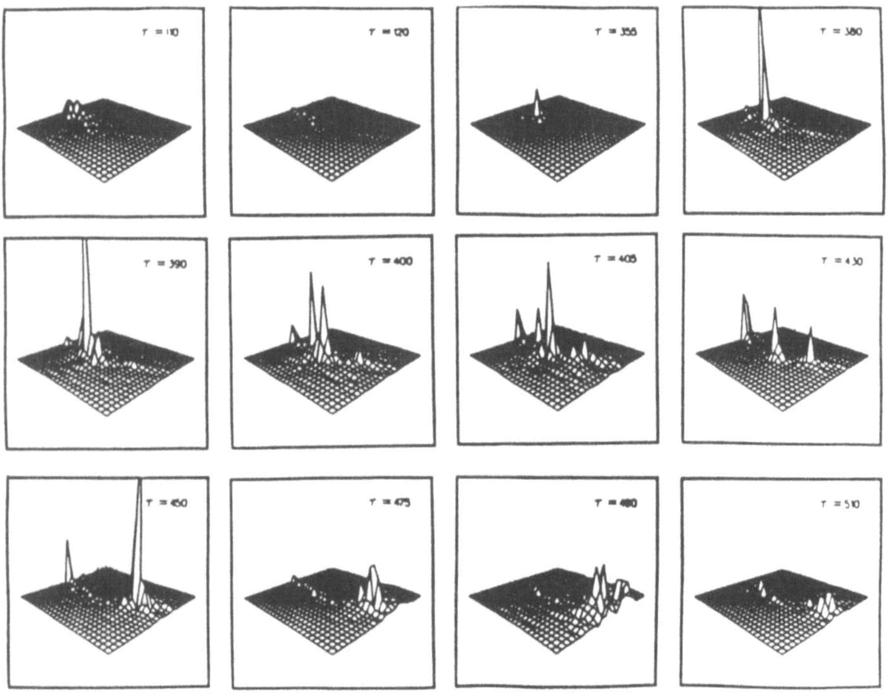

Fig. 8. Snapshots of the normal current in the horizontal bonds vs. positions in the 25 x 25 array of Fig. 7 for $i > i_c$ for various times τ.

and seemed to bifurcate near the right-hand edge of the sample. This initial vortex path is the breakdown path across which the first voltge appears and is entirely analogous to the fracture line in mechanical failure. The vortices as pictured in Fig. 8 fluctuate enormously in amplitude because we are plotting the normal current rather than the supercurrent around the vortices; these fluctuations make the characteristic noise observed in this voltage above the critical current as the vortices jump from defect cluster to defect cluster. This percolation of critical regions as well as the vortices themselves, of course, are a direct result of the non-linearites in the Josephson equations. Similar behavior must exist near the fracture stress in non-linear materials which emit dislocation lines at the tips of spreading cracks.

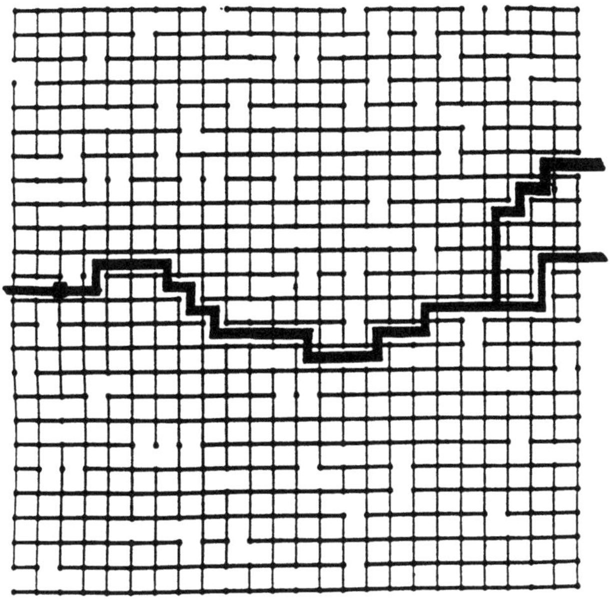

Fig. 9. Approximate vortex path of breakdown for $i > i_c$ in the 25 x 25 array of Fig. 7. The solid square in the third column marks the point of the first vortex motion which spreads across the sample in an irregular manner.

Finally, Leath and Xia[2,3] numerically studied the statistics of the critical current to compare with the linear cases and found that, since this is essentially governed by the statistics of defect clusters, there is little qualitative difference from the failure distributions in linear systems. The failure distribution $F(i)$ (probability of exceeding the critical supercurrent) for samples in an ensemble of 800 randomly diluted 25 x 25 arrays at p = 0.90 is shown in Fig. 10. The plot of

$$A(i) = -\ell n\{-\ell n[1 - F(i)]/L^d\} \tag{7}$$

versus $1/i$ is shown in Fig. 11a, as a test of the double-exponential distribution, and versus ℓn in Fig. 11b, as a test for the Weibull distribution. (This method of testing for the failure distributions was discussed previously at this conference[4]) The results of these numerical simulations suggested that, as in the linear case, the double-exponential distribution was best, especially in the high reliability tail of the distributions. Thus, although it will take numerical simulations on systems larger than 25 x 25 to definitively answer this question, there does not seem at this point to be any substantial qualitative differences between the linear systems and the non-linear ones as far as the failure distribution is concerned. This result would seem to be true because the qualitative behavior of the failure distribution is essentially determined by the defect cluster distribution and is relatively independent of the crack propagation dynamics.

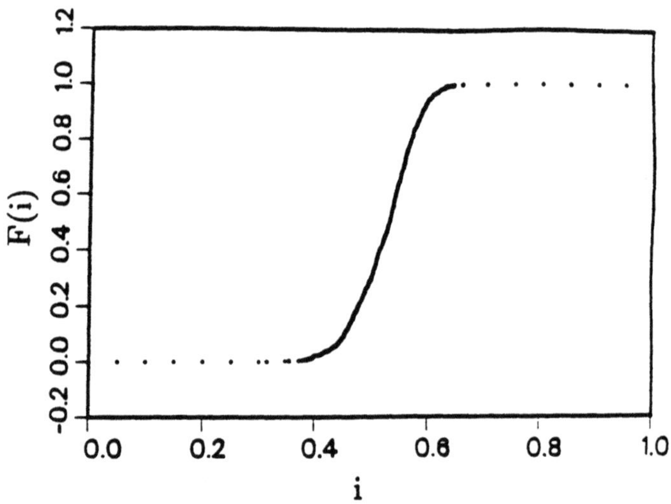

Fig. 10. Failure distribution F (i) vs. i for an ensemble of 800 random 25 x 25 arrays at p = 0.90.

3 Finite Temperature Fluctuations

Next we consider the effects of temperature on the behavior of perfect N x N arrays of Josephson Junctions[9] and finally we use this approach to numerically study the behavior near the Shapiro steps in Josephson junction arrays.

In 1969, Ambegaokar and Halperin[10] showed us how to include finite temperature effects into Josephson junctions. Following their approach we add a random noise current $\tilde{I}_{12}(t)$ to the current equation (1) for each single junction, which gives

$$J_o sin(\phi_1 - \phi_2) + \frac{h}{2eR}\frac{d(\phi_1 - \phi_2)}{dt} + \tilde{I}_{12}(t) = I_{12}, \tag{8}$$

where $\tilde{I}_{1,2}$ is a random variable on each bond (1,2), with the following characteristics

$$< I_{ij}(t) >= 0, \tag{9}$$

and

$$< \tilde{I}_{ij}(t+\tau)\tilde{I}_{kl}(t) >= \frac{2k_BT}{R_{ij}}\delta(\tau)\delta_{ij,kl}, \tag{10}$$

where < > denotes an ensemble average. Thus, the noise currents in different junctions are uncorrelated and the noise current in a single junction (bond) has

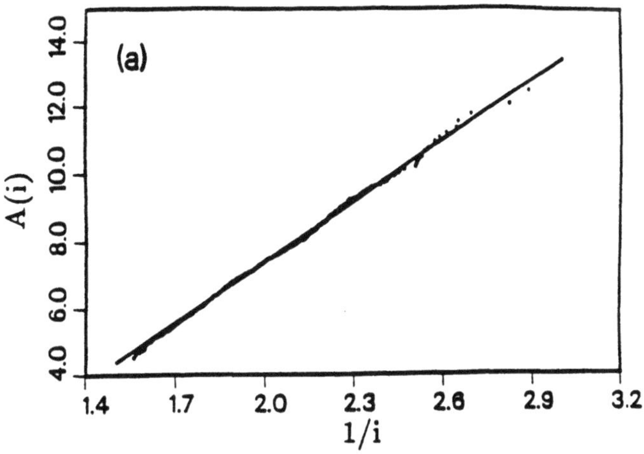

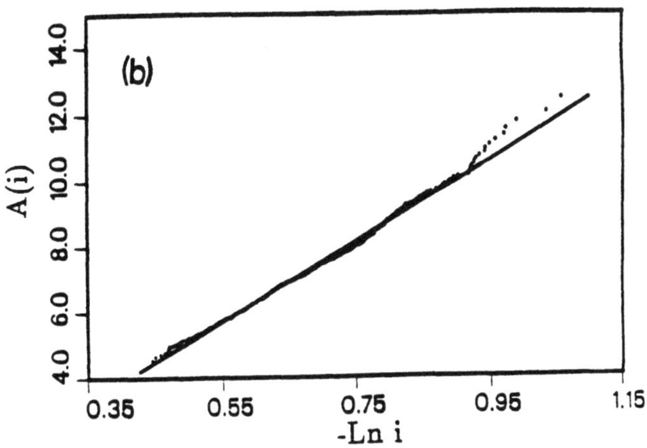

Fig. 11. Plot of A (i) as given by (7) vs $\frac{1}{i}$ [curve (a)] as a test of the double-exponential distribution, and A (i) vs. $\ell n\ i$ [curve (b)] as a test of the Weibull distribution.

zero correlation time corresponding to white noise (Josephson noise). Similarly, for an N x N array the coupled equations (2) have the random $i_{ij}(t)$ terms added for each bond.

For the case of a single junction (8) was solved exactly by Ambegaokar and Halperin[10] [as is reviewed and discussed in detail by Barone and Paterno[11]] who introduced a Fokker-Planck equation for the probability distribution of the phase $\phi(t)$ across the junction. Their result for the average voltage across the junction is

$$< V(t) >_T = TA(T)^{-1}(1 - e^{-i\pi/T}), \tag{11}$$

where

$$A(T) = \int_o^{2\pi} d\chi e^{-i\chi/2T} I_o(T^{-1} \sin(\chi/2)), \tag{12}$$

and where I_o is the zeroth-order Bessel function. It is easy to check that these equations reduce to (5) in limit $T \rightarrow 0$. Cai et al.[9] developed an algorithm, following the work of Chung, et.al,[8] to add the effects of temperature fluctuations numerically by the addition of a random current $i_{ij}(t)$ to each bond at each Monte Carlo time step. In order to check this algorithm they evaluated the average voltage versus T across a single Josephson junction numerically and compared with the Ambegaokar and Halperin formula as is shown in Fig. 12 for three values of the bias current.

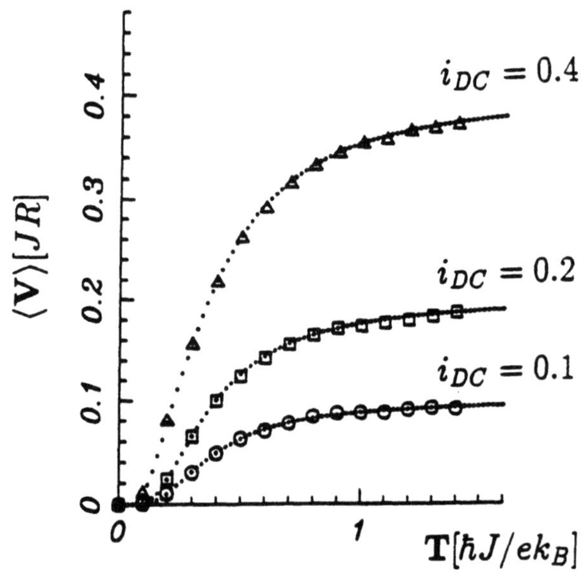

Fig. 12. Average voltage $< v >$ vs. T for a single Josephson junction for bias currents $i_{dc} = 0.1, 0.2$, and 0.4. The symbols represent the numerical solution of (8), the dotted curve represents the analytic solution in Eq (11).

Then Cai et al[9] performed numerical calculations on 16 x 16 arrays using the coupled equations(2) with the random $i_{ij}(t)$ current added to each bond. These arrays showed the Kosterlitz-Thouless (KT) phase transition as had been reported earlier by Mon and Teitel and by Chung et al[8]. In particular the behavior of the average voltage near the KT-phase transition is of the general form[6,9]

$$< V >= iA_1 exp[-A_2/(T - T_{KT})^{\frac{1}{2}}], \tag{13}$$

where A_1 and A_2 are parameters[9]. The comparison of the numerical simulations and (13) are shown in Fig. 13.

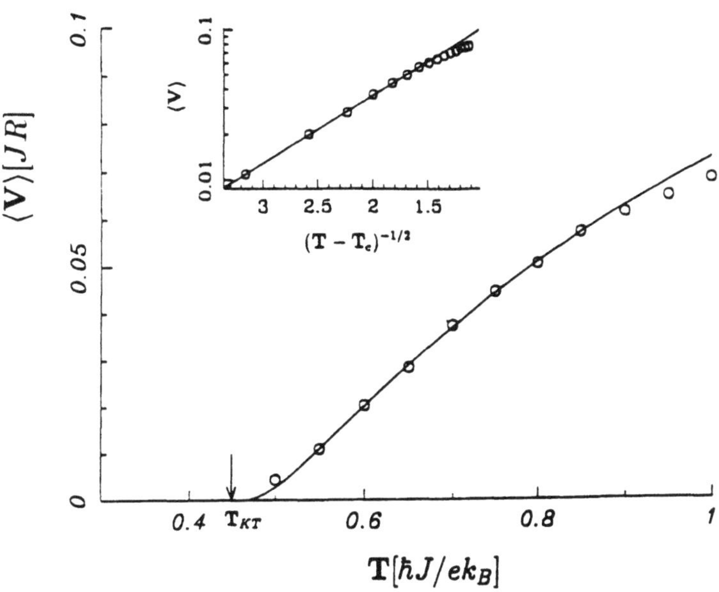

Fig. 13. Average voltage $< v >$ vs. T for a 16 x 16 perfect array of RSJ junctions. The circles are the numerical solution of the coupled Josephson equations, while the solid line is the fit to (13) with $T_{kT} = 0.45, A_1 = 3.01$, and $A_2 = 1.05$. The bias current is $i_{DC} = 01$. Inset: A semilog plot of $< v >$ vs. $(T - T_{kT})^{-\frac{1}{2}}$.

Furthermore this numerical simulation allows for the study of the creation of vortex-antivortex pairs at the KT phase transition. An example, of a single vortex-antivortex pair creation and subsequent annihilation is shown in the snapshots of Fig. 14. One might wonder how such pairs are actually created in the numerical simulations. Careful study of the events, such as that shown in Fig. 14 show that large temperature fluctuations (which appear as large, random values of the noise current $\tilde{i}_{ij}(t)$ at some location in the lattice, and at some time)are carrying essentially the local critical current J_o so that the applied bias current must detour around each side of the hot spot (illustrated schematically in Fig. 15) more or less as if it were a defect in the lattace. This distortion of the current flow creates a current enhancement on each side of the fluctuation and sometimes a pair of such enhancements depin and become a full fledged vortex-

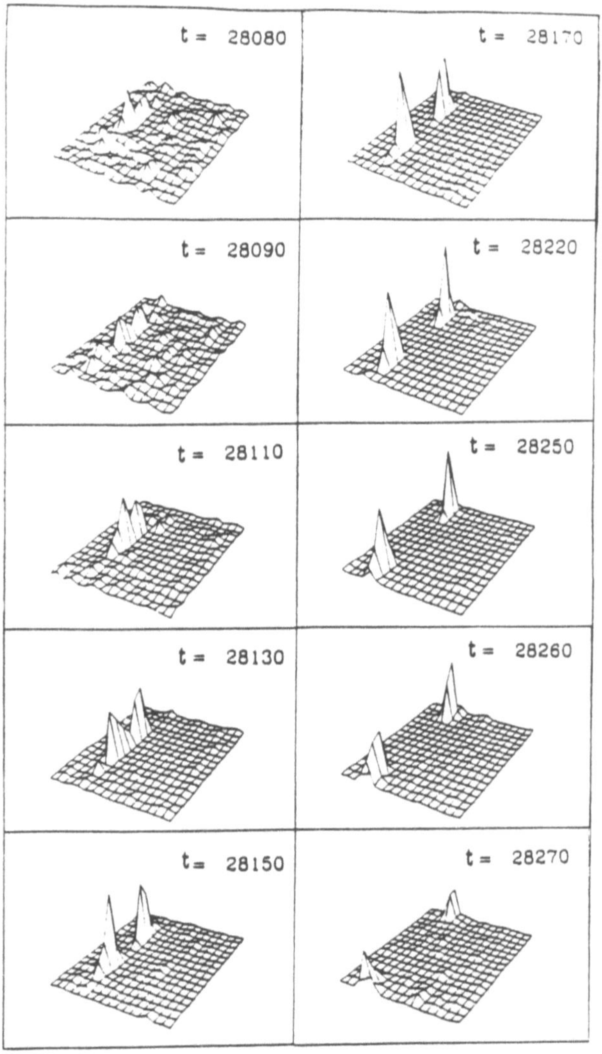

Fig. 14. Snapshots of the creation and annihilation of a vortex-antivortex pair in a perfect 16 x 16 array at T = 0.25, i_{DC} = 0.7, i_{ac} = 0.25, and f = 0.2. The Monte Carlo time steps are indicated on each picture as the temperature fluctuation produces a barrier to the supercurrent flow which causes the vortex and antivortex to be produced on each end of the fluctuation and to depin and travel across the sample annihilating at the periodic boundary conditions. The plot is of the magnitude of the supercurrent flowing in the bonds perpendicular to the external current.

antivortex pair which separates and moves across the sample perpendicular to the bias current. In the simulations these vortices can be counted and studied in great detail. With this new tool we began to study arrays near the Shapiro steps.

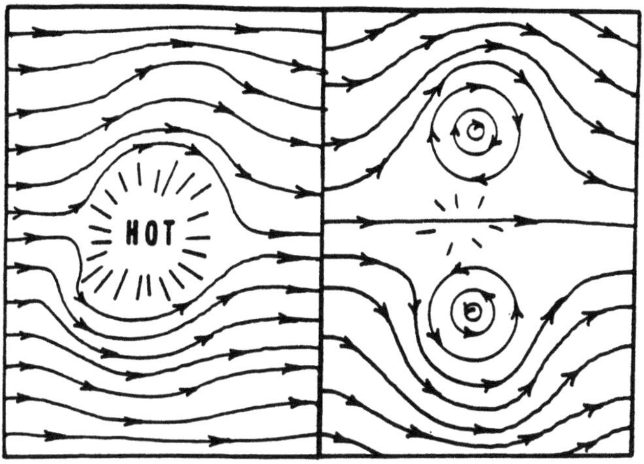

Fig. 15. A schematic picture of the creation of vortex-antivortex pairs by the temperature fluctuations, i_{dc}, and i_{ac}. The current flow produced by $i_{dc} + i_{ac}$ flowing past a high-temperature fluctuation is diverted around it, as if it were a defect, creating a vortex and antivortex on the opposite ends of the fluctuation which can be depaired by the applied current to move across the sample producing a voltage.

4 Risers Fluctuations Near Shapiro Steps

Finally, we give a preliminary report on our recent numerical studies[12] of the fluctuations that occur on the risers of the Shapiro steps as the system makes the transition from one step to the next. In this study we consider a perfect N x N array of Josephson junctions at finite temperatures, with each junction having the random noise current of (9-10), but with the external applied current at the sample boundaries consisting of both a dc-bias current and an ac-current, i.e.-

$$i^{ext} = i_{dc} + i_{ac} sin(\omega t). \tag{14}$$

The coupled Josephson equations of the array are then solved numerically as before with the external dc and ac currents inserted along the top row of the sample and removed along the bottow row. The results for the average voltage are shown in Fig. 16. In Fig 16a, we show the average voltage versus i_{dc} at T = 0. The voltage (per vertical bond) is zero up to the critical bias current at about $i_c = 0.64$ for $i_{ac} = 0.5$, and $f = \omega/(2\pi) = 0.2$. At the critical current i_c the

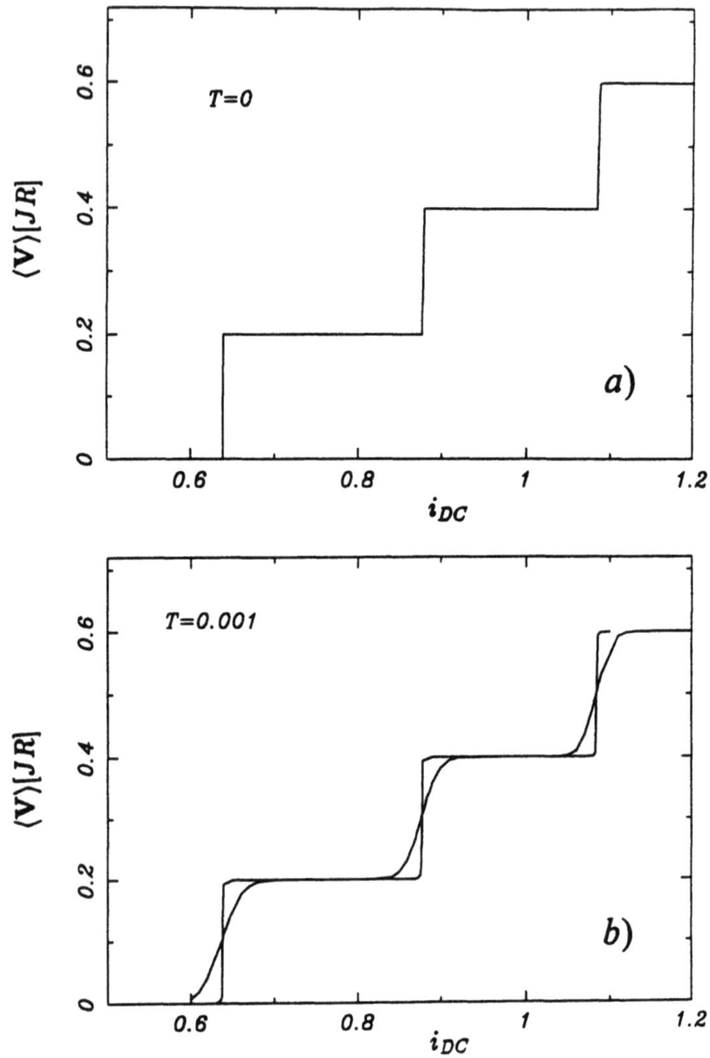

Fig. 16. **The average voltage per bond $< v >$ vs. i_{dc} for a 16 x 16 array with i_{ac} = 0.5, f = 0.2, and (a) T = 0, and (b) T = 0.001. The more rounded curve in (b) is for a single junction.**

entire array begins to oscillate, locked into the applied frequency f. As the bias current is increased still further the voltage remains constant (on the Shapiro step) since the frequency of the driven system remains locked into the applied frequency f until the natural frequency of the system becomes so high that the system can't remain locked into the applied frequency and therefore makes a transition to the first harmonic of the applied frequency to which it remains locked on the second step until making the transition to subsequent higher harmonic (steps). When the temperature is raised these steps become rounded (Fig. 16b) due to fluctuations as the system make the transitions between the various harmonics. It is evident (Fig 16b) that much less rounding occurs for an array than for a single junction. Clearly the cooperative behavior (due to non-linearity between the junctions) lessens the fluctuations. Associated with the rounding of the steps is a rise in the noise of the system, which can't be explained as ordinary thermal noise. There have been a number of experiments on single Josephson junctions whose reports included this anomalous noise rise[13,14] and explanations have been given for single junctions by Huberman et al[15], Kautz[16], and Pederson and Davidson[17]. These explanations involved numerical calculations for a single junction on essentially the same Josephson equations we have been using which are often called here the Stewart-McCumber model[18]. A clear analytic calculation of these effects for a single junction was given by Chiao et al[11] who considered small perturbations about the phase-locked oscillations and found that as the transition between harmonics is made there is an ac-induced null in the supercurrent. We have now looked closely at 16 x 16 arrays just at the Shapiro step risers where these transitions take place and find a dramatic increase in the number of vortex-antivortex pairs which are created in the sample just at the step risers. Figure 17 shows a plot of the number of vortices counted

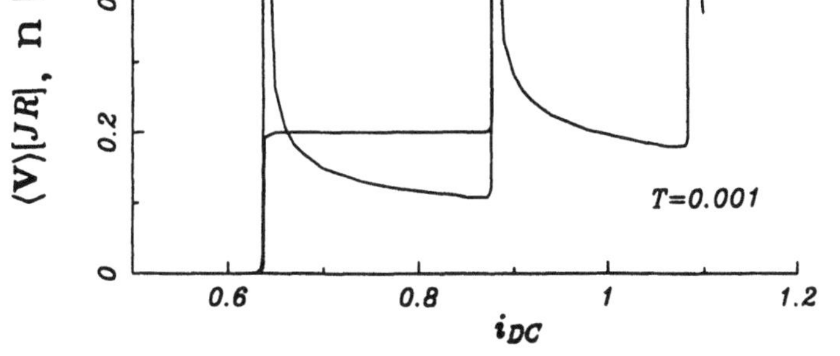

Fig. 17. Plots of the number of vortex-antivortex pairs (generated over an arbitary, but fixed, time) vs i_{dc}, and average voltage $< v >$ vs i_{dc} for $i_{ac} = 0.5$, $= 0.2$, and T = 0.001 in a 16 x 16 array.

verses bias current i_{dc} and shows that these increases occur just on and about the step risers.

We also calculated the noise power spectrum generated on a 5x5 array as the bias current was swept across a riser and this is shown in Fig. 18. The

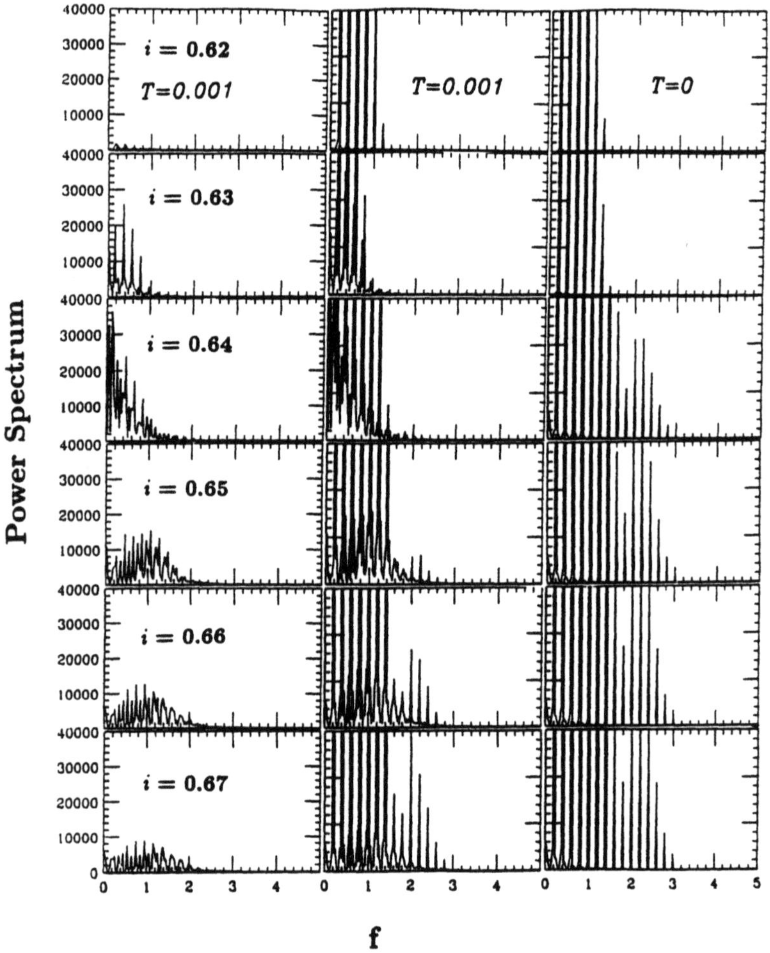

Fig. 18. The power spectrum (fourier components of the voltage squared vs. frequency) of a 5 x 5 RSJ array for $i_{dc} = 0.62$ - 0.67. The peak in the number of vortices occurs at the step riser at $i_{dc} = 0.64$. The right-hand column is for T = 0, the central column is T = 0.001, and the left-hand column is also at T=0.001, but with the harmonic peaks removed.

right-hand column of Fig. 18 is the power spectrum for T = 0, where there is no random noise, but only the various harmonics of the applied frequency. The central column is for T = 0.001, and the left-hand column is the same as the central column but with the harmonic peaks removed to better show the noise. Each row of Fig. 18 is for a different bias current i_{dc}, and the maximum noise occurs at $i = 0.64$, just on the riser. A log-log plot of this curve does give, perhaps, one decade over which this may be 1/f-noise, but the data are not conclusive. We also have looked at the voltage-voltage correlation functions across the sample as the bias current is swept across the step riser and see what are apparently transitions between power-law decay on each step as would be expected within a superconductor and exponential decay on each riser which would seem to indicate transitions out of and back into the superconducting state as the supercurrent is suppressed. These results are still preliminary but clearly there are very interesting and non-linear cooperative phenomena occuring at the Shapiro step risers.

REFERENCES

1 P.L. Leath and W. Tang, Phys. Rev. B39, 6485 (1989).
2 W. Xia and P.L. Leath, Phys. Rev. Lett. 63, 1428 (1989).
3 P.L. Leath and W. Xia, Phys. Rev. B44, 9619 (1991).
4 P.L. Leath and P.M. Duxbury, previous paper this conference.
5 J.G. Mooij and G. Schön, Physica B152,1(1988).
6 C.J. Lobb, D.W. Abraham and M. Tinkham, Phys. Rev. B27, 150 (1983).
7 K.K. Mon and S. Teitel, Phys. Rev. Lett. 62, 673 (1988).
8 J.S. Chung, K.H. Lee, and D. Stroud, Phys. Rev. B40, 6570 (1989).
9 Y. Cai, P.L. Leath, and Z. Yu, Phys. Rev. B49, 4015 (1994).
10 V. Ambegaokar and B.I. Halperin, Phys. Rev. Lett 22, 1364 (1969).
11 A. Barone and G. Paterno, "Physics and Applications of the Josephson Effect" (Wiley, New York, 1982).
12 P.L. Leath and Y. Cai, to be published.
13 R.Y. Chiao, M.J. Feldman, D.W. Peterson, B.A. Tucker, and M.T. Levinson, in "Future Trends in Superconductive Electronics", AIP Conf. Proc. 44, p. 259 (AIP, New York, 1978).
14 Y. Taur and P.L. Richards, J. Appl. Phys. 48, 1321 (1977).
15 B.A. Huberman, J.P. Crutchfield, and N.H. Packard, Appl. Phys. Lett. 37, 750 (1980).
16 R.L. Kautz, J. Appl. Phys. 52, 6241 (1981).
17 N.F. Pederson and A. Davidson, Appl. Phys. Lett. 39, 830 (1981).
18 W.C. Stewart, Appl. Phys. Lett. 12, 277 (1968); D.E. McCumber, J. Appl. Phys. 39, 3113 (1968).

Acknowledgements

We thank for their hospitality the Department of Physics and Astronomy of Michigan State University where the authors were visitors during the spring of 1993, and P.L.L. thanks Theoretical Physics and New College, Oxford University for their hospitality during the fall of 1993. The authors also gratefully acknowledge the CAIP center at Rutgers and the NCUBE corporation for the access to their NCUBE systems. And finally we acknowledge a vital grant of computing time on the Cray-YMP at the NSF Pittsburgh Supercomputing Center without which many of the computations reported here could not have been done.

Field-induced Transport in Random Media

Mustansir Barma [1], Ramakrishna Ramaswamy [2]

[1]Tata Institute of Fundamental Research, Homi Bhabha Road, Bombay 400 005, INDIA
[2]The Isaac Newton Institute for Mathematical Sciences, Cambridge University, Cambridge CB3 0EH, UK
[2]School of Physical Sciences, Jawaharlal Nehru University, New Delhi 110 067, INDIA

Abstract: We review the problem of particle transport in random media in the presence of an external field. The random medium is modeled by the infinite cluster above the percolation threshold. The field imposes a preferred direction of motion along which diffusing particles (random walkers) are more likely to move than against. Two kinds of traps occur – branches pointing in the direction of the field, and backbends, in which particles must move against the field. For noninteracting particles, the drift velocity is a nonmonotonic function of the biasing field, and the two kinds of traps make the current vanish above a threshold value of the bias. If there is hard-core repulsion between the particles, branches get filled up and eventually cease to be traps. Below the directed percolation threshold, transport is rate-limited by backbends, and the particle current flows predominantly along those paths on the percolation backbone on which the length of every backbend is bounded. The current is a nonmonotonic function of the biasing field. We also consider a different sort of interparticle interaction which leads to levels of particles equalising near backbend bottoms. The motion along a typical path is then described by 'drop-push' dynamics: between backbends, particles drop down, assisted by the field, and push those on the next backbend, possibly leading to a cascade of overflows. Drop-push dynamics has interesting connections with other lattice gas automata, and Monte Carlo simulations show that the model supports kinematic waves and exhibits interesting behaviour of time-dependent correlations.

1 Introduction

The general problem of random walks in disordered media is interesting since the stochastic motion of random walkers has a strong interplay with the random, frozen-in geometry of the medium itself [1-4]. This has some unusual and unexpected consequences. In the present article, we are primarily concerned with the nonlinear effects that arise from imposing a uniform external field which leads to a bias in the random walks. Our aim is not to model a specific experimental system in detail, but rather to discuss general characteristics that are brought in by the random geometry of the medium. Nevertheless, a concrete physical example we may keep in mind is that of sedimentation in a random, porous medium: particles (imagine heavyish pollen grains) perform biased random walks, more likely to take downward steps than upwards owing to the effect of gravity, but constrained to move only within the interstices or free pore space within the random medium.

The interest in this problem stems from the observation [5-11], first made in the context of disordered semiconductors, that the externally imposed bias brings into play two competing physical effects. On the one hand, a weak field aids transport as it leads to a finite drift velocity: this is a consequence of linear response theory. On the other hand, a strong enough field can actually inhibit transport, as local potential energy minima are set up in the network (e.g. points Q and H in figure 1). Such spots act as particle traps, as the particle must travel against the field in order to be able to leave. The eventual escape of a particle from such a region is a strongly nonlinear activation process, and is very slow if the field is large. The tendency of the field to promote both drift and trapping leads eventually to an unusual, nonmonotonic dependence of the drift velocity on the bias, and associated anomalous properties in a large region of parameter space.

A concrete theoretical model that has been considered [9-11] is that of biased random walkers on the infinite cluster in the percolation problem for occupation concentrations p above the percolation threshold p_c. The model is thus a natural generalization of the well-known 'ant in the labyrinth' problem pertaining to diffusion in random media [12]. The infinite cluster is well connected in the sense that it is not a fractal — unlike the cluster precisely at p_c — but there is a finite concentration of traps of the type shown in figure 1, and this is sufficient to make the transport properties unusual and interesting. Two types of random walkers have been considered in the past (a) noninteracting particles, or equivalently a single walker in the network [9,10] (b) particles with hard-core interactions which lead to mutual exclusion effects [11].

Hard-core interactions have a strong effect on field-induced trapping. To get a qualitative idea of the result, it is necessary to distinguish between two types of traps. The first type consists of branches which point predominantly in the direction of the field (e.g. PQ in figure 1). The second consists of backbends, which are those portions of the percolation backbone where the particle current flows against the direction of the field (e.g. HK in figure 1). Interparticle interactions effectively screen out deep branches, and the particle current is rate-limited only by backbends. Transport properties are governed not by the geometrically shortest, but by the temporally quickest paths. This leads to an interesting geometrical question in the percolation problem, *viz.* looking for the formation of infinitely long paths with the smallest possible backbend lengths [12,13].

The existence of field-induced drift and trapping leads essentially to the following picture. In strong fields, particles move rapidly between traps, but spend a

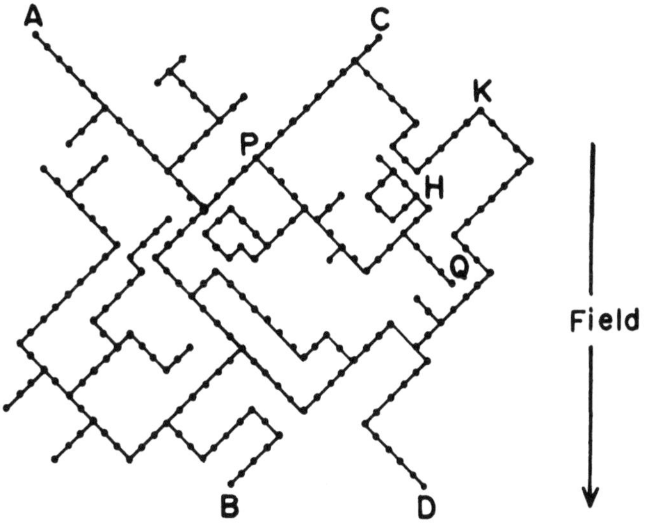

Fig. 1. A portion of the infinite percolation cluster is illustrated. A downward acting field sets up local potential energy minima which act as particle traps. Examples are the branch PQ and the backbend HK.

long time near and at traps, resulting in the establishment of a quasi-equilibrium state near each trap. The nature of this quasi-equilibrium state is evidently sensitive to the sort of interparticle interaction, and it is interesting to examine a variant — in particular, what if the trapped material is not compressible (like the 'pollen grain lattice gas' described by the models discussed above) but rather, incompressible, as in a liquid — imagine beads of mercury, say. In the latter case, levels equalize around backbending trap bottoms, and the typical state consists of puddles around such traps. The dynamics then has elements both of activation and of overflow from already-full traps to traps at lower heights. With this picture in mind, we propose a new stochastic lattice-gas model which we call drop-push dynamics, which mimics some aspects of the process described above.

Besides reviewing the status of the biased random walk problem in random media, we present here some new results on drop-push dynamics. The plan of this article is as follows. Section 2 deals with noninteracting particles, for which trapping effects are the strongest, in fact, strong enough to make the drift velocity vanish above a threshold value of the field. In Section 3, we turn to the effects of hard-core interactions, in which case the current is rate-limited by the length of the smallest backbend required to form infinitely long connections. In Section 4, we turn to a discussion of drop-push dynamics in a one-dimensional model, and show that the steady state fluctuations are transported by kinematic waves. Section 5 is the conclusion which includes an account of the history of the problem, including a discussion of several controversies.

2 Noninteracting Particles

We consider the transport properties of a system of noninteracting particles moving through a randomly connected network in the presence of a biasing field. We will show that the drift velocity is a nonmonotonic function of the bias and vanishes above a threshold value. Above this value, transport is sublinear (distance $\sim t^x$ with $x < 1$), and the exponent x varies continuously with the bias.

2.1 Model

For definiteness, we model the random medium by the infinite percolation cluster (IC) on a simple cubic lattice with $p > p_c$. Suppose that the field is in the $\hat{e}$ direction, where $\hat{e}$ is a unit vector in the $(\bar{1}\bar{1}\bar{1})$ direction, so that it has an equal component along every bond of the network. The motion of a particle is modeled by a continuous time random walk on the sites of the IC. The effect of the field is incorporated as a bias in the random walk, making steps along the field more likely than against it. If $P_n(t)$ is the probability that the particle is at site n at time t, we have [9]

$$\frac{dP_n}{dt} = \sum_m (W_{mn} P_m - W_{nm} P_n) \tag{1}$$

where the summation runs over the occupied nearest neighbours of site n. The transition probability per unit time to hop from site m to site n is

$$W_{mn} = W(1 \pm g) \tag{2}$$

Here g is the bias, and satisfies $0 \leq g < 1$; the plus sign holds for hops along the field (when site m is above n), and the minus sign for hops against the field (when site m is below n). The relationship between g and experimentally accessible quantities is given in Eqs. 5 and 6. The limit $g = 0$ corresponds to the unbiased or purely diffusive case. In the other extreme $g = 1$, the field is infinitely strong and hops against the field are not allowed.

Equation (1) describes a set of coupled, linear differential equations, and can, in principle be solved for any given initial condition — difficult in practice, because of the random geometry. However, we can learn a good deal by studying the nature of the the time-independent steady state which the system approaches after a long enough time.

2.2 Steady State

Suppose there are $\mathcal{N}$ noninteracting particles in the system. In steady state, let ρ_m be the particle density at site m, the product of $\mathcal{N}$ and the steady state value of P_m. The corresponding current from site m to an adjacent site n is given by

$$J_{mn} = W_{mn}\rho_m - W_{nm}\rho_n. \tag{3}$$

Now, the IC is made up of the backbone (which consists of those sites which cannot be disconnected from the IC by removing any other single site), and the remainder, which consists of branches attached to the backbone. In steady state, the current through any site in a branch is evidently zero, implying that the ratio of densities on adjacent sites is $(1+g)/(1-g)$. Thus, if j is a site in a branch which is attached to the backbone at site k, we have

$$\rho_j = \rho_k e^{\Delta_{jk}/L(g)} \tag{4}$$

where Δ_{jk} is the difference of heights (along $\hat{e}$) between sites i and k, and $L(g)$ is a bias-induced length in the problem, defined by

$$L(g) = a \ln \left(\frac{1+g}{1-g}\right) \tag{5}$$

where a is the lattice spacing. The bias induced length $L(g)$ describes the exponential growth of density near a branch bottom, familiar as the law of exponential atmospheres in a gravitational field G at temperature T. This leads to the identification

$$L(g) = \frac{k_B T}{mG}. \tag{6}$$

Thus, changing the temperature is a simple, effective way of varying $L(g)$, thereby scanning all the regimes of interest.

The average particle density $\bar{\rho}$ in the system can be found by summing over all backbone sites k, counting the particles on k and sites in branches attached to k. This is accomplished by writing

$$\bar{\rho} = \frac{1}{V}\sum_{jk} \rho_k \eta_{jk} e^{\Delta_{jk}/L(g)} \tag{7}$$

where η_{jk} is an indicator function which is 1 if k belongs to the backbone, and j belongs to a branch attached to the backbone at site k, and is zero otherwise. Now ρ_k itself may be quite large (for instance if k is a backbend bottom), but even if it is not, there can be a large density pile-up in branches attached to k. Replacing ρ_k in Eq. (7) by a 'typical' value ρ_{bb} and performing a configuration average (justified by the self-averaging property of a large system), we get

$$\bar{\rho} \approx \rho_{bb} \sum_{r} e^{\Delta_{r,0}/L(g)} \langle \eta_{r,0} \rangle. \tag{8}$$

Since $\langle \eta_{r,0} \rangle$ is a two-point correlation function, on general grounds, it is expected to decay exponentially for large separations:

$$\langle \eta_{r,0} \rangle \sim exp(-r/\xi(p)) \tag{9}$$

where $\xi(p)$ is the percolation correlation length in the $\hat{e}$ direction.

Now the macroscopic drift velocity v is related to the steady state current J (the average flux per unit area per second) and the density $\bar{\rho}$ by

$$J = \bar{\rho} v \tag{10}$$

For small bias g, the drift velocity $v = Kg$ where K is proportional to the undirected diffusivity of the percolation network. As the strength of the field is increased, the bias-induced length $L(g)$ decreases. Once $L(g) < \xi(p)$, the mean density $\bar{\rho}$ diverges, and correspondingly the drift velocity v vanishes. The drift velocity is thus a nonmonotonic function of the bias (figure 2) and reaches zero at a threshold value of the bias determined by $L(g) = \xi(p)$.

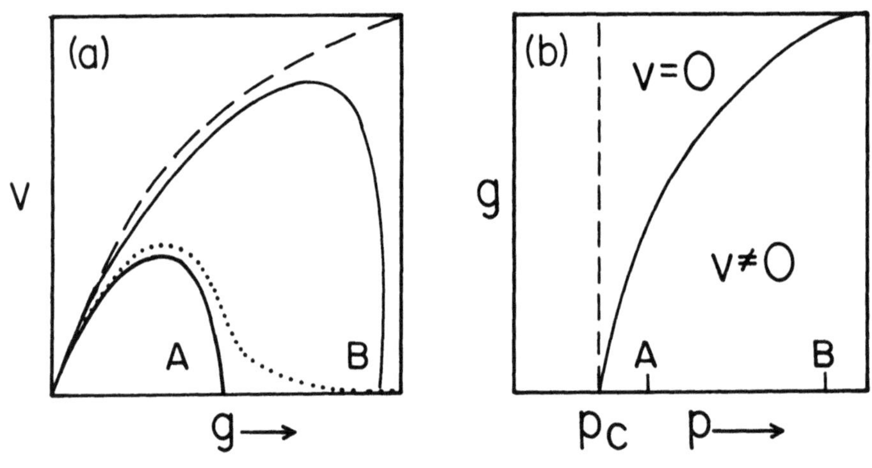

Fig. 2. (a) The variation of the drift velocity with bias g is shown schematically. The dashed curve is the variation for $p = 1$ (undiluted lattice). The solid curves (A and B) refer to noninteracting particles for values of p depicted in (b). The dotted curve shows the variation of v vs. g when hard core interactions are included. (b) Depiction of the regions in the concentration-bias plane where noninteracting particles have nonvanishing and vanishing drift velocity.

The result is quite robust and holds for other types of random media as well, provided only that the probability of having long dead-ends decreases exponentially with their length. Examples of other structures with this property are random combs [10]. For percolation systems, a study of branch statistics bears out the exponential decay form. Further, the results have been explicitly checked by a calculation in the steady state on the randomly diluted Bethe lattice [10]. On the random comb, a full time-dependent solution [15] of the problem bears out the conclusion that the drift velocity vanishes above the threshold value of the bias.

It is interesting to ask whether a finite drift velocity could be sustained if all branches were removed, and we considered flow only along the backbone. The existence of backbends along backbone paths again leads to trapping, and a calculation of transport through a single random path along the backbone shows that the drift velocity vanishes above a threshold value of the field [9,16,17]. As with branch trapping, this can be traced to two competing exponential factors: in a strong field, there is an exponentially large pile-up of density at the bottoms of backbends, but long backbends are exponentially rare.

2.3 Distance-time Laws in Different Regimes

There is a simple physical picture of transport of a single particle through the random network. Motion between traps is relatively rapid, while there is a long residence time in traps, be they branches or backbends. In order to escape from a trap of depth Δ, a particle must overcome a potential energy barrier $mG\Delta$. This is the classical Kramers escape problem [18], and the mean residence time is τ given by

$$\tau(\Delta) \sim \tau_0 \exp\left[\frac{mG\Delta}{k_B T}\right] = \tau_0 \exp\left(\frac{\Delta}{L(g)}\right). \tag{11}$$

On the other hand, the probability of occurrence of a trap of depth Δ follows Eq. (9), so Eqs. (6) and (11) together imply that the probability of finding a trap with residence time t is

$$\tilde{p}(t) = \frac{A}{(1+Bt)^{1+x}} \tag{12}$$

where A and B are constants and the exponent

$$x = \frac{L(g)}{\xi} \tag{13}$$

is the ratio of the two fundamental lengths in the problem.

Consider the mean transit time T_N of the particle as it moves through a large distance during which it encounters N traps. N is proportional to the distance moved, and T_N is approximately given by the sum of N residence times t_i drawn from the probability distribution $\tilde{p}(t)$ [10]. The distribution $\tilde{P}$ of T_N is determined by the limit theorems of probability theory [1,19]. The answer for $\tilde{P}$ depends strongly on the value of x, and we discuss three regimes separately.

If $2 < x < \infty$, the second moment $\langle t^2 \rangle \equiv \int dt\, t^2 \tilde{p}(t)$ is finite; both the law of large numbers and the central limit theorem apply. The resulting distribution $\tilde{P}(T_N)$ is centered around the mean value $N\langle t \rangle$ and the drift velocity is given by Eq. (10). Furthermore, the probability distribution of the scaled variable $(T_N - N\langle t \rangle)/\sqrt{N}$ tends to the normal distribution with probability density $e^{-\frac{1}{2}y^2}/\sqrt{2\pi}$.

If $1 < x < 2$ the mean value $\langle t \rangle$ is finite, but $\langle t^2 \rangle$ is infinite. Thus, the law of large numbers holds, but the central limit theorem is not applicable. The drift velocity v continues to follow Eq. (10). However, unlike in the regime $x > 2$, the distribution $\tilde{P}$ is not given by the normal distribution, and its width does not scale as $\sqrt{N}$.

The range $0 < x < 1$ defines the strong field regime $g > g_c$. The mean value $\langle t \rangle$ over the probability distribution of Eq. (12) is infinite. The law of large numbers ceases to hold, and the drift velocity v is zero. In terms of the scaled variable $y =$ constant $T_N/N^{1/x}$, the probability distribution approaches the stable (Lévy) form $\tilde{P}_x(y)$ whose Laplace transform is $\exp(-s^x)$ [19]. The typical (most probable) value of T_N scales as $N^{1/x}$. This marks a breakdown of the customary linear relationship between distance and time, and clarifies the sense in which the drift velocity is zero. Both this regime and the previous one $(1 < x < 2)$ are characterized by abnormally large fluctuations.

3. Particles Interacting via Hard-Core Exclusion

In this section we explore the effects of inter-particle hard-core repulsion on transport properties. We show that such interactions reduce the potency of branches as traps, and particle transport is rate-limited by backbends in the percolation backbone. The current flows primarily through a sub-network of paths on the IC, each of which has backbends no longer than a certain value ζ. For a given value of p, finding the smallest value of ζ which leads to infinitely long paths defines a geometrical problem which is a variant of the percolation problem. This geometrical information, together with an estimate of the largest current through a backbend of length ζ, determines the way transport properties depend on p.

3.1 Model

The effects of hard-core repulsion are incorporated in the lattice model by limiting the allowed occupancy n_i of each site to 0 or 1. The evolution of a configuration $\{n\}$ occurs through interchanges of particle-hole pairs (Kawasaki dynamics). The time dependence of the probability $P(\{n\})$ of configuration $\{n\}$ follows

$$\frac{d}{dt} P(\{n\}) = \frac{1}{2} \sum_{ij} [W_{ji} n_j \bar{n}_i P(\{n\}_{ij}) - W_{ij} n_i \bar{n}_j P(\{n\})] \tag{14}$$

where configuration $\{n\}_{ij}$ is obtained from $\{n\}$ by replacing a particle (hole) in $\{n\}$ at site i by a hole (particle) and performing the same operation at a neighbouring site j. Here $\bar{n}_i \equiv 1 - n_i$ is the number of holes on site i, and factors like $n_j \bar{n}_i$ arise as a particle can hop only onto a vacant site. Eq. (14) cannot be reduced to a single-particle equation like Eq. (1), and describes the asymmetric exclusion process [20] on a random network.

3.2 Effect on Branch Trapping

In steady state, a current flows through the backbone, but not through branches. Thus particles in a branch achieve thermal equilibrium with respect to the Hamiltonian

$$\mathcal{H}_k = mG \sum_j{}' \Delta_{jk} n_j \tag{15}$$

where the prime indicates that the sum is over sites in a branch attached to the backbone at site k. Introducing a fugacity z, the branch grand partition function can be found, and also the mean occupancy

$$\rho_j \equiv \langle n_j \rangle = \frac{z e^{\Delta_{jk}/L(g)}}{1 + z e^{\Delta_{jk}/L(g)}} \tag{16}$$

The fugacity z is determined by the value of ρ_k, which in turn is determined by solving the steady state problem on the backbone alone. Eliminating z, we find

$$\frac{\rho_j}{1-\rho_j} = \frac{\rho_k}{1-\rho_k} e^{\Delta_{jk}/L(g)} \tag{17}$$

which generalizes the noninteracting particle result Eq. (4).

If $\rho_k << 1$, the density ρ_j rises approximately exponentially until it reaches values of order unity. This happens at a depth approximately given by

$$\Lambda(\rho_k, g) = L(g) \ln\left(\frac{1-\rho_k}{\rho_k}\right) \tag{18}$$

For $\Delta_{jk} >> \Lambda$, the density is close to the saturation value of one particle per site. The number of particles in very long branches with depth >> the screening length $\Lambda(g)$ is roughly proportional to the number of sites in the branch — a much slower growth than the exponentially large density found for noninteracting particles. As a result, the mean density $\bar{\rho}$ is always bounded, and hence (in view of Eq. (10), the drift velocity is nonzero for all values of the bias.

The trapping effect of branches is reduced strongly despite the fact that a particle at the bottom of a branch takes much longer to escape (compared to the noninteracting case) as it has to wait for the unlikely event of all other particles between itself and the backbone having escaped as well. The expected time for this to happen can be estimated as the ratio of the branch grand partition function $Z(z,g)$ to a constrained grand partition function $Z_j'(z,g)$, where the constraint is that there is at least one clear path (devoid of particles) between site j and the backbone. This yields [21]

$$\tau_k \sim \exp(ka/L(g)) \text{ if } \Delta \ll \Lambda \tag{19}$$

$$\tau_k \sim \exp(k^2 a/2L(g)) \text{ if } \Delta \gg \Lambda. \tag{2}$$

Escape from a branch bottom is thus an extremely rare occurrence. As against that, it is very difficult for a fresh particle from the backbone to reach the branch bottom, as it is almost always full. The combination of extreme inaccessibility and extremely long trapping times (in the rare event that a fresh particle does get trapped) leads to a near cancellation, and eventually to a finite mean time of residence in a branch for all values of the bias.

3.3 Motion in a Single Backbend

Let us now focus on the motion of particles on the backbone alone and disregard the effect of branches. The backbone connects opposite faces of the system but, for $p_c < p < p_d$, even the shortest connecting paths have backbends which act as traps and serve to limit the current. We first study transport of mutually excluding particles through a single backbend, of length la (e.g. HK in figure 1) [11]. There is a steady source of particles from the top to H and particles are continually drained off at K. In order to get an upper bound on the current, we are led to consider biased diffusion of hard core particles in the segment $[0, l]$ of a 1D lattice, with the boundary conditions $\rho(0) = 1$ and $\rho(l) = 0$. The effects of boundary conditions on the steady state of the asymmetric exclusion process have been investigated recently [22]. In our problem, the boundary conditions force a current opposite to the bias — a case which has not been treated explicitly earlier, although it seems likely that the matrix methods developed in [22] can find application. The master equation that describes transport is invariant under interchange of particles and holes and simultaneous relabelling of sites in reverse order, i.e. $n_j \rightarrow \bar{n}_{l-j}$ The boundary conditions respect this symmetry, implying that the steady-state density $\rho(j)$ at site j satisfies $\rho(j) = 1 - \rho(l - j)$. Thus in the steady state the number of particles in the backbend is $l/2$ irrespective of the strength of the bias g. The principal effect of increasing g is to sharpen the region which marks the transition from the particle-rich half of the backbend to the hole-rich half. The steady-state profile approaches a step function centered at $j = l/2$ as $g \rightarrow 1$.

The current in the steady state is the number of particles crossing site l in unit time. Results of a Monte Carlo study are consistent with the large-l asymptotic behaviour

$$J_l \sim \exp\left(-\frac{1}{2}\, la/L(g)\right). \tag{21}$$

To see this, we write the current within a mean field approximation (which neglects two-site density correlations) as

$$J = W(1+g)\rho_j(1-\rho_{j+1}) - W(1-g)\rho_{j+1}(1-\rho_j) \tag{22}$$

The problem is then to find the value of J for which the boundary conditions $\rho_0 = 1$, $\rho_l = 0$ hold. This problem can be solved by going to the continuum limit $(\rho_{n+1} - \rho_n \rightarrow d\rho/dx)$ and setting up a differential equation for ρ. The solution is [11]

$$(g^2 + 2jg)^{1/2} = g \coth \frac{1}{2}\, l(g^2 + 2jg)^{1/2}. \tag{23}$$

For long backbends ($l \gg 1$), J is small and is given by $J \approx 2g\, e^{-lg}$. This agrees with Eq. (21) when g is small ($L(g) \approx 1/2g$), a necessary condition for the validity of the continuum approximation made above.

One can understand the result — in particular the factor $\frac{1}{2}$ in the exponent in Eq. (21) — as follows. The transport of a single particle through the backbend can be viewed as occurring in two (approximately) causally independent steps. In the first step, the topmost particle (located at site $k \approx l/2$ in large fields) is activated a distance $la/2$ from the top of the backbend. The energy cost is $\frac{1}{2}$ $mGla$ and the associated activation time is $\tau_{1/2} \sim \exp\left(\frac{1}{2}\ la/L(g)\right)$. In the second step, the consequent hole that remains in the steady-state distribution moves to the bottom and is filled up, by moving each of $l/2$ particles up a distance a. The energy cost is $\approx \frac{1}{2}$ Ela and the time required is $\tau_{1/2}$ again. The current J is thus proportional to $\tau_{1/2}^{-1}$, and consequently follows Eq. (21).

3.4 Percolation with Limited Backbend Lengths

Since the largest current that can flow through a backbend is exponentially small in its length, the largest current through a single path on the IC is determined by the largest backbend encountered on that path. Hence it becomes important to answer the following purely geometrical question: are there infinitely long paths which span the lattice, along which all backbends are bounded by a fixed value [14]?

To be definite, consider paths — self-avoiding walks (SAW) whose steps lie on the percolation network — that connect opposite faces perpendicular to the field direction $\hat{e}$. Along a particular SAW, if R_n is the location of the walk after the nth step, then $r_n \equiv (R_n \cdot \hat{e})$ is the displacement along $\hat{e}$. Let the maximal displacement up to and including step n be

$$M_n = \max_{n' \leq n}(r_{n'}) \tag{24}$$

The backward displacement (from maximal) is then

$$l_n = M_n - r_n. \tag{25}$$

On each path $\mathcal{P}$ between opposite faces, let $L(\mathcal{P})$ be the maximum value attained by l_n; it is the length of the longest backward excursion, or backbend, on path $\mathcal{P}$.

At fixed p, in the thermodynamic limit, let ζ be the *smallest* value of L such that there exist infinitely long paths which connect opposite faces. Clearly, ζ depends on the fraction p of occupied sites. In lattice models of the type under discussion, $\zeta(p)$ is the smallest integer such that there are infinitely long paths satisfying the constraint that *every* backbend on the path is *at most* of length $\zeta(p)$.

It is helpful to consider familiar limits. Let p_c and p_d denote the critical probabilities for ordinary and directed percolation. For $p > p_d$, directed paths (which do not have any backbends) proliferate, so that $\zeta(p) = 0$ in the range $p_d < p \leq 1$. At the other extreme, for p slightly larger than p_c, connections are very tortuous and involve large backbends. As p approaches p_c, $\zeta(p)$ diverges as $(p - p_c)^{-\nu}$ where ν is the critical exponent which characterizes the divergence of the correlation length.

$\zeta(p)$ can be calculated on the randomly diluted Bethe lattice of coordination number $2m$, with m in-pointing and m out-pointing arrows. (An arrow on a bond represents the component of $\hat{e}$ along that bond). Since the definition of a backbend along any path involves the maximum displacement up to that point, the upper bound on backbend lengths is a non-Markovian constraint as it depends on the full history of the path. However it is possible to handle the constraint by keeping track of both the height and the backward excursion coordinate along the path, and thus determine the variation of $\zeta(p)$ with p [14]. The result is shown in figure 3. Near p_c, we find $\zeta \sim (p-p_c)^{-1/2}$, consistent with the expectation that ζ scales like the correlation length ξ near p_c.

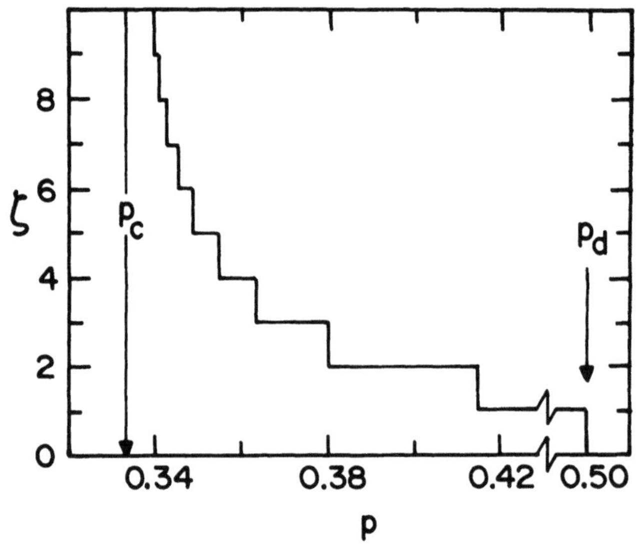

Fig. 3. The minimal backbend length ζ as a function of p for a Bethe lattice with coordination number 4. Near p_c (which is 1/3 on this lattice) ζ diverges as $(p-p_c)^{-1/2}$.

Returning to the question of transport of hard-core particles through the IC [11], it is evident that 'ζ-paths' are very important as along any such path, every backbend is bounded by $\zeta(p)$ and the current $\sim \exp(-\zeta/2L(g))$. The union of all such 'ζ paths' forms a finite fraction of the backbone and would be expected to carry most of the current in strong fields; portions with backbends of length much greater than $\zeta(p)$ contribute comparatively little to the current and may be neglected, in the first instance, in the strong-field limit.

For $p > p_d$, we have $\zeta(p) = 0$ and the primary current-carrying network is the directed backbone. On the other hand, as p approaches p_c, the minimal backbend length $\zeta(p)$ is proportional to the percolation correlation length ξ and diverges. We then expect that as $p \to p_c$,

$$J \sim \exp\left(-\frac{bmG}{k_B T(p-p_c)^\nu}\right) \tag{26}$$

where the exponent ν characterises the divergence of ξ in that limit, and b is a constant of order unity.

4 Drop-Push Dynamics

As has been elaborated in the preceding sections, the physical picture of lattice gas transport in strong fields on the percolation backbone is as follows. At a given value of p, with $p_c < p < p_d$, the paths that are most relevant for transport have backbends of a given *maximum* length, $\zeta(p)$, with $\zeta(p)$ diverging as $p \to p_c$. These are the quickest (as opposed to shortest) paths. The dynamics of transport is dominated by the backbends rather than the portion of the backbone between successive backbends, since on these the direction of the field is along the path rather than against it. We may take this latter motion to be ballistic, and thus the rate-determining aspect reduces to a consideration of the dynamics on successive backbends alone.

A single backbend consists of a long inlet and a shorter outlet. Even when there is a long column of particles in the inlet, particles in the outlet column rise only halfway up the outlet; levels of particles do not equalize. As discussed in the last paragraph of Section 3.3, this feature is ultimately responsible for transit times growing exponentially with backbend lengths in the case of the lattice gas.

However, we can imagine other physical situations in which levels do equalize — for instance, if we are considering the transport of a liquid rather than a lattice gas. In such a case, if a stochastic move (figure 4) results in dropping some liquid into a 'lake' at a lower level which happens to be full, there is an overflow which feeds into the next valley. A cascade of overflows may result, as illustrated in the figure. The situation is very similar to the motion of fluid on a tilted, rough inclined plane, considered recently by Narayan and Fisher [23].

We model this process by a one-dimensional asymmetric exclusion model of particles hopping in a preferred direction. Each site of this 1-d lattice corresponds to a backbend on the percolation backbone. Each site i can hold z_i particles, $z_i = 0, 1, \ldots (z_c - 1)$ particles, where $z_c = \zeta(p)$. Particle have hard core interactions. (A model that attempts a more detailed description of the backbend might involve two variables at each lattice site corresponding to the number of particles on the left and right portions of the backbone at a backbend. To begin with we take the model defined as above with z_i the "effective" number of particles on the backbend).

The dynamics of particles in this system is activated, i.e. the topmost particle at each site i hops to the right (say), with probability $exp(-\beta(z_c - z_i))$, with $\beta = 1/L(g)$. If a particle should land at a site j where $z_j = (z_c - 1)$, then it moves to the next site k on the right such that $z_k < (z_c - 1)$, effectively bouncing to the next site available for occupation. Alternatively, one can imagine that when the particle moves from site i to $i+1$ with $z_{i+1} = (z_c - 1)$, it drops to the

Fig. 4. Illustrating drop-push dynamics, which applies if levels in a backbend equalize. An activated hop (bold arrow) can lead to a cascade of overflows (dashed arrows). The dynamics is modeled by a lattice gas with upto z_c particles per site, with activated rightward hopping. $z_c = 3$ in the figure above.

bottom of the backbend and immediately pushes the topmost particle at that site to the next site: hence the terminology, ***drop-push*** dynamics. This model captures the elements of activation and overflow of the motion of liquid through a backbending path, but is non-random.

One limit of this model, which has been studied earlier is the case of $z_c = 2$, which is a 1-d lattice gas equivalent to the low-noise dynamics of the Toom interface model [24,25]. In the Toom model of Ising spins on a two-dimensional lattice, evolution proceeds according to a stochastic north-east-center(NEC) majority rule [26,27]. In the low-noise limit, the dynamics of interfaces (between regions of up and down spins) can be recast in terms of a one-dimensional lattice gas model, with horizontal links in the interface mapping into holes and vertical links to particles. The stochastic NEC majority rule for spins becomes a stochastic rightward interchange: a particle (hole) is exchanged with the nearest hole (particle) on the right with probability $p(q)$. Setting $q = 0$ and $p = exp(-\beta)$ explicitly gives the $z_c = 2$ case of the drop-push model.

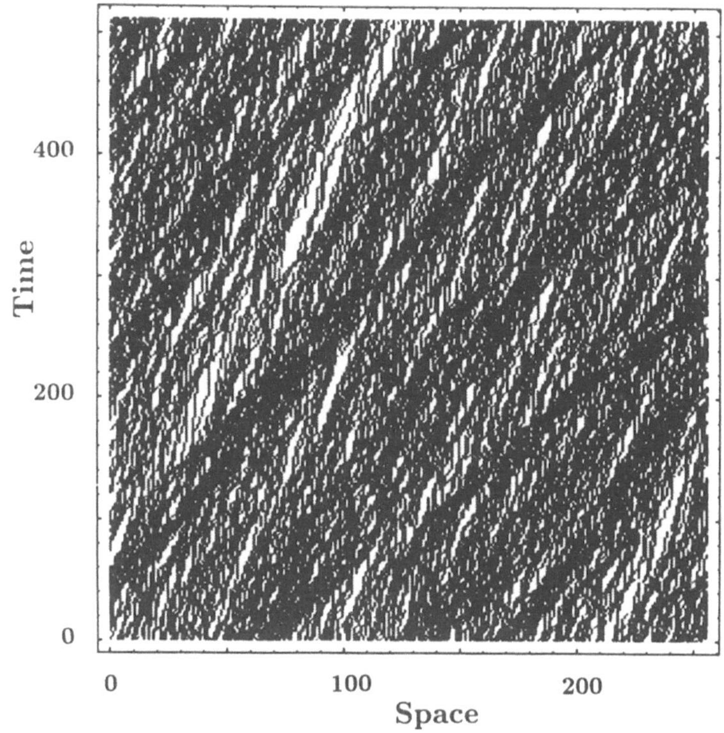

Fig. 5. Space-time plot of the evolution of a lattice-gas under drop-push dynamics with $z_c = 3$. Sites with 0, 1 and 2 particles are respectively white (blank), grey and black. The simulation, with periodic boundary conditions, is for a lattice with $N_s = 256$ sites and $N_p = 256$ particles, with $\beta = 1$.

In this Toom-interface limit, the properties of this model are well known, and in particular, the steady state has product invariant measure. The current, J, is a nonlinear function of the density ρ, which therefore allows for kinematic waves [28,29] that travel through the system with velocity $v_K = dJ/d\rho$. The average particle velocity is, of course, $v_P = J/\rho$.

For general $z_c(\geq 3)$, the steady state is somewhat more complicated. With the possibility of more than one particle at a site, each microscopic configuration can be reached from several others with very different kinetic rates. No explicit expression for the invariant measure is known. However, the dependence of the current on the density remains nonlinear, and thus these general drop-push models continue to support kinematic waves.

Here we report results from numerical (Monte Carlo) simulations of the model with $z_c = 3$, on a lattice with periodic boundary conditions. Shown in figure 5 is a space-time plot of the lattice gas with $\rho = 0.5, \beta = 1.0$ on a lattice of size $N_s =$

256, with black, grey and white corresponding to occupation numbers 2,1 and 0 respectively: the vertical and horizontal directions are time and space (particles are moving to the right). Visual inspection of this figure strongly suggests the presence of density waves (a more quantitative measure is given below shortly) inasmuch as there are stripes of black and white that run diagonally across. Indeed the two have different slopes, which further suggests that there may be more than one sort of wave in the system, at least over short times.

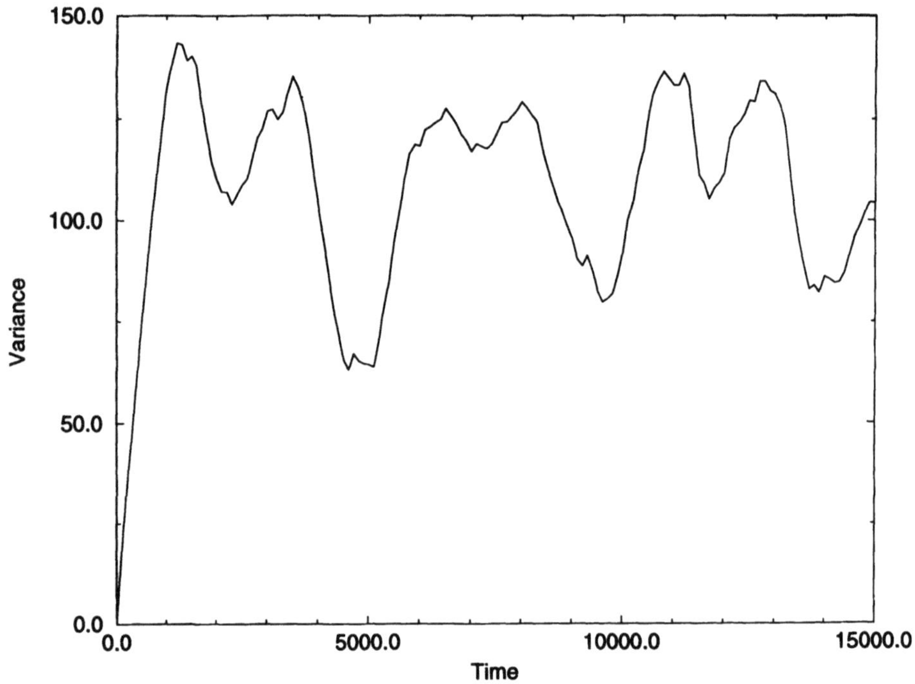

Fig. 6. Time dependence of the variance σ_0^2. The simulations are for a lattice of size $N_s = 10000$ with $N_p = 17000$ and $\beta = 1.5$. Fluctuations have been averaged over 10 samples in order to reduce the noise arising from size-effects.

To study the wave motion, it is convenient to examine the following correlation function which measures the spread of displacements of the individual particles,

$$\sigma_0^2(t) = \langle [\ell(j,t) - \ell(j,0) - v_P t]^2 \rangle \tag{27}$$

where $\ell(j,t)$ is the location of the jth particle at time t, v_P is the mean particle velocity, and $\langle\rangle$ denotes an average over the particles $j = 1, 2, \ldots, N_p = \rho N_s$. Shown in figure 6 is the variation of σ_0^2 with time for a system with $N_s = 10000$ and $N_p = 17000$, $\beta = 1.5$, when the particle velocity $v_P = 0.365$. The velocity of the kinematic waves, $v_K \approx 2.2$ is quite different and this shows up as oscillations in the variance σ_0^2 [30,31]. A detailed study of this system will appear elsewhere [32].

In recent years the study of driven nonequilibrium systems such as the present model, has attracted considerable attention [33]. This interest derives in part because these systems are seen to exhibit generic scale invariance, *i.e.* power-law decay of spatial and temporal correlation functions [33,34]. Temporal correlations under drop-push dynamics can also be long-ranged, and this can be seen if one defines a local energy E_i, in addition to the occupation number z_i at a site i. Then each configuration of the system is characterised by a total energy

$$E(t) = \sum_{i=1}^{N_s} E_i = \sum_{i=1}^{N_s} \sum_{k=1}^{z_i} [z_c - k]. \tag{28}$$

The signature of long-ranged temporal correlations can be effectively detected by an examination of the power-spectrum of E(t). In particular, for sufficiently high particle density the spectrum has a power law character, decaying as $1/f^{\alpha}$ where f is the frequency, and $\alpha \approx 1$, as for example in figure 7.

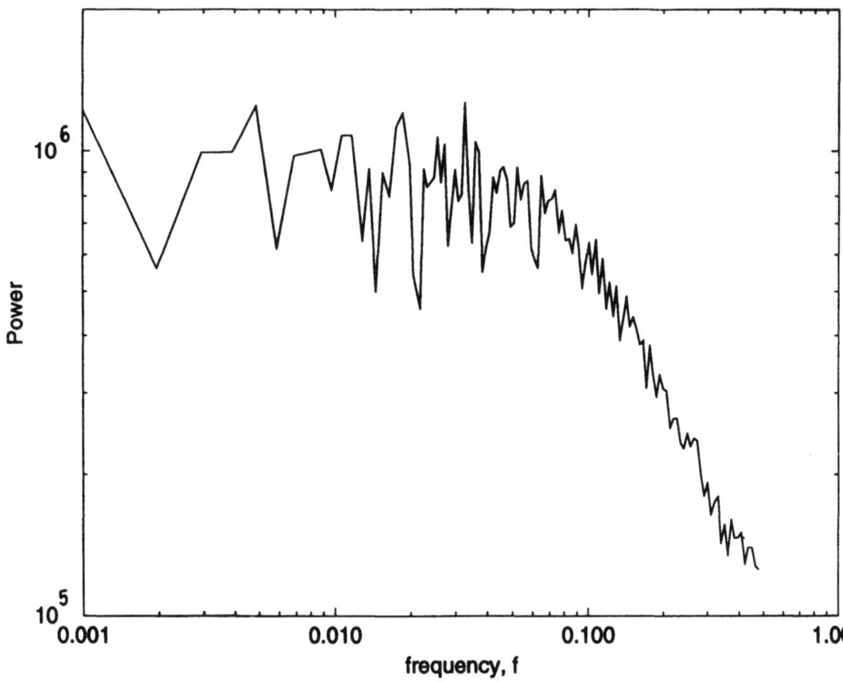

Fig. 7. Power spectrum of energy fluctuations for the same system as in figure 6.

5 Discussion

The different regimes of behaviour in the problem of biased random walkers in random media can be understood in terms of the different lengths which enter the problem. For noninteracting particles, there are two length scales of importance — the bias-induced length $L(g)$ and the percolation correlation length ξ. For weak fields ($L(g) \gg \xi$), the drift velocity is a linearly increasing function of the bias. The strong-field regime ($L(g) < \xi$), however, is marked by anomalous transport properties, with sublinear growth of distance with time. In this regime, the exponent which characterizes this sublinear growth is given by the ratio of the two lengths (Eq. 13). As discussed in section 2.3, there is also an intermediate regime in which the drift velocity is finite, but the dispersion is anomalously large.

Hard core interactions bring a new length into the problem, namely the branch screening length Λ. Branch depths $\gg \Lambda$ are effectively screened out, and so no longer potent as traps. The geometrical length which is important in the interacting case is ζ, the minimum backbend length that is required in order to have infinite connections. The definition of ζ involves a nonlocal constraint, which however can be handled analytically on the Bethe lattice. ζ diverges as $p \to p_c$ proportionally to ξ, and vanishes at the directed percolation threshold; its value determines the current through the network.

That a strong applied field can slow down, rather than speed up, motion was first pointed out in a discussion of transport processes in disordered semiconductors. The notion that a field in conjunction with random connections can set up traps was discussed by Zvyagin [5] and by Böttger and Bryskin [6]. These latter authors envisage transport through paths in which all backbends are bounded, like the ζ-backbone of section 3.3. However, in view of the conclusions of Sections 2 and 3, it would seem that their results hold for a finite density lattice gas, and not in the extreme low density limit. As we have seen, noninteracting particles do not move only along the optimal ζ-paths, and so encounter backbones and branches of all depths in the course of their motion. It is only hard core interactions which effectively screen out very deep branches and backbends, and ultimately channel the current through the 'ζ-backbone' of bounded backbend paths. In an interesting paper, Lien and Shklovskii [8] considered the way density of carriers influences strong-field transport in disordered semiconductors. They concluded that the current is low for low densities (noninteracting particle limit) and highest at half-filling — in agreement with our results. An important difference between disordered semiconductors and the model considered here is that there are no absolute dead ends in the former case, as forward tunnelling is always possible; as a consequence, the drift velocity is always nonzero in the former case.

For biased diffusion of a single particle on the percolation IC, our results are consistent with those of Ohtsuki and Keyes [35] on the domain of validity of linear response theory, viz. $L(g) \gg \xi(p)$. In the nonlinear regime, as the analysis of Section 2 shows, the drift velocity is nonmonotonic and actually vanishes beyond a critical value. This result has evoked mixed reactions. Gefen and Goldhirsch [36] questioned the nonmonotonicity of the response. They defined a velocity $\tilde{v}$ on a random comb (a straight backbone with branches of random lengths) through the relation $J = \tilde{v}\bar{\rho}_{bb}$ where $\bar{\rho}_{bb}$ is the mean particle density on the backbone; $\tilde{v}$ is monotonic in the bias g. However, the velocity $\tilde{v}$ is not the same as the macroscopic drift velocity v in Eq. 10, which is defined as the large time limit of the ratio distance/time. In particular, consideration of the steady state on the random comb implies that v is a nonmonotonic function of the bias and vanishes above a critical value [10]; this is confirmed by Balakrishnan's calculation of the full time dependence [15].

There have been several numerical studies of biased diffusion, without and with interparticle interactions. Pandey [37] made a Monte Carlo study of biased diffusion of noninteracting particles on a site-diluted simple cubic lattice. He observed a nonmonotonic behaviour of the drift velocity on the bias in conformity with theoretical predictions, but questioned the vanishing of the drift velocity in strong fields. In our opinion, neither his data nor that of later workers [38,39] clearly confirm nor preclude a zero velocity regime, nor do they seem incompatible with it, as has been explicitly pointed out by Dhar [40]. Roux and Guyon [41] used a numerical transfer matrix method [42] for biased diffusion on a two-dimensional percolation network, eliminating the need to average over random walk histories. In their study, they obtained a clear signature of a trapping regime. At $p = p_c$, Monte Carlo simulations by Stauffer [43] reveal a growth of distance with time that is slower than any power law, in agreement with the theoretical expectation [10,40] $x = 0$ (see Eq. 13). Simulations of interacting particles [44] show that hard-core interactions lead to reduced trapping effects, as a consequence of branches filling up, in agreement with the theoretical expectation.

Harder, Bunde and Havlin [45] studied the motion of noninteracting random walkers subject to an oscillating bias at $p = p_c$ and found evidence for scaling in frequency and amplitude. Bouchaud and Georges [46] extended the scaling arguments to the off-critical case, and pointed out that the AC response can be understood in terms of the steady state results, in particular the existence of a critical bias.

Bunde and coworkers [47] considered a variant of the problem, namely the effect of topological bias which acts uniformly outward from a central injection point, in contrast to the effect of an external field like gravity. Not surprisingly, the effects of trapping by branches are similar in the two situations. However, backbends play no role in the topological bias problem. Hence the behaviour of interacting random walkers would be expected to be quite different in the two cases.

The nonmonotonic dependence of drift velocity on the field may have some practically useful consequences. In particular, Roux and Guyon [41] point out that the fact that optimal transport occurs at intermediate values of the field may have important ramifications for filtration and separation processes.

In the transport of biased random walkers with hard core interactions on a lattice, backbend traversal times grow exponentially with backbend lengths. By contrast, in other physical situations (where particle levels in a backbend equalize, as in a liquid), overflows may occur, resulting in much more rapid transport. This was modelled by drop-push dynamics in Section 4. This model has interesting points of correspondence with earlier work. It resembles the Narayan-Fisher model of fluid flow on a rough inclined plane [23]. However, the focus is somewhat different, as these authors are interested in the geometrical characteristics of connected lakes near a threshold tilting angle, whereas we are interested in time-dependent fluctuations in the flowing steady state. Another connection, as has already been elaborated above, is with the lattice gas version of the Toom interface model.

A third class of models related to drop-push dynamics in the limit of z_c large ($\rightarrow \infty$) are zero-range processes [20]. These are lattice gases with no hard-core repulsion (a site can be occupied by any number of particles), but the jump rate depends on the number of particles already at that site. For finite z_c, there are obvious similarities between the two models — in that the number of particles at a site can determine the hopping rate, but also equally obvious differences, in that particles can hop arbitrarily long distances. In addition, zero-range processes are not asymmetric. However, an important and interesting feature of the latter processes is that there are invariant product measures, which allow for a complete and simple characterization of the steady state [48].

A fourth connection is between drop-push models and a long-range nonequilibrium lattice model studied [49] in the context of self-organized criticality [33]. These are 1-d sandpile models with local rules for particle addition and particle transport. Depending on boundary conditions, these can be considered as two-state models (lattice sites are singly occupied or not), with particles hopping to the nearest vacant sites. Again there are similarities (most closely to Toom interface dynamics [24,25]) and significant differences, especially in the microscopic details — for instance, the transition rates in [49] satisfy detailed balance, whereas in drop-push dynamics they do not.

In conclusion, many open problems remain in the general area of transport of interacting particles in random media. As we have seen in this paper there are several situations where the steady state current in the random medium is well characterized and shows interesting nonmonotonicities. Relatively little is known, however, about time-dependent correlation functions in such steady states. Even in the absence of randomness (as in the drop-push model) dynamical correlation functions show a wealth of interesting behaviour such as kinematic waves and $1/f$ noise. The study of analogous phenomena in random media is likely to reveal new and interesting effects.

References

1. J.-P. Bouchaud, A. Georges: Phys. Rep. **195**, 127 (1990)
2. J.W. Haus, K.W. Kehr: Phys. Rep. **150**, 263 (1987)
3. S. Havlin, D. Ben Avraham: Adv. Phys. **36**, 695 (1987)
4. H. Böttger, V.V. Bryskin: Phys. Stat. Sol. (b) **113**, 9 (1982)
5. I.P. Zvyagin: Phys. Stat. Sol. (b) **95**, 227 (1979)
6. H. Böttger, V.V. Bryskin: Phys. Stat. Sol. (b) **96**, 3219 (1979)
7. H. Böttger, V.V. Bryskin: Phil. Mag. B **42**, 297 (1980)
8. Nguyen Van Lien, B.I. Shklovskii: Sol. St. Comm. **38**, 99 (1981)
9. M. Barma, D. Dhar: J. Phys. C **16**, 1451 (1983)
10. S.R. White, M. Barma: J. Phys. A **17**, 2995 (1984)
11. R. Ramaswamy, M. Barma: J. Phys. A. **20**, 2973 (1987)
12. P.G. de Gennes: Le Recherche **7**, 916 (1976)
13. H. Böttger, D. Wegener: Phys. Stat. Sol. (b) **121**, 413 (1984)
14. M. Barma, R. Ramaswamy: J. Phys. A **19**, L605 (1986)
15. V. Balakrishnan: Proc. Solid State Symp., Nagpur, **28C**, 77 (1985)
16. H. Kesten, M.V. Kozlov, F. Spitzer: Comp. Math **30**, 145 (1975)
17. B. Derrida, Y. Pomeau: Phys. Rev. Lett. **48**, 627 (1982);
    B. Derrida: J. Stat. Phys. **31**, 433 (1983)
18. P. Hänggi, P. Talkner, M. Borkovec: Rev. Mod. Phys. **62**, 251 (1990)
19. W. Feller: *An Introduction to Probability Theory and its Applications, Vol. 2*, (Wiley, New York, 1971)
20. F. Spitzer: Adv. Math., **5**, 246 (1970);
    T.M. Liggett: Ann. Prob. **1**, 240 (1973), **5**, 795 (1977)
21. M. Barma, R. Ramaswamy: J. Stat. Phys., **43**, 561 (1986)
22. B. Derrida, M.R. Evans: J. Physique **3**, 311 (1993);
    B. Derrida, M.R. Evans, D. Mukamel: J. Phys. A. **26**, 4911 (1993);
    B. Derrida, M.R. Evans, V. Hakim, V. Pasquier: J. Phys. A **26**, 1493 (1993);
    G. Schütz, E. Domany: J. Stat. Phys. **72**, 277 (1993)
23. O. Narayan, D.S. Fisher: preprint (1993)
24. B. Derrida, J.L. Lebowitz, E.R. Speer, H. Spohn: Phys. Rev. Lett. **67**, 165 (1991);
    J. Phys. A. **24**, 4805 (1991)
25. M. Paczuski, M. Barma, S.N. Majumdar, T. Hwa: Phys. Rev. Lett. **69**, 2735 (1992)
26. A.L. Toom: *in Multicomponent Random Systems*, edited by R.L. Dobrushin, Ya.G. Sinai (Marcel Dekker, New York, 1980)
27. C.H. Bennett, G. Grinstein: Phys. Rev. Lett. **55**, 657 (1985)
28. M.J. Lighthill, G.B. Whitham: Proc. Roy. Soc. A**229**, 281 (1955); **229**, 317 (1955)
29. G.B. Whitham: *Linear and Nonlinear Waves* (J. Wiley, New York, 1974)
30. S.N. Majumdar, M. Barma: Phys. Rev. B **44**, 5306 (1991); Physica A **177**, 366 (1991)
31. P.-M. Binder, M. Paczuski, M. Barma, Phys. Rev. E **49**, 1174 (1994)
32. M. Barma, R. Ramaswamy, in preparation
33. P. Bak, C. Tang, K. Wiesenfeld: Phys. Rev. A **38**, 364 (1988)
34. G. Grinstein: J. Appl. Phys. **69**, 5441 (1991)
35. T. Ohtsuki, T. Keyes: Phys. Rev. Lett. **52**, 1177 (1984)
36. Y. Gefen, I. Goldhirsch: J. Phys. A **18**, L1037 (1985)

37. R.B. Pandey: Phys. Rev. B **30**, 489 (1984)
38. E. Seifert, M. Suessenbach: J. Phys. A. **17**, L703 (1984)
39. G. Michel: J Phys. A **19** 2461 (1986)
40. D. Dhar: J. Phys. A **17**, L257 (1984)
41. S. Roux, E. Guyon, Europhys. Lett. **4**, 175 (1987)
42. S. Roux, C. Baudet, E. Charlaix, C.D. Mitescu: J. Phys. A **19**, L687 (1986)
43. D. Stauffer: J. Phys. A **18**, 1827 (1985)
44. M. Przyborowski, M.V Woerkom: Eur. J. Phys. **6**, 242 (1985)
    L. Heupel: J. Stat. Phys. **42**, 541 (1986)
45. H. Harder, A. Bunde, S. Havlin: J. Phys. A **19**, L609 (1986)
46. J.-P. Bouchaud, A. Georges: J Phys. A **23**, L1003 (1990)
47. A. Bunde, S. Havlin, H.E. Stanley, B. Trus, G.H. Weiss, Phys. Rev. B **34**, 8129 (1986)
48. H Spohn: *Large Scale Dynamics of Interacting Particles*, (Springer Verlag, Berlin, 1991)
49. J.M. Carlson, J.T. Chayes, E.R. Grannan, G. Swindle: Phys. Rev. Lett. **65**, 2547 (1990);
    J.M. Carlson, E.R. Grannan, G.H. Swindle: Phys. Rev. E **47**, 93 (1993);
    J.M. Carlson, E.R. Grannan, C. Singh, G.H. Swindle: Phys. Rev. E **48**, 688 (1993)

Percolation and Tunneling in the Quantum Hall Effect

Alex Hansen

Institutt for fysikk, Norges tekniske høgskole, N–7034 Trondheim, Norway and Groupe Matière Condensée et Matériaux, URA CNRS 804, Université de Rennes 1, F–35042 Rennes Cedex, France

Abstract: The question of localization in the quantum Hall system is very close to a classical percolation problem. The dominating quantum correction to this picture is tunneling. We show how this effect changes the universality class of the localization problem from that of percolation to a different one. Our result is in agreement with both experimental and numerical results.

1 Introduction

A quick check of the *Science Citation Index* shows that the Quantum Hall Effect is still — fourteen years after its discovery [1] — a very active field of research: There are now about 150 papers published each year on this subject. Even though many of the "grand" outstanding questions have been answered by now, quite a few still linger on [2].

The quantum Hall effect is seen in quasi two-dimensional electron systems pierced by a strong perpendicular magnetic field. It manifests itself through the conductance matrix: Applying a potential across the sample will not produce a current in the orthogonal direction. Thus, the longitudinal conductance $\sigma_{\parallel}$ is zero. The transversal conductance $\sigma_{\perp}$, on the other hand, is quantized in multiples of e^2/h. (However, at very strong fields, fractional multiples occur. This is the fractional quantum Hall effect [3]. We will not deal with this effect here.) Furthermore, plotting $\sigma_{\perp}$ as a function of $\rho h/eH$, where ρ is the density of carriers, reveals that plateaus of height ne^2/h form centered at integer values of $\rho h/eH$. Halfway between these integer values, the value of $\sigma_{\perp}$ jumps with a derivative that diverges with the size of the sample [4]. A similar plot of the longitudinal conductance $\sigma_{\parallel}$, shows that at the values of $\rho h/eH$, halfway between integers, where the perpendicular conductance moves from one plateau to the next, $\sigma_{\parallel}$ forms spikes whose sharpness increases with the size of the system. It is these spikes that we will be dealing with here.

The absence of longitudinal conductance for most values of $\rho h/eH$ signals that the electrons are localized. When $\rho h/eH$ passes through half-integer values the system undergoes at localization-delocalization transition. The critical exponent $\tilde{\nu}$ associated with the divergence of the localization length ξ at these values have been measured both experimentally [5] and numerically [6]. The most accurate experimental value is today $\tilde{\nu} = 2.3 \pm 0.1$ [4], and the best numerical value is $\tilde{\nu} = 2.34 \pm 0.04$ [7].

There have been attempts at deriving this exponent [8, 9]. As we will discuss in the next section, the delocalization transition in the quantum Hall system has much more in common with the percolation transition [10] than with localization in other two-dimensional electronic systems [11, 12], where it is thought that any amount of disorder induces localization [13].

The problem with the percolation picture is that it predicts a localization exponent which is equal to the percolation correlation length exponent which has been conjectured to be exactly $\nu = 4/3$ [14] — thus far away from the observed value quoted above. Thus, at best the percolation picture is incomplete. What is lacking — according to References [8] and [9] — is *tunneling.* This is different from "standard" localization where backscattering plays the leading rôle. In Section 3 we develop a simple renormalization argument [9] that shows how tunneling changes the localization exponent from the percolation one ν to $\tilde{\nu} = \nu + 1 = 7/3$. Furthermore, we show how all the other geometrical percolation exponents are changed due to this effect, thus defining a new universality class for percolation in this system.

In Section 4 we present our conclusions.

2 The Percolation Picture

Our starting point is a gas of *non-interacting* electrons moving on a plane pierced by a strong perpendicular magnetic field. The one-electron Hamiltonian is

$$H = \frac{1}{2m}\left(\boldsymbol{p} - \frac{e}{c}\boldsymbol{A}\right)^2 + V(\boldsymbol{x}) , \tag{1}$$

where $V(\boldsymbol{x})$ is a potential originating from the disorder in the system, including impurities, surface roughness,... This potential may be split into two components: (i) One originating from "hard scatterers" which form well-localized spikes in the potential, and (ii) a smoothly varying potential with a length scale a associated with it. In the following we will assume that V conists only of the smooth component. The vector potential $\boldsymbol{A}$ is related to the magnetic field by the relation $\boldsymbol{H} = \nabla \times \boldsymbol{A}$, and we may choose a gauge where $\boldsymbol{A} = (-yH/2, +xH/2)$. Setting the potential $V(\boldsymbol{x})$ equal to zero for a moment and assuming that the system has size $L \times L$ leads to wave functions [15]

$$\varphi_{nX}(\boldsymbol{x}) = \frac{1}{\sqrt{L}} e^{i(xy - 2Xy)/2l^2} \chi_n(x - X) , \tag{2}$$

where

$$\chi_n(x) = \frac{1}{\sqrt{2^n n! \pi^{1/2} l}} e^{-x^2/2l^2} H_n(x/l) \,. \tag{3}$$

H_n is the nth order Hermite polynomial, and $l = \sqrt{e\hbar/cH}$ is the magnetic length. n and X are two quantum numbers having the following interpretation: In a classical system, the perpendicular magnetic field forces the electron to move in a circle with radius r — the *Larmor radius.* The center of this circle defines the ***guiding center coordinates*** (X, Y). Quantizing this system, *i.e.* returning to the problem set by equation (1) (but still with $V = 0$), leads to a quantization of the circular motion, with the appearance of a quantum number n, enumerating the nth *Landau level.* Furthermore, the guiding center position is characterized by the quantum number X. The energy levels corresponding to the solutions (2) and (3) are

$$E_n = \hbar\omega_c \left(n + \tfrac{1}{2}\right) \,, \tag{4}$$

where $\omega_c = eH/mc$.

Let us now turn the smooth potential V "back on" again. We write the new wave functions $\psi(\boldsymbol{x})$ as linear combinations of the zero-potential wave functions,

$$\psi(\boldsymbol{x}) = \sum_{nX} C_n(X) \varphi_{nX}(\boldsymbol{x}) \,. \tag{6}$$

If we now assume that magnetic field is strong enough so that no transitions occur between different Landau levels n, the original Schrödinger equation $H\psi = E\psi$ is transformed into

$$\tilde{V}_n(X, Y) C_n(X) = (E - E_n) C_n(X) \,, \tag{7}$$

where E_n is defined in equation (4) and $C_n(X)$ in equation (6). Furthermore,

$$\tilde{V}_n(X, Y) = \frac{i^n}{\sqrt{2\pi} l} \int d\xi d\eta e^{-i\xi\eta/l^2} \chi_n(\xi) V(X+\xi, Y+\eta) \chi_n(\eta) = V(X, Y) + \mathcal{O}(l^2) \,. \tag{8}$$

The center coordinates do not commute. Rather, they obey

$$[X, Y] = il^2 \,. \tag{9}$$

Thus, X and Y are canonical conjugates, and the plane (X, Y) forms a ***phase space*** for the motion of the guiding center. We see that equation (8) is a ***one-dimensional*** Schrödinger equation with Hamiltonian V (in the limit $l \to 0$, *i.e.* strong magnetic field). The corresponding classical problem that this Scrödinger equation describes (which is not the classical version of the original problem stated in (1)) is one whose phase space trajectories are given by

$$V(X, Y) = E - E_n = constant \,, \tag{10}$$

and the corresponding wave functions ψ are localized to within a distance l of these classial paths.

Thus, at this stage, the picture to have in mind is a disordered energy landscape varying on a length scale a which is much larger than the magnetic length

l. The wave functions of the electrons in this landscape are localized to within a distance l of the equipotential lines in this landscape [15].

Let us now discuss how this picture leads to a localization-delocalization transition [11, 12]. We turn the potential V off again in our sample measuring $L \times L$, where $L \gg l$. Then, in the interior of the sample the electrons are localized; they simply swirl around each stationary guiding center. However, along the edges they drift [16, 17]. Classically, the picture to have in mind is an electron which keeps bouncing off the edge, so that its guiding center always moves in the same direction along the edge. This edge current is a supercurrent, *i.e.* not dissipative, since there is no mechanism present that can transport the electrons at the Fermi level across the sample from one edge to the opposite one where the current is going in the opposite direction [17]. As a result, the longitudinal resistance *and* conductance are both zero (since there is a non-zero transversal conductance, which must be taken into account when inverting the conductance or resistance matrix).

We now include the smoothly varying potential V in the discussion. The edge currents are still be present. However, the potential V opens for the possibility of scattering across the sample from one edge to the opposite one, thus leading to a non-zero longitudinal conductance $\sigma_{\|}$. The mechanism is as follows: The wave functions are localized along the classical paths defined by the equipotential curves $V = E - E_n = constant$. Most such paths form closed loops in valleys or around mountains of V. However, paths may be connected across saddle points, thus forming clusters of paths. At some energy level $E_c - E_n$ there will be a cluster of paths spanning the system from edge to edge, in analogy with percolation, thus providing a mechanism for transport across the sample. When the Fermi level of the electron system passes through this value (by adjusting *e.g.* the magnetic field), conductance appears, as there now is scattering across the sample. For each Landau level, there will be a value of the magnetic field for which the Fermi level passes through the percolating value of V. The Fermi electrons thus undergo a percolation delocalization transition, thereby explaining the occurence of multiple peaks in the $\sigma_{\|}$ *versus* $\rho h/eH$ plot. Furthermore, the regular spacing along the $\rho h/eH$ is due to the quantization of the Landau levels.

Percolation theory [10] predicts that the size of the largest cluster of connected paths in the V landscape diverges as $|E - E_c|^{-4/3}$. However, both real and numerical experiments [5, 6] lead to a different value for this exponent, namely $\simeq 2.3$. What is missing? First of all, we have thrown out the "hard scatterers." This seems to be realistic for the experiments, both numerical and real, even though a potential dominated by "hard scatterers" also gives rise to a percolation-type delocalization problem [18]. Thus, retaining the idea of a potential which is smooth on the scale of l, leaves one to suspect that the percolation picture breaks down at the saddle points where the equipotential curves meet to form clusters. There are two mechanisms at the saddle points which may change the classical picture of paths overlapping and therefore connecting: (i) Tunneling before the paths are within a distance l apart and thus overlap, and (ii) backscattering. However, we see that these two processes tend in opposite directions: Tunneling makes delocalization appear stronger than classical percolation as the paths connect before. On the other hand, backscattering softens delocalization in comparison to percolation, as it favors localization. Since the delocalization exponent measured in the quantum Hall system is larger than the percolation exponent, thus demonstrating that delocalization in the quantum Hall system is stronger than what is predicted by percolation, we draw the conclusion that it is tunneling wich is the dominant effect. We discuss in the next Section how tunneling changes the picture presented so far.

3 Tunneling

Suppose now that we have adjusted the magnetic field so that the Fermi level is exactly at $E = E_c$. Moving a little away from this value will disconnect two paths at some saddle point, and as a result, classically the system is no longer percolating. If we now continue to tune E away from E_c, a new pair of paths will become disjoint at a second saddle point, thus breaking the earlier percolating cluster into three disjoint clusters, and so on. The typical distance between these "disconnecting" saddle points (barriers) is proportional to the percolation correlation length, and thus behaves as

$$\frac{\xi(E)}{l} \sim |E_c - E|^{-\nu}, \quad (\nu > 0) \tag{11}$$

near E_c.

We now add the possibility of quantum tunneling to the above picture. This does not change the idea that the percolation properties of the system close to threshold are determined by a few barriers in the potential energy landscape. However, it does change how effective these are in disconnecting paths. In the above picture, the blocking length is l: Wave functions are connected if they overlap. Since they localized over a distance l, overlap must happen within this distance.

When E is very near the percolation threshold, whether the system percolates or not is determined only by what is happening at one particular saddle point. As E approaches E_c the probability of tunneling through the barrier converges to unity. In the semiclassical limit this convergence may be associated with a *tunneling length* λ. If we call the saddle point under consideration $V(\boldsymbol{x})$, then the probability that a particle with energy $E - E_n < V(\boldsymbol{x})$ will tunnel through the potential barrier is given in the semiclassical WKB-approximation, by

$$\wp(E, V) = e^{-1/\lambda(E,V)} , \tag{12}$$

where

$$\lambda(E, V)^{-1} = \int_{\mathcal{P}(\mathbf{x}_i,\mathbf{x}_f)} dx \,\mathrm{Im}\, k(E, U; x) \,. \tag{13}$$

Here $\mathcal{P}(\boldsymbol{x}_i, \boldsymbol{x}_f)$ is the easiest path through the barrier, here parametrized by x. When E is sufficiently close to criticality we can usually (i.e., when $V(\boldsymbol{x})$

is not too pathological) without loss of generality replace this path by a small symmetrical domain around the maximum $E_c - E_n = V(0)$ of the potential barrier. The canonical momentum along the path of least resistance, $k(E, V; x)$, is determined by the semiclassical dispersion relation.

Our idea [9] is now that the tunnel length takes over the rôle of l in equation (11),

$$\frac{\tilde{\xi}(E,V)}{\lambda(E,V)} \sim |E_c - E|^{-\nu} . \tag{14}$$

This is so since by including tunneling, it is λ and not l that determine whether two paths are connected or not. Furthermore, since percolation is a statement about the statistical distribution of barriers, the correlation exponent remains the cassical one equal to 4/3. As we will demonstrate below,

$$\frac{\lambda(E,V)}{l} \sim |E_c - E|^{-\mu}, \quad (\mu > 0) . \tag{15}$$

Combining this equation with (14) gives

$$\frac{\tilde{\xi}(E,V)}{l} \sim |E_c - E|^{-(\nu+\mu)} , \tag{16}$$

and we see that the effective quantum percolation correlation length exponent $\tilde{\nu}$ is

$$\tilde{\nu} = \nu + \mu . \tag{17}$$

There is no danger in using the WKB approximation in the above derivation. This approximation is good when the tunnel probability is small, which is exactly the condition that defines the tunnel length λ.

Next, we ask how the other critical exponents associated with percolation, such as the specific heat exponent α, the order parameter exponent β or the susceptibility exponent γ, are changed when including tunneling. In order to calculate these exponents, it is convenient to define the size of a cluster, s. This is a fundamental concept in percolation theory. Imagine a square lattice with lattice constant a — the length scale characterizing V — superposed on the two-dimensional plane in which the electrons move. Count the number of boxes of size $a \times a$ that contains a path of interest. This number is s.[1] The cluster size distribution $N(s, E)$, gives the density of clusters of size s per site [10].

$$N(s,E) \sim s^{-\tau} F(R(s)/\xi(E)) , \tag{18}$$

where $R(s)$ is the average radius of gyration of clusters of size s, and F is a scaling function which approaches a constant for small arguments and falls off faster than any power law for large arguments. The radius of gyration behaves as a power law,

$$R(s) \sim s^{\nu\sigma} , \tag{19}$$

[1] Why not use l instead of a? This makes no difference with respect to the scaling properties, but the choice made is conceptually simpler.

which allows us to write the cluster size distribution as $N(s,E) \sim s^{-\tau}G(|E-E_c|s^\sigma)$, where G is a scaling function with the same behavior as F. All other critical exponents may be expressed in terms of τ and σ. In two dimensions, $\tau = 187/91$ and $\sigma = 36/91$. The "usual" critical exponents may all be expressed in terms of τ and σ: $\alpha = 2-(\tau-1)/\sigma$, $\beta = (\tau-2)/\sigma$, $\gamma = (3-\tau)/\sigma$, and $\delta = 1/(\tau-2)$.

Taking tunneling into account in the semiclassical limit, $N(s,E)$ is transformed into $\tilde{N}(s,E)$ by substituting the tunneling length λ for l. However, as the percolation problem is identical to a classical one measured relative to the tunneling length scale λ, neither the cluster-size distribution exponent τ of (18) nor the exponent relating the radius of gyration to cluster size, $\nu\sigma$ of equation (19), are changed, since they only measure *relative* quantities. Thus, we have that $\tilde{\tau} = \tau$ and $\tilde{\nu}\tilde{\sigma} = \nu\sigma$, *i.e.* $\tilde{\sigma} = \sigma\nu/(\nu+\mu)$. Hence, when $\mu = 1$ — which we will show it is in the quantum Hall system — $\tilde{\tau} = 187/91$, and $\tilde{\sigma} = 144/637$. This gives

$$\tilde{\alpha} = 2-(2-\alpha)\frac{\mu+\nu}{\nu}\,, \quad \tilde{\beta} = \beta\frac{\mu+\nu}{\nu}\,, \quad \tilde{\gamma} = \gamma\frac{\mu+\nu}{\nu}\,, \quad \tilde{\delta} = \delta\,. \tag{20}$$

We find that the scaling relations $2-\tilde{\alpha} = \tilde{\gamma}+2\tilde{\beta} = \tilde{\beta}(\tilde{\delta}+1) = 2\tilde{\nu}$ are satisfied.

Finally, in order to obtain the missing exponent $\tilde{\eta}$ we proceed via the correlation function, which is given by

$$g(r,E) = l^{-2}\left(\frac{r}{l}\right)^{-\eta} e^{-r/\xi}\,, \tag{21}$$

so that at percolation threshold it simply behaves as $g(r,E_c) \sim l^{-2}(r/l)^{-\eta}$. Forming the ratio $g(br,E_c)/g(r,E_c) = b^{-\eta}$ shows how the correlation function scales relative to a length scale r.

In order to take tunneling into account we again substitute the tunneling length λ for a, to find:

$$\tilde{g}(r,E) = \lambda^{-2}\left(\frac{r}{\lambda}\right)^{-\eta} e^{-r/\tilde{\xi}} = l^{-2}\left(\frac{r}{l}\right)^{-\tilde{\eta}} e^{-r/\tilde{\xi}}\,. \tag{22}$$

As in the classical case, from the ratio $\tilde{g}(br,E_c)/\tilde{g}(r,E_c) = b^{-\eta} = b^{-\tilde{\eta}}$ we immediately conclude that

$$\tilde{\eta} = \eta\,. \tag{23}$$

We can thus verify the last scaling relation $(2-\tilde{\eta})\tilde{\nu} = \tilde{\gamma}$.

So far, we have only assumed equation (15), stating that λ diverges as $E \to E_c$. We now show that this is so for the quantum Hall system. In addition, we show that μ is equal to one in this case.

The classical dispersion relation for an electron moving in a strong magnetic field in the viscinity of a saddle point has been obtained by Mil'nikov and Sokolov [8]. It is

$$E - E_n = V(x, k(x)l^2)\,. \tag{24}$$

Assuming that the coordinate system is oriented such that the x-axis is the "easy" direction, and that the saddle point is quadratic and located at the origin of the coordinate system so that $E_c = U(0,0)$, we find from (24) that

$$k(x)^2 = -\frac{2}{l^4 V_{yy}} \left[(E_c - E) + \frac{V_{xx}}{2} x^2 \right] , \tag{25}$$

where $V_{xx} \equiv \partial^2 V(0,0)/\partial x^2 < 0$ and $V_{yy} \equiv \partial^2 V(0,0)/\partial y^2 > 0$.

Explicitly, performing the integral in (13) after inserting (25) gives[2]

$$\lambda(E,V)^{-1} = \frac{\pi}{l^2} \sqrt{\frac{1}{|V_{xx} V_{yy}|}} (E_c - E) . \tag{26}$$

We see that $\mu = 1$, so that the quantum Hall delocalization exponent $\tilde{\nu}$ in the quantum Hall system, is $\tilde{\nu} = \nu + 1 = 4/3 + 1 = 7/3$. This value was first obtained in Reference [8], but by a very different method. It is in excellent agreement with both experimental and numerical estimates.

---

[2] For those who still distrust the WKB approximation, one can go back to the one-dimensional Schrödinger equation (7), which in the strong magnetic field limit reads

$$-\frac{l^4}{2} V_{yy} \frac{d^2 C_n(X)}{dX^2} + \frac{1}{2} V_{xx} X^2 C_n(X) = (E - E_n - V(0,0)) C_n(X) . \tag{A}$$

The transmission coefficient through this potential for E near $E_n + V(0,0)$ may be found by studying tunneling in the one-dimensional potential $V_0 / \cosh^2(x/a)$, as this potential has the same quadratic behavior near the maximum as our original problem. The transmission coefficient is [19]

$$T = \frac{e^{-2\pi/l^2 \sqrt{1/|V_{xx}V_{yy}|}(E_n + V(0,0) - E)}}{1 + e^{-2\pi/l^2 \sqrt{1/|V_{xx}V_{yy}|}(E_n + V(0,0) - E)}} . \tag{B}$$

The WKB approximation gives from (26)

$$T_{WKB} = e^{-2\pi/l^2 \sqrt{1/|V_{xx}V_{yy}|}(E_n + V(0,0) - E)} , \tag{C}$$

so that

$$T = \frac{T_{WKB}}{1 + T_{WKB}} . \tag{D}$$

As we are interested in the length at which $T \approx 1/e$, we see that WKB is sufficient.

4 Conclusions

We have in this paper reviewed a recent attempt at clarifying the mystery of the delocalization length exponent in the quantum Hall system. It is believed that delocalization in the quantum Hall system is a percolation effect. In a strong magnetic field, charge carriers with energy E are localized to within a band of size l, the magnetic length, of the curve defined by $V(\boldsymbol{x})+\hbar\omega_c(n+1/2) = E$. For a particular energy $E = E_c$, curves coalesce to form a connected network spanning the sample. When the Fermi level passes through this energy, the system is delocalized, and a longitudinal conductance appears as a result. The problem with this picture is that it predicts that the localization length diverges with the percolation correlation length exponent $\nu = 4/3$, in stark disagreement with both numerical and experimental values results, which set it at $\simeq 2.3$.

The idea that has been put forward, is that globally the percolation picture is correct. However, the key to the problem lies in how the paths determined by $V(\boldsymbol{x}) + \hbar\omega_c(n + 1/2) = E$ connect across saddle points. Rather than assuming that curves are connected when they are within a distance l of each other, tunneling has to be taken into account. This means that l should be replaced to the tunneling length λ, as the distance which defines connectivity across saddle points. However, the tunnel length itself diverges at the same energy E_c as the percolation transition. Thus, we have a classical percolation problem relative to a smallest length scale which itself diverges. Transforming into a "non-dilating reference frame," changes the universality class of the percolation transition, and in particular the correlation length exponent becomes $\tilde{\nu} = \nu + 1 = 7/3$. Interpreting this exponent as the localization length exponent leads to a result which is in accordance with both numerical and experimental measurements.

Let me end by posing the following problem: It has recently been shown that the classical problem of charged particles moving in a plane pierced by a strong magnetic field among "hard scatterers" undergo a percolation-type localization-delocalization transition [18]. In particular, the localization length exponent is thus predicted (and numerically shown) to be 4/3. Will a quantum treatment of this problem give rise to a $\tilde{\nu} = 7/3$?

Acknowledgements

I first and foremost thank C. A. Lütken for our collaboration on this problem. I also thank M. Engineer, E. H. Hauge, J. S. Høye F. Maaø and J. Støvneng for several very clarifying discussions. This work has been partially supported by the Groupement de Recherche CNRS "Physique de Milieux Hétérogènes Complexes."

References

1. K. v. Klitzing, G. Dorda and M. Pepper, Phys. Rev. Lett. **45**, 494 (1980).
2. R. E. Prange and S. M. Girvin, editors, *The Quantum Hall Effect* (Springer Verlag, Berlin, 1987).
3. D. C. Tsui, H. L. Störmer and S. C. Gossard, Phys. Rev. Lett. **48**, 1559 (1982).
4. S. Koch, R. J. Haug, K. v. Klitzing, and K. Ploog, Phys. Rev. Lett. **67**, 883 (1991).
5. H. P. Wei, D. C. Tsui, and A. M. M. Pruisken, Phys. Rev. B **33**, 1488 (1985); H. P. Wei, D. C. Tsui, M. A. Paalanen, and A. M. M. Pruisken, Phys. Rev. Lett. **61**, 1294 (1988); L. Engel, H. P. Wei, D. C. Tsui, and M. Shayegan, Surf. Sci. **229**, 13 (1990); S. Koch, R. J. Haug, K. v. Klitzing, and K. Ploog, Phys. Rev. Lett. **67**, 883 (1991).
6. J. T. Chalker and P. D. Coddington, J. Phys. C **21**, 2665 (1988); B. Huckestein and B. Kramer, Phys. Rev. Lett. **64**, 1437 (1990); D. K. K. Lee and J. T. Chalker, Phys. Rev. Lett. **72**, 1510 (1994).
7. B. Huckestein and B. Kramer, Phys. Rev. Lett. **64**, 1437 (1990).
8. G. V. Mil'nikov and I. M. Sokolov, JETP Letters **48**, 536 (1988).
9. A. Hansen and C. A. Lütken, preprint, 1993.
10. D. Stauffer and A. Aharony, *Introduction to Percolation Theory* 2. ed. (Taylor and Francis, London, 1992).
11. R. F. Kazarinov and S. Luryi, Phys. Rev. B **25**, 7626 (1982).
12. S. A. Trugman, Phys. Rev. B **27**, 7539 (1983).
13. E. Abrahams, P. W. Anderson, D. C. Licciardello and T. V. Ramakrishnan, Phys. Rev. Lett. **42**, 673 (1979).
14. M. P. M. den Nijs, J. Phys. A: Math. Gen. **12**, 1857 (1979); J. L. Black and V. J. Emery, Phys. Rev. B **23**, 429 (1981).
15. M. Tsukada, J. Phys. Soc. Japan, **41**, 1466 (1976).
16. B. I. Halperin, Phys. Rev. B **25**, 2185 (1982).
17. M. Büttiker, Phys. Rev. B **38**, 9375 (1988).
18. F. Maaø, A. Bobylev, A. Hansen and E. H. Hauge, preprint, 1994.
19. D. ter Haar, *Selected Problems in Quantum Mechanics* (Infosearch, London, 1964).

MIX
Papier aus verantwortungsvollen Quellen
Paper from responsible sources
FSC® C105338

If you have any concerns about our products,
you can contact us on
ProductSafety@springernature.com

In case Publisher is established outside the EU,
the EU authorized representative is:
Springer Nature Customer Service Center GmbH
Europaplatz 3, 69115 Heidelberg, Germany

Printed by Libri Plureos GmbH
in Hamburg, Germany